THE CLIMATE DEBATE

Updated and Modified
November 2012

DONALD RAPP

THE CLIMATE DEBATE
DONALD RAPP

TABLE OF CONTENTS

1. THE CLIMATE DEBATE 1
2. CLIMATE OVER PAST FEW THOUSAND YEARS 7
2.1 Use of proxies to estimate historical temperatures 7
2.2 Proxies and climate 8
2.2.1 Processing proxy data 8
2.2.2 Challenges in using proxies 9
2.2.3 Combining multiple proxies 9
2.3. The Medieval Warm Period and the Little Ice Age 11
2.3.1 Proxies for Specific Sites and Regions 11
2.3.2 Combining Multiple Proxies for Global Averages 17
2.3.3 Proxy Analysis by Mann, Bradley, and Hughes – & Others ... 20
2.3.4 Fallacies in Reconstruction of Millennial Temperatures 23
2.3.4.1 *The Fallacy of Choosing the Wrong Mean* 24
2.3.4.2 *Hiding the Decline* 41
2.3.4.3 *Sparse Data Set* 45
2.4 The Alarmist Cabals 46
2.4.1 Paleoclimatic Cabal 46
2.4.2 The Climate Alarmism Cabal 57
2.5 The Climate Debate Revisited 64
3. GLOBAL SCARES, SUBJECTIVE SCIENCE & CLIMATOLOGISTS 69
3.1 Global Scares 69
3.2 Dealing With Subjective Science 72
3.2.1 Nature of Subjective Science – Emergence of Consensus 72
3.2.2 Consensus to Orthodoxy 74
3.2.3 Counting Adherents to the Orthodoxy 76
3.3 Hyperbole on Impacts of Global Warming 78
3.4 Spreading the Gospel 80
3.5 Climatologists 83
3.6 The Lunatic Fringe 93
3.7 The Climate Debate Revisited 93
4. EARTH TEMPERATURES OVER THE PAST ~ 200 YEARS 95
4.1 Introduction 95
4.2 Quality and Reliability of Land Temperature Networks 97
4.3 Quality and Reliability of Ocean Temperature Measurements 108
4.4 Measured Global Average Temperature 113

4.4.1 U. S. Temperatures 114
4.4.2 Global Temperatures 118
4.5 Tropospheric Temperatures 125
4.6 Arctic and Antarctic Temperatures 131
4.6.1 Antarctic Temperatures 132
4.6.2 Arctic Temperatures 135
4.7 The NH temperature dip, 1940–1978: effect of aerosols 139
4.8 Warming in the Early 20th Century 140
4.9. Climate in the 20th Century and El Niños 146
4.10 Warming of the Oceans 156
4.10.1 Introduction 156
4.10.2 Brief Overview of Model 159
4.10.3 The Initial Response to Forcing at the Ocean Surface 160
4.10.4 Establishment of a New Equilibrium 163
4.10.4.1 Zero'th Order Model 164
4.10.4.2 First Order Model 166
4.10.4.3 Analysis by Ramanathan (1981) 169
4.10.4.4 Summary of Models 171
4.10.5 Excess back radiation IR due to rising CO_2 173
4.10.6 Some Basic Quantities 175
4.10.7 Measured Heat Content of the Oceans 176
4.10.8 Effective Back Radiation at the Ocean Surface 181
4.11 The Climate Debate Revisited 186
5. CARBON DIOXIDE THROUGH THE AGES 189
5.1 Recent and Future Times 190
5.2 CO_2 and the Greenhouse Effect 199
5.3 CO_2 and Climate 205
5.4 CO_2 and Ancient Climates 209
5.5 Climate Models for the 21st Century 215
5.5.1 The Good, the Bad and the Ugly 215
5.5.2 Water Vapor and Cloud Feedbacks 225
5.5.3 Model of Spencer and Braswell 237
5.6 The Climate Debate Revisited 241
6. THE EARTH'S HEAT BALANCE - GREENHOUSE EFFECT 243
6.1 Estimates of Average Heat Balance and Effect of Rising CO_2....243
6.2 An Estimate of Heat Balance of Earth from 1950 to 2005 247
6.3 Solar Power Input to Earth... 250
6.3.1 Total Solar Irradiance Measurements 250
6.3.2 Appearance of the Sun... 252
6.3.3 Reconstructing Total Solar Irradiance (TSI) in the Past........ 256
6.3.3.1 *Constant Quiet Sun Models* 258
6.3.3.2 *Solar Cycles 23 and 24* 262
6.3.3.3 *Other Reconstructions* 262
6.3.4 TSI Reconstructions Based on Cosmogenic Isotope Proxies 266
6.4 The Albedo of the Earth 273

6.5 Effect of Black Carbon Deposition on Ice and Snow 277
6.6 The Dynamic Earth..282
6.7 The Climate Debate Revisited 286
7. ENERGY AND CLIMATE IN THE 21ST CENTURY...................288
7.1 World Energy Requirements..288
7.2 Constraints on CO_2 Production due to Limits of Fossil Fuels 295
7.3 The Climate Debate Revisited 297
8. IMPACTS OF GLOBAL WARMING 299
8.1 Sea Level Rise 299
    8.1.1 Measurement of Sea Level 299
    8.1.2 Historic Sea Level Change 301
    8.1.3 Recent Sea Level Change 310
    8.1.4 Global Warming and Future Sea Level Change 317
    8.1.5 Evidence from Previous Deglaciations..................................324
8.2 Sea Ice Extent 327
8.3 Future Increases in Global Temperature..336
8.4 Changes in Precipitation: Floods, Drought and Storms..............339
    8.4.1 Drought..339
    8.4.2 Floods 340
    8.4.3 Storms..341
    8.4.3.1 *Tropical Hurricanes* 341
    8.4.3.2 *Tornados*..347
    8.4.3.3 *Extreme Weather* 349
8.5 Species Extinction..350
8.6 Vegetation..351
8.7 Coral Reefs 352
8.8 Food Production...353
8.9 The Climate Debate Revisited 355
9. GLOBAL CLIMATE CHANGE AND PUBLIC POLICY 357
9.1 The Kyoto Protocol..357
9.2 Economics: Will it cost more to do nothing?................................361
    9.2.1 The Stern Report..361
    9.2.2 Critiques of the Stern Report..365
9.3 Investment opportunities in climate change 372
9.4 U.S. Congress: meeting the climate change challenge 372
9.5 Renewable Energy 379
9.6 The Climate Debate Revisited 383
10. FINAL REMARKS...385
11. REFERENCES..393

LIST OF FIGURES

2.1. Concept of calibration period for a proxy 8
2.2. Ice core records showing LIA and MWP 13
2.3. Mg/Ca analyses as a measure of sea surface temperatures 13
2.4. Mean temperature anomaly for 18 non-tree ring proxy series 14
2.4a. Calibration curve for northern Scandinavian tree ring data 15
2.4b. 2000-year history of No. Scandinavian temperatures 16
2.4c. 2000-year history of No. Scandinavian temperatures 16
2.5. Estimates of Historical Temperatures in Asia 17
2.6. Cartoon of hypothetical dependence of variable Y on X 18
2.7. Cartoon of hypothetical dependence of temperature on date range 19
2.8. Reconstructed temperatures since 1400 22
2.9. Temperature anomaly vs. year since AD 1000 23
2.10. Hypothetical set of proxies 25
2.11. Two tree ring temperature anomaly series from the MBH data set26
2.12. Some of the proxies used by Mann *et al.* (2008). 29
2.13. MBH centering vs. using centering across the whole time span ...33
2.14. Fifteen individual proxies from various locations 34
2.15. Derived NH temperature anomalies 35
2.16. "Spaghetti chart" of individual proxies 39
2.17. Simple average of proxy data 39
2.18. Temperature anomalies from MBH 40
2.19. Eight reconstructions of northern summer temperatures 42
2.20. Comparison of reconstructions 42
2.21. Comparison of tree ring density with temperature for NH 43
3.1. Measured cloud feedback vs. Earth surface temperature 90
3.2. Cloud feedback plot with four-month time lag. 91
4.1. Locations of GHCN mean temperature station locations................101
4.2. Locations of GHCN max/min temperature station locations 102
4.3. World temperature measurement stations 103
4.4. Distribution of temperature stations in the U. S. 104
4.4a. Smoothed monthly sea surface temperatures for 2 hemispheres..111
4.4b. Global smoothed monthly sea surface temperatures....................112
4.5. U.S. mean temperature anomalies 114
4.6. Historical variation of U.S. heat wave index 115
4.6a. Number of daily high U. S. T_{max} records (1895-2011) 117
4.6b. Number of daily U. S. T_{max} and T_{min} records (1895-2011) 117
4.6c. Fraction of monitored area that exceeds 3σ threshold..................118
4.7. Global temperatures 119
4.8. Area-weighted surface temperatures over latitude bands 119
4.9. Amplification of air temperature as a function of latitude............122

4.10. BEST ten-year moving average of global land temperatures.......122
4.10a. BEST ten-year moving average of global land temperatures.....123
4.10b. BEST yearly data compared to volcanic eruptions 124
4.11. Tropospheric temperature over three decades...........................126
4.12. Global temperature anomaly compared to the Nino 3.4 index 128
4.13. Satellite temperatures vs. Nino 3.4 index and volcanic eruptions 129
4.14. Trend of tropospheric temperature 129
4.14a. Measured global temperatures compared to Nino3.4 index 132
4.14b. Arctic-wide annual average surface air temperature anomalies.137
4.15. Global sulfur dioxide emissions..141
4.16. Reconstructed sea surface temperatures 142
4.17. Long-term SOI index 149
4.18. SOI Index since 1950 150
4.19. Integral of SOI anomaly since 1950..151
4.20. Integral of Douglass' modified El Niño index 155
4.21. El Niño index vs. surface and tropospheric temperatures...........155
4.22 Schematic temperature profile in a tropical ocean 159
4.23 Schematic temperature profile in a tropical ocean 160
4.24 Forcing due to doubling CO_2 as a function of altitude.................174
4.25. Percent of world ocean volume below any depth 176
4.26. Effective back radiation 182
5.1. Rough estimates of carbon storage and annual carbon fluxes........191
5.2. Atmospheric CO_2 concentration over the past 2,000 years 192
5.2a. Annual and long-term variation of CO_2 concentration 193
5.2b. CO_2 emissions and concentration and Niño index 194
5.3. Annual emissions of carbon for future scenarios 196
5.4. Buildup of CO_2 in the atmosphere 197
5.5. Absorption factor for CO_2 vs. wavelength..................................199
5.5a. Lindzen's picture of how the greenhouse effect works...............201
5.5b. Lindzen's picture of how the greenhouse effect works; part 2 202
5.6. Estimated forcing of climate due to changes in CO_2 203
5.7. Hypothetical single curve relating T_G to CO_2 concentration..........207
5.8. Hypothetical curves relating T_G to CO_2 concentration..................208
5.9. Range of CO_2 concentration for 21st century climate change 208
5.10. Hypothetical variation of T_G with CO_2 concentration.................209
5.11. Variation of ΔT_G with latitude..211
5.12. Inclusion of LGM in relationship of CO_2 to T_G 211
5.13. Model used by Roe and Baker 224
5.14. Dependence of ΔT on f...224
6.1. Solar power distributed uniformly around the Earth.....................244
6.2. Solar Power Input to the Earth 245

6.3. Heat transfer between the Earth, the atmosphere and space. 245
6.4. Spectra of irradiance from the surface and top of atmosphere.......246
6.5. Factors contributing to heat balance of the Earth...........................248
6.6. Net forcing due to greenhouse gases vs. ocean heat content 249
6.7. Redundant, overlapping satellite TSI monitoring experiments 251
6.8. Composite of TSI measurements 251
6.9 Group sunspot numbers since 1600 253
6.10. Historical record of sunspot activity 254
6.11. Relation between TSI and sunspot number in a CQSM...............260
6.12. Modeled TSI through the MM up to the present..........................261
6.13. Measured sunspot numbers for solar cycle 23 263
6.14. Reconstructed TSI based on sunspot number or cycle duration ..265
6.15. Cosmo-nuclide production as percent of present production.......267
6.16. Modeled TSI during the period 1850 to the present 268
6.17. Radionuclide fluxes and ice-rafted debris....................................272
6.18. Comparison of various models for the albedo of the Earth..........274
6.19. History of BC emissions by region since 1850 279
6.20. Snow-albedo reduction attributed to BC......................................282
7.1. Projection of world population made by IS92a..............................289
7.2. Energy mix for generation of electric power in the 21st century ..290
7.3. "carbon factor" in the 21st century 290
7.4. Carbon emissions to stabilize CO_2 concentration 291
7.5. Estimated cumulative production of hydrocarbons and coal 295
7.6. CO_2 emissions and IPCC projections for future CO_2. 296
7.7. MIT estimates of carbon emissions for various levels of control ..296
8.1. Sea-level rise since the Last Glacial Maximum.............................303
8.2. Reconstruction of relative sea level since 1700 303
8.3. Comparison of estimates of sea level rise for the 20th century 307
8.4. Calibration curve for sea level model 309
8.5. Ice sheet mass balance 314
8.5a. Projection of future sea level rise from ground water 317
8.6. Projections of future sea level rise 320
8.7. Relative temperatures of the last four interglacials........................325
8.8. Northern Hemisphere seasonal sea ice extent 330
8.9. Sum of changes in sea ice extent in four Arctic seas 332
8.10. Ice extent in the Greenland, Barents, and Kara Seas 332
8.11. Ice extent in the Laptev, East Siberian, and Chukchi Seas 333
8.11a. Summer Arctic sea ice extent in recent years............................335
8.11b. Average August sea ice extent (1980-2012). 336
8.12 Atmospheric methane concentrations 339
9.1. Distribution of sources of world energy use in 2008 381

LIST OF TABLES

1.1. Dimensions of the climate debate 5
2.1. Number of proxies vs. earliest date...21
3.1 Personality Traits 84
3.2. Personality Traits of Climate Scientists vs. general public 85
3.3. Alarmist view of climate change vs. global scares 94
4.1 Change in total heat loss as a function of T_S when the atmosphere is unchanged 166
4.2 Change in total heat loss as a function of T_S 168
4.3 Clear sky forcing at various altitudes for various changes in greenhouse gas concentrations 174
4.4 Clear sky forcing at various altitudes for various changes in greenhouse gas concentrations 175
4.5 Rate of heat absorption for the 0-700 m ocean layer 178
4.6 Heat gains by time period and ocean layer 179
4.7 Total temperature rise by time period and ocean layer 179
4.8 Rate of heat absorption by time period and layer 179
4.9 Clear Sky Formulas for Back Radiation 185
4.10. Simple formulas for back radiation including clouds 185
4.11 Alarmist view of climate change vs. 20th century observations...149
5.1 Parameters for analyzing LGM – pre-industrial transitions............212
5.2. Evidence for the climate debate 241
6.1. Dimensions of the climate debate 286
7.1. Dimensions of the climate debate 297
8.1. Estimates of acceleration of the of sea level rise 308
8.2. Projected rise in sea level 318
8.3. Predicted rise (cm) in sea level by 2100 319
10.1. Dimensions of the climate debate 392

Donald Rapp

1 THE CLIMATE DEBATE

In late 2011, Richard Muller and co-workers released preliminary versions of reports that measured global temperatures over the past 200 years showing a significant increase over that period. A number of Internet sites responded with blaring headlines such as *"Richard Muller, Global Warming Skeptic, Now Agrees Climate Change Is Real"*.

What is the climate change debate about? In the film "Karate Kid" Miyogi said:

> "Answer only matter if ask right question".

The climate debate is not about whether the Earth has warmed in the last 200 years. That is settled. It was settled long before Muller *et al.* (2011) released their findings. *The Earth has warmed in the last 200 years.* Or put differently, it was colder 200 years ago than it is today. There has been a small climate change. That is fact. It is not debatable.

People ask me "do I believe in climate change"? I respond: "Sure, the Earth's climate has gone through wild gyrations over millions of years."

Before attempting to define the boundaries of the climate debate, some background needs to be established.

According to convention, weather represents the short-term variability of temperature, wind, precipitation, humidity, cloudiness, etc. and climate represents the long-term variability of average weather. Weather varies at any location diurnally, from season to season, and year to year. It varies even more with location. It is not clear how to define the global climate at any time by averaging the weather at myriad points across the surface of the Earth. Indeed, the whole concept that there even is a global climate may not have much utility, except in discussing the extremes such as ice ages and hothouse

Earth. Only for such large changes are the worldwide parameters widely in conformance. For small variations in climate, the dividing line between weather variability and climate variability is usually obscure.

Generally, variations from year to year can be construed to be changes in weather, and persistent universal changes over many millennia may be construed to be changes in climate. But for intermediate periods, such as decades, or perhaps even as much as a century or two, it is difficult to resolve whether an unusually cold or warm period represents a statistical fluctuation in weather within a stable climate, or a climate change. If it is a climate change, it is difficult to resolve whether it is part of a cycle or a unidirectional trend.

Today, we are concerned about potential climate changes that might occur in the future. To seek background data, we search for variability in past climates. However, random changes in weather from year to year and decade to decade are so large as to mask any underlying trends in climate change (if there is such a thing as a global climate).

There is ample geological evidence that over hundreds of millions of years, the Earth's climate has ranged from very warm in which there was no polar ice or mountainous glaciers, to very cold with extensive glaciation, possibly down to equatorial latitudes. The widely held view amongst geologists and climatologists alike, is that the primary cause of long-term climate change is variability of CO_2 concentration due to imbalances between CO_2 degassing at spreading centers and the conversion of atmospheric CO_2 to mineral carbon through long-term silicate weathering and oceanic carbonate formation. The argument goes (more or less): If it wasn't CO_2, what else could it have been? Paleoclimatologists have been trying for decades to establish a relationship between climate and CO_2 concentration over many millions of years. However, a detailed examination of the data shows that while CO_2 is probably one of several major factors in long-term climate change, other factors are also important such as the placement of the continents on Earth, the functionality of ocean currents, the past history of the climate, the orientation of the Earth's orbit relative to the Sun, the luminosity of the Sun, the presence of aerosols in the atmosphere, volcanic action, land clearing, biological evolution, etc. Nevertheless, it is widely accepted that the Earth's climate tends to go up and down with the CO_2 concentration.

The Earth was enveloped in an Ice Age 20,000 years ago in which gigantic ice sheets covered high northern latitudes, soaking up so much of the Earth's water that the oceans were over 100 meters lower than they are today. The Earth began warming about 18,000 years ago, and reached a relatively warm plateau roughly similar to today's climate about 7,000 years ago. Since that time, the Earth's climate has been relatively stable, meandering up and down through moderate variations, the magnitude of which are uncertain. The CO_2 concentration has also been relatively constant over the past few thousand years at around 280 ppm.

Starting late in the 19th century, and continuing through the 20th century into the 21st century, emissions of CO_2 by human activity (factories, power generation, cement production, deforestation, ...) have upset the delicate balance between emission and absorption of CO_2 by the Earth ecosystems, resulting in a significant increase in atmospheric CO_2 concentration in the 20th and 21st centuries.

One aspect of the climate debate is this:

Alarmists believe that the natural variations in climate over the past several thousand years were minor and slowly changing while CO_2 concentrations held fairly constant. The temperature rise over the past century is regarded as a significant change from the past, particularly the rate of change. The fact that over the past century, CO_2 concentrations and global temperatures both increased suggests that the rising CO_2 concentration was the primary cause of the temperature increase.

Extreme skeptics believe that natural variations in climate over the past several thousand years were significant enough that the temperature rise over the past century may be considered to be just another natural fluctuation having little to do with rising CO_2.

Moderate skeptics believe that natural variations in climate over the past several thousand years were significant enough that the temperature rise over the past century is probably partly due to natural fluctuations and partly due to rising CO_2. The fact that the pattern of rising CO_2 only very roughly matches the pattern of temperature rise during the 20th century supports this view. The proportion of temperature rise in the 20th century attributable to rising CO_2 is unknown.

Everyone must agree that in a business-as-usual scenario for the future, as the Earth's population expands, and industry and technology spreads in developing nations, human use of fossil fuels will emit ever more CO_2, resulting in even greater increases in atmospheric CO_2 concentration as the 21st century proceeds forward.

A second aspect of the climate debate is this:

Alarmists believe that the increases in CO_2 projected for the 21st century in a business-as-usual scenario will warm the Earth significantly, resulting in a large number of serious impacts to the world, the global economy, and various populations, as documented in IPCC publications. Perhaps the most serious consequence of projected global warming would be higher sea levels, which would threaten large population centers adjoining the oceans. The only remedy for this is an immediate draconian reduction in fossil fuel usage that they believe could be accommodated with expanded use of renewable energy. It should be emphasized that many alarmists think this has not been proven beyond any doubt, but feel that the probability is high enough that we should act on it.

Extreme skeptics believe that rising CO_2 in the 21st century will have little or no effect on global temperatures, and the Earth's climate will be dictated by other factors. They also believe that even if the Earth warms, the putative deleterious impacts described by alarmists are grossly exaggerated, and there may even be benefits associated with global warming. They believe that an immediate draconian reduction in fossil fuel usage would plunge the world into economic depression with major dislocations in energy supply due to inadequacies and costs of renewable energy.

Moderate skeptics accept that rising CO_2 in the 21st century may contribute to further increases in global temperatures, but they remain dubious about specific predictions of future runaway warmth made by alarmists based on climate models. They also believe that even if the Earth warms, the putative deleterious impacts described by alarmists are grossly exaggerated, and there may even be some benefits associated with moderate global warming. They agree with extreme skeptics that an immediate draconian reduction in fossil fuel usage would plunge the world into economic depression with major dislocations in energy supply due to inadequacies and costs of renewable energy.

These ideas are summarized in tabular form in Table 1.1.

Table 1.1. Dimensions of the climate debate.

Aspect	Alarmists	Extreme Skeptics	Moderate Skeptics
Variability of climate over past few thousand years	Minor	Significant	Significant
Temperature rise over past century compared to natural fluctuations	Significant difference in 20th century	Within past variations	Probably within past variations
Rising CO_2 was the cause of rising temperatures in 20th century	Agree	Disagree	Rising CO_2 was probably one of several factors
CO_2 concentration will rise further in 21st century in business-as-usual	Agree	Agree	Agree
Climate models provide reasonable estimates of future warming in business-as-usual scenario	Agree	Disagree	Uncertain; doubtful
Impacts of future global warming are known to be disastrous	Agree	Disagree	Exaggerated
Only remedy is immediate draconian reduction in fossil fuel usage	Agree	Disagree	True if predicted impacts are accurate; but predictions are dubious
Draconian reduction in fossil fuel usage can be accommodated by renewable energy supply	Agree	Disagree	Disagree

2 VARIABILITY OF CLIMATE OVER THE PAST FEW THOUSAND YEARS

A recurring theme in most of Isaac Asimov's science fiction novels is the search for historical roots in the fog that obscures the past. These books have been very successful and are very readable, partly because they are well conceived and well written, but also because they evoke an empathetic response from our natural desire to understand our origins and roots. Similarly, the history of climatic variations leaves behind a fog that is difficult to penetrate. Many incredibly ingenious proxy methods have been devised to peer into the past. However, none is entirely satisfactory, and many uncertainties remain. In addition to proxy data, there are many anecdotal accounts in historical records that indirectly infer information about past climate (e.g., extent of glacier expansion and contraction in the Swiss Alps, paintings showing skaters on lakes that presently don't freeze, etc.). All of these have been utilized in the imperfect attempt to estimate past climates.

2.1 Use of proxies to estimate historical temperatures

In the context of historical temperatures, proxies are residual data from processes that occurred in the past, when the processes were dependent on local temperatures at the times they took place, and the evidence is preserved in the present in an accessible form. In all cases, extraction of implied past temperature data from confounding influences requires considerable analysis and manipulation. As a result, the credibility and reliability of such proxy data vary widely from data set to data set, as well as in the eye of the beholder. For example, there are trees that are several thousand years old. The growth (width and density) of tree rings depends on the temperature prevailing during the growth period. By examining old tree rings corresponding to historical times, one can infer past temperatures. However, tree growth is also affected by other factors (water availability, humidity, wind, cloudiness, CO_2 content in the atmosphere, nutrients, etc.). These add noise to the temperature signal. Hence, it is not a simple matter to extract accurate historical temperature data from tree rings (or other proxies, for that matter). Some prominent proxies include: tree rings (width, density, stable isotope composition), ice cores (oxygen isotope ratios, gas content in bubbles), ocean sediments (isotope ratios), pollen, boreholes and corals.

2.2 Proxies and climate

2.2.1 Processing proxy data

The common approach to historical climate reconstruction from proxies is to establish a relationship between actual temperature measurements and the variability of the proxy over this recent period of overlap (calibration period). This provides a transfer function that enables the proxies to be used to infer the past climate in historical times when proxy data are available but direct temperature measurements are not. Ideally this works out as shown in Figure 2.1.

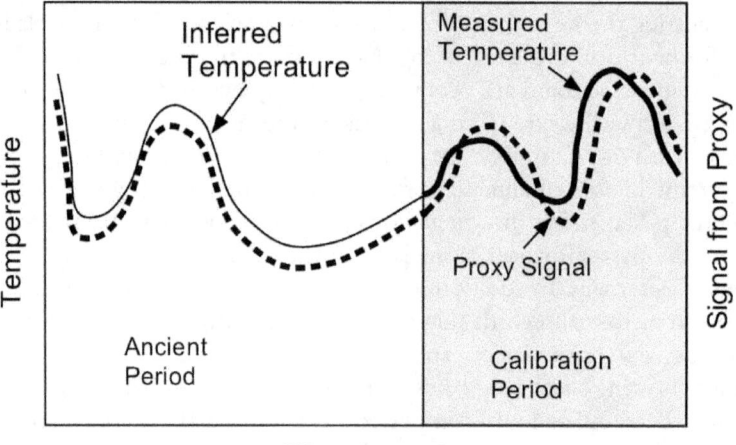

Figure 2.1. Concept of calibration period for a proxy.

During the calibration period measurements of the temperature are available for the locality where the proxy is located. A mathematical connection is made between the proxy signal and the temperature during this period. The proxy signal extends back in time prior to the calibration period. It is assumed (without proof) that the same relationship between proxy and local temperature holds in the past, and from this, past temperatures are estimated prior to the calibration period.

However, in many cases, the "fit" between the proxy and the measured temperature during the calibration period is not nearly as good as that depicted in Figure 2.1. There is usually a wide scatter in the data points and the putative relationship between proxy and temperature is not always clear. This is because many factors other than temperature often affect the proxy signal. Furthermore, in many published papers, the details of the proxy-temperature relationship during the calibration period are either not presented, or are so complex as to defy simple evaluation.

Over the past several decades, quite a number of scientists have analyzed data from one or several proxies and derived estimates of climate history in various localities and regions. Since much of the available data are from the northern hemisphere (NH), a good deal of this analysis pertains mainly to the NH.

2.2.2 Challenges in using proxies

Ogilvie and Jonsson (2001) noted that essentially all current calibrations of proxies to large-scale instrumental measurements have been made over periods of rising temperature. They raised the concern that a different calibration response might arise when the procedure is extended to an untested climate regime associated with a persistent cooling phase. They also raised other issues as well.

Jones *et al.* (1998) presented an extensive review of proxies. The spatial and timescale constraints on proxies were described in some detail. It was concluded that each reconstruction

"... is probably limited in its ability to reproduce past temperature variations faithfully on the longest of timescales. This limitation varies from proxy to proxy and it is virtually impossible to quantify the degree to which this has occurred because instrumental series are not long enough."

It was found that for most proxies, the correlation with instrumental data during the calibration period was not as good as reported in the journal articles.

Rapp (2010) also provided an extended review of proxies.

2.2.3 Combining multiple proxies

Most proxies provide estimates of ancient temperatures at specific localities or regions. Climatologists are interested in the overall global climate, which necessitates developing a synoptic view of temperatures around the globe. The challenge is to find ways to combine temperature estimates from multiple proxies at various locations, over variable time periods.

Starting as early as the late 1970s and following through the 1980s and 1990s, culminating in a pair of very influential papers in 1998 and 1999 (Mann, Bradley, and Hughes, 1998, 1999), and continuing to this day, a loosely allied cadre of scientists has attempted to statistically combine large numbers of proxies (indeed, all the proxies that were available) into a reconstruction of global (or at least NH) average temperatures for the past millennium or two. They have assembled as many as a thousand proxies into a database. These proxies include a variety of geographical locations, ranges of time, and degrees of credibility. A major question is how should these proxies be combined? Measurements at different locations, particularly different latitudes, will have different absolute temperatures and different temperature trends with time. Another major question is how to assign weights to various proxies based on their degree of credibility. Equally challenging is the question of how (or whether) to include documentary information that is typically discontinuous and often anecdotal in nature. These scientists typically employed sophisticated statistical approaches for combining proxy data sets into reconstructions of past NH or global average temperatures. The underlying basis for these approaches is the assumption that each proxy supplies an estimated temperature as a function of time (T_E) that contains a temperature signal (T_S) plus noise (T_N).

$$T_E = T_S + T_N$$

It is further assumed that if one utilizes a collection of proxies, the signals T_S will have similar trends for the various proxies while the trends for the noise T_N will vary randomly from proxy to proxy, sometimes plus and sometimes minus. Hence, if one adds up a number of proxies, the signals will tend to reinforce while the noise will tend to cancel out, leading to an estimate of T_S with less noise. Since the noise is typically quite large, sophisticated statistical mathematical schemes have been utilized to extract the signal according to this hypothesis.

Scientists who process proxies with complex statistical procedures have produced a steady stream of journal articles that justify their procedures (e.g., Rutherford *et al.*, 2005). These articles seem to rarely show the actual original proxies during the calibration period, but only the result of feeding them into their analysis machines. The end-result is typically a relatively flat meandering curve of temperature over the past millennium with a sudden rise in the 20th century (the so-called "*hockey stick*"). In the latest in this series of self-justifying reports, Jones *et al.* (2009) provided a very lengthy review of the use of proxies to unravel the climate of the past millennium. The review covered (1) high-resolution proxy disciplines (trees, corals, ice cores, and documentary evidence), (2) various approaches for combining multiple climate proxy records to provide estimates of past hemispheric climates, and (3) use of climate model simulations of the past millennium. The end-result for each proxy is a wiggly line representing a plot of temperature vs. year for a location. They then faced the problem of combining a large number of wiggly lines into a regional or global climate representation. The major stumbling blocks in combining multiple wiggly lines are (1) the spatial and temporal diversity and sparseness of the data, (2) the fact that most wiggly lines are heavily laden with local variations, and (3) large areas of the globe are underrepresented in the database. As a result, when large numbers of independent wiggly lines from various regions and time periods are simply summed and averaged democratically, the result tends to average out differences due to the wide variety of phases and amplitudes of the wiggles. Jones *et al.* (2009) describe a number of sophisticated approaches for "reconstructing the underlying spatial patterns of past surface temperature changes at global scales" and "assimilating proxy records into reconstructions of the underlying spatial patterns of past climate change." Basically, this means extracting T_S from $T_S + T_N$. However, as several critics have shown, the resultant regional or global climate reconstruction does indeed depend upon the method used for reconstruction, and depending on the method used, almost any result can be obtained. Furthermore, the net result of combining many wiggly lines with variable amplitude and phase tends to be a relatively flat profile. When this flat profile for the past millennium is combined with a measured upward trend of temperature vs. time in the 20th century, the inevitable result is a hockey stick type of figure (relatively flat for the prior thousand years with a sudden upturn in the 20th century). It is also remarkable that Jones *et al.* (2009) made no mention of major criticisms of methods used to combine multiple wiggly lines but simply pretended that such criticisms do not exist. While it is true that most of these criticisms do not appear in the published literature, the reason for this is that it is difficult to get climate papers published that do not support the alarmist position.

2.3. The Medieval Warm Period and the Little Ice Age

2.3.1 Proxies for Specific Sites and Regions

Historic proxy studies have distinguished two periods of particular interest in the past millennium. One is the putative *Medieval Warm Period* (MWP) centered near 850–1050, which was supposedly an unusually mild climate. The other was the so-called *Little Ice Age* (LIA) from perhaps 1600 to about 1850 (depending on location) when temperatures were unusually cold. Apparently, the Earth was not uniformly warm during the MWP or uniformly cold during the LIA at all locations at all times. The same is true for the warming of the 20th century in which 1/3 of the measurement stations report that their regions cooled during the 20th century while the other 2/3 warmed (Muller *et al.*, 2011a). There has not been any period over the past 2,000 years in which all regions of the Earth warmed or cooled in lock-step.

The website: *http://co2science.org/data/mwp/mwpp.php* continually seeks new peer-reviewed scientific journal articles pertaining to the MWP, and provides brief summaries of the findings of each paper. The locations of these studies are plotted on an interactive map of the globe. Studies are categorized by the degree to which they support the notion of a MWP as well as by geographical location (Africa, Antarctica, Asia, Australia/New Zealand, Europe, North

America, Northern Hemisphere. Oceans, and South America). An amazingly large number of studies exist. The majority of studies find good evidence for a pronounced MWP although a few studies find no evidence that the MWP was as warm as present day temperatures. Many of these studies also provide evidence on the LIA. In addition, the studies by Idso, Carter and Singer (2011) and Idso (2008) provide extensive evidence in favor of the existence of significant MWP and LIA. Grove (1998) provided 1,000 pages of evidence for the LIA. Grove (2001) asserted that a great amount of information about the LIA can be gleaned from the Swiss Alps where historical records are unusually rich and moraine dating is good. Many ice fronts extended below the current tree line and were in full view of settlements for hundreds of years, or even abutted onto farmland. Written records, paintings, and drawings made by both local observers and visitors are plentiful. Identification of the calendar dates at which many in situ trees were killed by advancing ice, together with their ages at death, has been made possible by multiple dendrochronological analyses.

A number of independent proxy studies show evidence of distinct MWP and LIA. For example, the GISP2 ice core record shows evidence of a MWP and a LIA (Rapp, 2009). Thorsteinsson showed evidence for the MWP and LIA in the Camp Century ice core. Dansgaard (2005) claimed that the MWP and the LIA "were recognizable" and "stand out clearly" in the Camp Century ice core. He also presented the data shown in Figure 2.2. More recently, Vinther et al. (2010) revisited the matter of using Greenland ice cores to infer temperature variations over the past few thousand years. (This paper was partly based upon the Ph. D. dissertation "Greenland and North Atlantic climatic conditions during the Holocene - as seen in high resolution stable isotope data from Greenland ice cores" by Bo Møllesøe Vinther at the University of Copenhagen in 2006.). "These authors worked with 20 ice core records from 14 different sites, all of which stretched at least 200 years back in time, as well as near-surface air temperature data from 13 locations along the southern and western coasts of Greenland that covered approximately the same time interval (1784–2005), plus a similar temperature dataset from northwest Iceland (said by them to be employed —in order to have some data indicative of climate east of the Greenland ice sheet)" (NIPCC, 2011). Vinther et al. (2010) proceeded to demonstrate that "Greenland winter temperatures are much more variable than summer temperatures and thus dominate the annual average variability". So, they utilized winter ice core measurements of $\delta^{18}O$ at three sites on the Greenland ice sheet to examine the variability of climate from year 600 to year 2000. They found that the winter climate was highly variable with rather wild swings and even with 50-year smoothing, the oscillations were large. Nevertheless, temperatures from about years 800 to 1000 were comparable to those of today, and temperatures from about 1400 to the late 19th century were demonstrably lower. This provides further support for the notions of a MWP and a LIA.

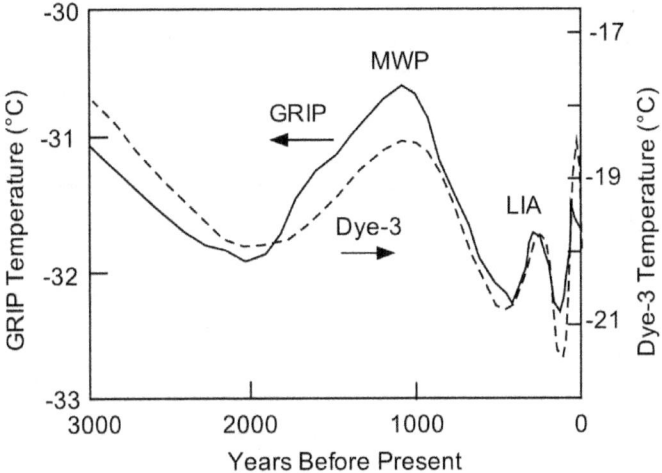

Figure 2.2. Ice core records showing LIA and MWP (Dahl-Jensen *et al.* 1998).

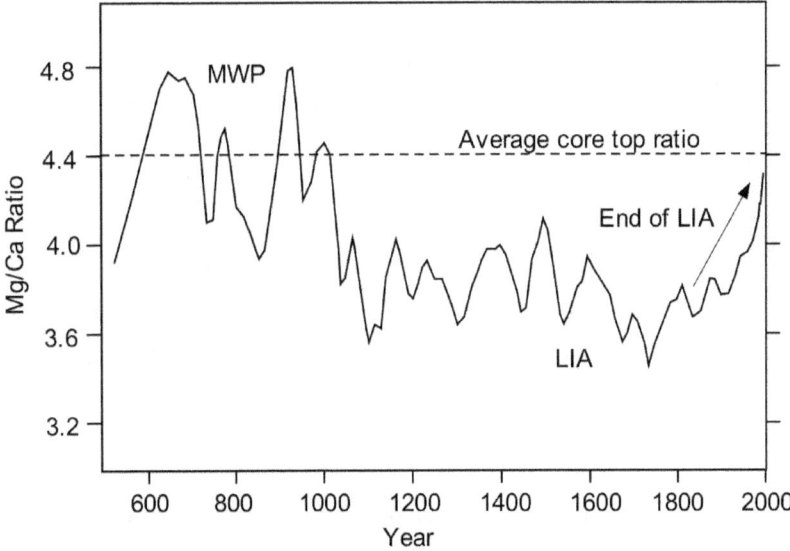

Figure 2.3. Mg/Ca analyses in the white variety of the planktic foraminifera delta, which were obtained from the northern Gulf of Mexico as a measure of sea surface temperatures (Richey *et al.*, 2007).

Yang *et al.* (2009) updated previous results for arid central Asia with new data over the last 2000 years. They reported:

"The most striking features are the existence of the *Medieval Warm Period* (MWP) and the *Little Ice Age* (LIA). The MWP was recorded in the 9–12th centuries and was accompanied by an anomalously dry climate, whereas the LIA extended from the 15–18th centuries and was accompanied by pluvial conditions."

Their result is shown in Figure 2.5.

Richey *et al.* (2007) used Mg/Ca analyses in the white variety of the planktic foraminifera delta, which were obtained from the northern Gulf of Mexico as a measure of historical sea surface temperatures. The results are shown in Figure 2.3.

Loehle (2007) produced a reconstruction of past temperatures that avoided the use of tree ring proxies. His results are shown as Figure 2.4. The MWP and LIA are clearly delineated and the MWP is indicated to be warmer than the present.

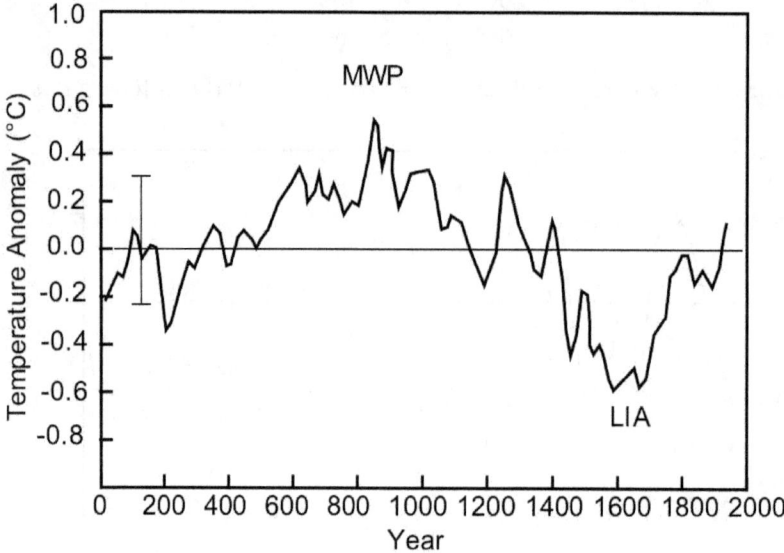

Figure 2.4. Mean temperature anomaly for 18 non-tree ring proxy series (Loehle, 2007).

Esper *et al.* (2012a,b) developed a 2,000-year summer temperature reconstruction based on 587 high-precision maximum latewood density (MXD) series from northern Scandinavia. The record was "developed over three years using living and subfossil pine (Pinus sylvestris) trees from 14 lakes and 3 lakeshore sites at latitudes > 65°N, making it not only longer but also much better replicated than any existing MXD time series". The reconstruction was calibrated against regional June-July-August (JJA) temperature (1876 – 2006) and spanned from the 138 BC to 2006 AD. The calibration curve is shown in Figure 2.4a. It is to the credit of this team that unlike most paleoclimatologists, they provided revealing data on the calibration period.

The resultant estimate of JJA temperatures for the past 2,000 years showed considerable noise from year to year with variations of ±1°C being typical. Using a 100-year filter, they obtained the curve shown in Figure 2.4b. These results indicate that past temperatures during Roman times and during the MWP were higher than the present, and there is a recognizable LIA. Esper *et al.* (2012a,b) chose to fit a straight line to the data, thus showing a fairly constant millennial scale cooling trend. However, this interpretation is affected by the relatively high temperatures that occurred at the starting point in Roman times. Had the starting point been say, 400 AD, the trend would have been oscillatory with a positive lobe from 400 AD to 1200 AD and a negative lobe from 1200 AD to 1900 AD. The interpretation of a fairly constant millennial scale cooling trend might suggest a sudden change in the 20th century due to human influence, whereas the cyclical interpretation might suggest a new positive lobe beginning near 2000 AD (see Figure 2.4c).

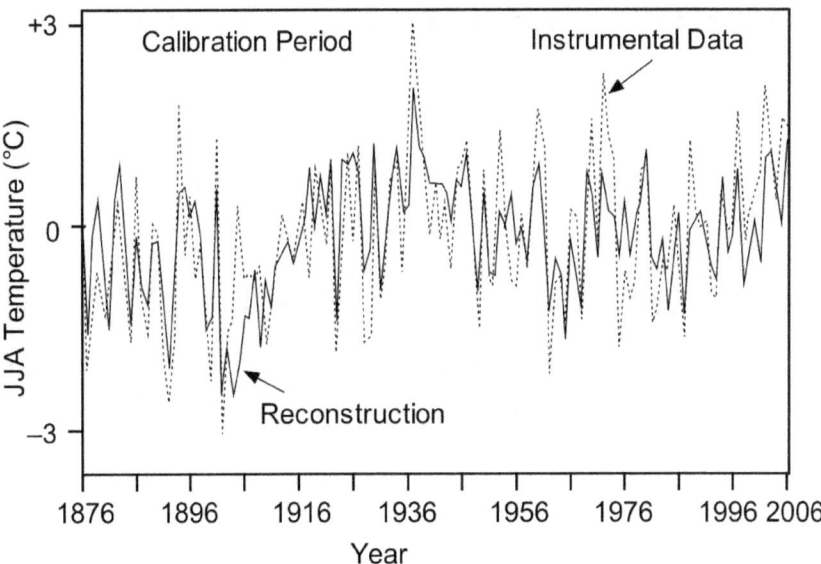

Figure 2.4a. Calibration curve for northern Scandinavian tree ring data (Esper *et al.* (2012a,b)).

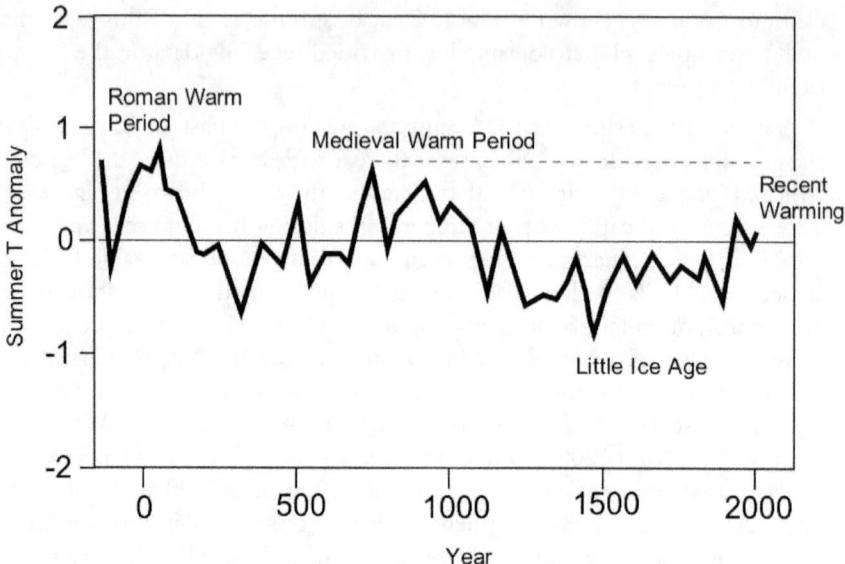

Figure 2.4b. Reconstruction of 2000-year history of Northern Scandinavian temperatures from tree ring data (Esper *et al.* (2012a,b)).

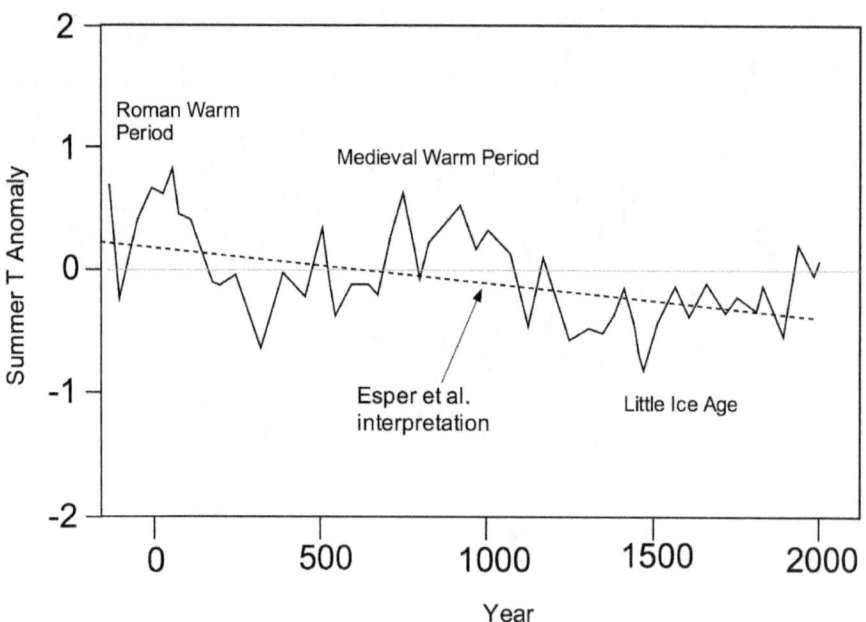

Figure 2.4c. Reconstruction of 2000-year history of Northern Scandinavian temperatures from tree ring data with linear interpretation by Esper *et al.* (2012a,b).

2.3.2 Combining Multiple Proxies for Global Averages

The proxy data shown in Figures 2.2 to 2.5 are for specific regions. A number of studies have attempted to combine such proxy data for many sites and regions into a global average. Since 70% of the Earth is covered by ocean, this necessitates assembling proxies for as many land and ocean sites as possible and developing methods for combining data over different locations with variable time periods.

Each of the proxy data sets has variable geographical distribution. The task is to combine these into a uniform function that best expresses the putative single global average temperature over a long time span. The processes used for data reduction are typically too complex to discuss in detail here. Typical analyses work with variances from the mean ("temperature anomalies"), rather than actual temperatures. The data from any proxy is a table with years in one column and proxy measurements in a second column. The common procedure is to standardize the temperature data by subtracting the mean of the proxy data column from each proxy entry and dividing the result by the standard deviation from the mean. This re-centers the anomalies to a mean of zero and re-scales the data to measure deviations from the mean in units of the standard deviation.

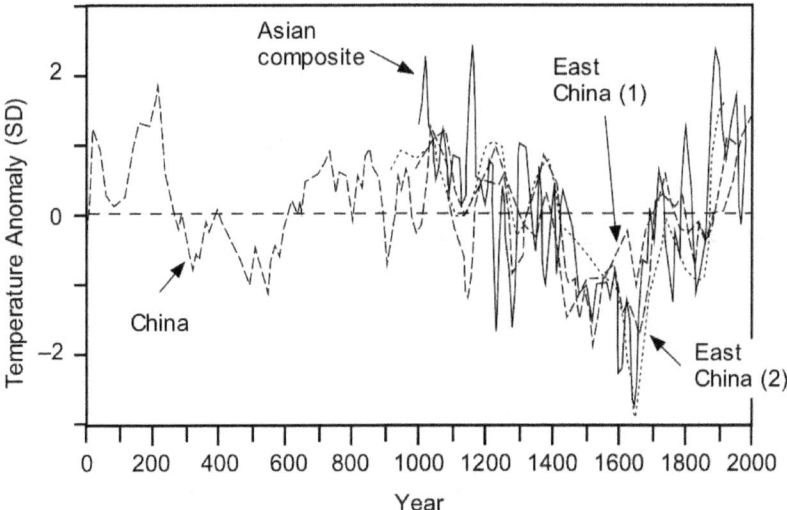

Figure 2.5. Estimates of Historical Temperatures in Asia (adapted from Yang *et al.*, 2009).

Statistical interpretation of large noisy data sets is a specialized niche of mathematics that is beyond the scope of this book. However, it is useful to consider some simplified examples. Figure 2.6 shows hypothetical measurements of variable Y for various values of variable X (the x's). Many measurements are made, but the results show wide scatter. It is possible to produce a so-called "best fit" of a straight line to the data using a "least squares" procedure, resulting in the straight line shown in the figure. This does not prove that the dependence of Y on X is linear. It merely shows that *if* one assumes the dependence is linear, that is the best straight line fit for the data that are available. However, there is considerable uncertainty associated with that estimate of the straight line as evidenced by the gray area in the figure. If more data were taken, it is possible that the best-fit straight line would rotate anywhere within the gray area. One can fit a straight line to noisy data but the uncertainty in assessing this straight line as a good representation of the relationship between Y and X may be significant.

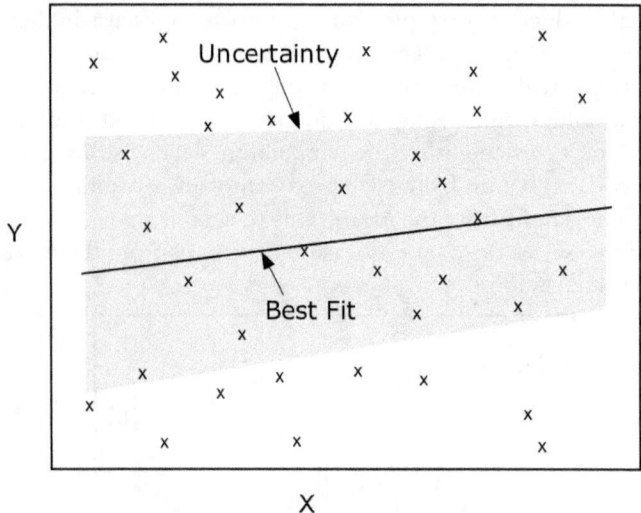

Figure 2.6. Cartoon showing hypothetical dependence of variable Y on variable X for many measurements (x's).

Figure 2.7 shows an oversimplified rendition in which hypothetical curves of temperature anomaly vs. year are assembled, and from this, a "best" approximation to the dependence of average temperature on year can be derived ("trend") by a statistical analysis technique. The gray area represents the uncertainty in the trend. The uncertainty in this example is very large.

The simplest thing to do would be to treat all proxies equally and simply take an average of the whole set. As we discussed previously, the hope would be that each proxy contains a signal (T_S) reflecting the true long-term trend of temperature plus noise (T_N) reflecting short-term weather effects. In adding up many proxies, one could hope that the signals would reinforce one another, while the noise would sometimes be positive and sometimes negative, ultimately cancelling out. This would improve the signal-to-noise ratio. Unfortunately, the signals also seem to vary from positive to negative and it is difficult to arrive at anything other than a meandering path centered on zero for the plot of temperature anomaly vs. year from such a procedure.

A widely used statistical procedure for deriving trends from noisy data is *principal components analysis* (PCA). PCA is typically used to analyze a noisy data set in which the trend may not be apparent due to excessive scatter. Visual observation of the data may suggest that there is no correlation between variables (in our case temperature anomaly vs. year).

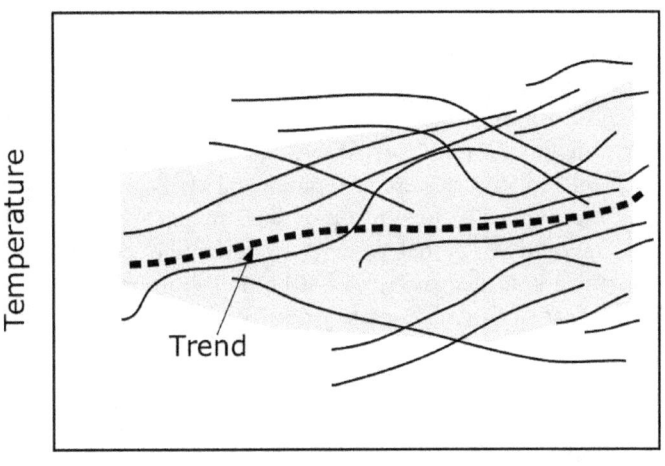

Year

Figure 2.7. Cartoon showing hypothetical dependence of temperature on date range for many measurements.

Instead, sophisticated analysts of long-term climate data have utilized some form of PCA procedure. PCA applies weights to the various component proxies in proportion to how much trend is contained in each proxy. The purpose is to focus on those proxies that generate the greatest amount of trend. McKitrick (2005) summarized PCA as follows:

"Principal components analysis involves replacing a group of data series with a weighted average of those series, where the weights are chosen so that the new vector (called the principal component or PC) explains as much of the variance of the original series as possible. This leaves a matrix of unexplained residuals, but this matrix can be reduced to a PC as well. In that case the original PC is called the first PC (PC1), and the PC of the residuals is called the second PC, or PC2. And there will be residuals from it too, yielding PC3, PC4, etc. The higher the number of the PC, the less important is the pattern it explains in the original data. PC1 is the dominant pattern, PC2 is the secondary pattern, etc. In many cases a large number of data series can be summarized with relatively few PCs."

2.3.3 Proxy Analysis by Mann, Bradley, and Hughes – and Others

Mann, Bradley, and Hughes (1998, 1999) attempted a comprehensive analysis of the history of global average temperatures using PCA to analyze a multi-proxy network consisting of "widely distributed high-quality annual-resolution proxy climate indicators, individually collected and formerly analyzed by many paleoclimatic researchers." The network included annual resolution dendro-climatic, ice core, ice melt, and long historical records previously assembled, combined with other long instrumental records. This was intended to integrate as many proxy sources as possible into a single comprehensive view of how a single global average temperature (or Northern Hemisphere [NH] average temperature) varied over the past millennium. A number of subsequent related studies were also published by the same group, as well as other collegial groups. The final result of such an analysis is typically a reconstruction of a single NH or global average temperature over the past one or two millennia with a so-called *hockey stick* structure: a rather flat meandering profile for most of the millennium prior to the 20th century, with a significant rise in the 20th century.

The paper by Mann, Bradley, and Hughes (1998) is sometimes referred to as "MBH" after the names of the three authors of the principal paper. Subsequently, Mann, Bradley, and Hughes (1999) extended the period of analysis from 1400 back to 1000, and Mann and Jones (2003) added an additional millennium back to year 200. Mann *et al.* (2008) updated previous results. There are also a number of other relevant papers.

The paper by Mann, Bradley, and Hughes (1998) is compact, full of jargon, and difficult to follow. However, this is a characteristic shared by many papers that deal with large data sets for historic Earth temperatures. Wegman, Scott, and Said (2006) said:

> "The papers of Mann *et al.* in themselves are written in a confusing manner, making it difficult for the reader to discern the actual methodology and what uncertainty is actually associated with these reconstructions. Vague terms such as 'moderate certainty' give no guidance to the reader as to how such conclusions should be weighed. While the works do have supplementary websites, they rely heavily on the reader's ability to piece together the work and methodology from raw data. This is especially unsettling when the findings of these works are said to have global impact, yet only a small population could truly understand them."

Wegman, Scott, and Said (2006) also said: "The description of the work in Mann, Bradley, and Hughes (1998) is both somewhat obscure and as others have noted, incomplete."

The reference period for calibration of proxies with actual temperature data was 1902 to 1980. The various proxies tended to become more numerous in recent times and less numerous in the more distant past. The number of proxies vs. earliest date is shown in Table 2.1.

Table 2.1. Number of proxies vs. earliest date according to Mann, Bradley, and Hughes (1998, 1999).

Earliest date	Number of proxies
1000	12
1400	22
1450	24
1600	57
1700	74
1763	93
1820	112
1854	219
1902	1,082

The final result from Mann, Bradley, and Hughes (1998) is shown in Figure 2.8. Note that the mean is the mean for the calibration period 1902–1980, and therefore most of the data (1400–1920) lie below the mean. We will have more to say about this in the following sections. This figure is the first rendition of a series of so called *hockey stick* figures published in subsequent papers with a relatively flat profile prior to 1900 and a sudden rise after 1900.

The final result from Mann, Bradley, and Hughes (1999) is shown in Figure 2.9. This extended their 1998 model back to year 1000. Note that the "X" at the far right of this figure, which is meant to be the current temperature, is 1.1°C higher than the 1895 temperature, whereas it is widely believed that this temperature differential is more like 0.7°C. This exaggerated the shape of the hockey stick.

Mann and Jones (2003) extended the work of Mann, Bradley, and Hughes (1998, 1999) back to year 200. Their result is similar to that shown in Figure 2.9 with the addition of essentially no change in temperature from year 200 to year 1000. Taken at face value, these figures would suggest: (1) there was no *Medieval Warm Period*, (2) there was a very minor *Little Ice Age*, (3) Earth temperatures have been remarkably stable for 2,000 years, and (4) the only significant change in Earth temperature took place in the 20th century with a sudden and decisive sharp rise after 1900. However, MBH chose the mean for the calibration period (1902–1980) rather than the mean for the entire data set. As we shall see, this had major repercussions regarding the form and credibility of the result.

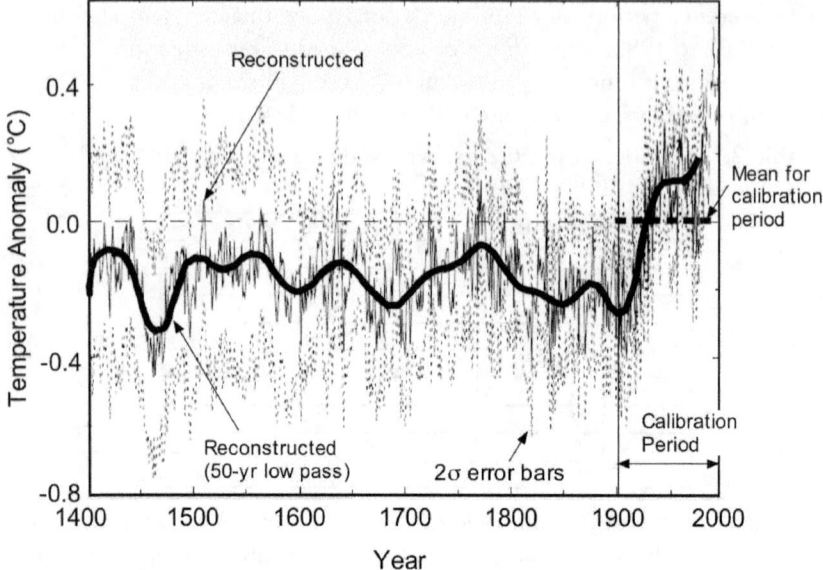

Figure 2.8. Reconstructed temperatures since 1400 (Mann, Bradley, and Hughes, 1998). Note that the mean is for 1902–1980. Also note that the 2σ error bars are so wide that they could hide almost any imaginable temperature curve.

A number of other groups published reconstructions of historical temperatures using similar methods. For example, Jones, Osborn, and Briffa (2001) obtained similar results using similar data processing schemes. It is particularly noteworthy that Jones, Osborn, and Briffa (2001) decided not to show some of the proxy data late in the 20th century because it ticked sharply downward and conflicted with the desire to emphasize recent global warming (as we will discuss in a later section). Esper *et al.* (2005) discussed differences between various reconstructions based primarily on tree rings and presented a comparison. There is considerable variation in amplitude of the MWP and the LIA from study to study.

Moberg *et al.* (2005) indicated that although differences in the amplitude of centennial temperature variability have been discussed in the literature, the picture with relatively small variability prior to the 20th century (i.e., the *hockey stick*) "is arguably best known by a wider audience. One reason for this is the prominent role that the multi-proxy reconstruction by MBH had in the latest IPCC report and in public media." However, they went on to point out that recent findings suggest that considerable underestimation of centennial Northern Hemisphere temperature variability may result when regression-based methods (like those used by MBH) are applied to noisy proxy data with insufficient spatial representation. Moberg *et al.* (2005) also referred to well-documented difficulties in reliably reproducing multi-centennial temperature variability based on tree ring proxies. Note the emphasis on data that "are best known" due to promotion by the IPCC and the media.

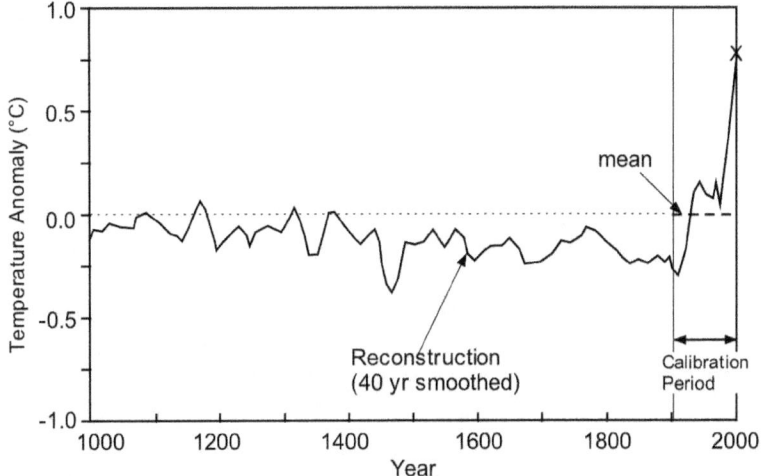

Figure 2.9. Temperature anomaly vs. year since AD 1000. Adapted from Mann, Bradley, and Hughes (1999). The X at the far right is their estimate for 1998. The measured temperature rise from 1895 to 2008 is around 0.7°C, whereas Point X suggests a rise of 1.1°C. Note that the mean is for 1902–1980.

von Storch *et al.* (2004) used a coupled atmosphere–ocean model simulation of the past 1,000 years to test empirical reconstructions of historical temperatures, specifically those of MBH. They found that centennial variability of the NH temperature is underestimated by the MBH regression-based methods. Their results also suggest that actual centennial variability may have been at least twice as large as the variability obtained in the MBH studies. Juckes *et al.* (2006, 2007) provided an extensive survey of a number of recent temperature reconstructions based on proxies. Juckes *et al.* (2006) presented a number of graphs of reconstructions of historical temperatures. However, all of these were based on MBH-type models in which the mean was chosen only for the calibration period (20th century) and as a result, almost all of the temperature data for the past millennium (except for the 20th century) lie below the mean.

2.3.4 Fallacies in Reconstruction of Millennial Temperatures

Generally, the published papers on reconstruction of millennial temperatures tend to be very terse and full of jargon. The MBH papers are particularly bad in this respect. These papers present their results in small graphs with poor resolution but provide little insight into the calibration periods of specific proxies. Comparisons of proxies with temperatures during the calibration period are rarely provided and probably for good reason; the comparison is likely to be poor.

The only way to really understand what was done is to go back to their original data and follow the original procedures. As more and more of these reconstructions appeared in the literature with their typical hockey stick results, McKitrick (2005) and McIntyre and McKitrick (2005, 2006, 2007) took it upon themselves to review these reconstructions by working with the original data in detail. These original studies by McIntyre and McKitrick continue to this day in the form of sporadic entries on the blog: *climateaudit.org*. The first obstacle they ran into was obtaining the data from the authors. Publications in journals are highly compressed and do not provide adequate means for others to reproduce the claimed results of the paper. Some journals require that authors archive the detailed data for access by the public but this is rarely enforced. Sensing that McIntyre and McKitrick (M&M) were antithetical to the hockey stick results, authors of papers on reconstruction of millennial temperatures resisted providing M&M with data and script from their work. Evidently, they were defensive about their work and did not cooperate in allowing their work to be checked. When M&M utilized the *Freedom of Information Act* (FOIA) in an attempt to obtain data generated by government-funded work in the U.S. and England, the authors of papers enlisted help from politicians to pervert and circumvent the FOIA on specious grounds.

2.3.4.1 The Fallacy of Choosing the Wrong Mean

After much perseverance, M&M succeeded to a considerable degree in penetrating the MBH data and procedure. In the course of doing this, they uncovered several major errors in the MBH approach. The principal problem was summarized by McKitrick (2005). A paraphrased rendition of some of his remarks is given in the next paragraph.

In a conventional PCA, the temperature data are standardized by subtracting the mean of entire data set and dividing by the standard deviation of the entire data set. This re-centers and re-scales all the data to a mean of 0 and measures deviations from the mean in units of the standard deviation. In the MBH program, a scaling was applied, but rather than subtract the mean of the entire data set over all years, they subtracted the mean of the 20th-century portion used for calibration, and then divided by the standard error of the 20th century portion. While this may appear at first glance as innocuous, it has important consequences for the results derived from this procedure. The overwhelming majority of individual proxy series do not have the form of hockey sticks, but appear as random noise, and since they don't change much in the 20th century, this procedure did not make much difference for them. For these proxies, the mean of the calibration period is roughly the same as the mean of the whole series (as is the standard error) so either way of standardizing yields more or less the same result. But a few of the proxy series trend upward in the 20th century. For these, the MBH method has a huge effect. Since the mean of the 20th century portion is higher than the mean of the whole series, subtracting the 20th-century mean de-centers the series, shifting it off a zero mean. This, in turn, inflates the deviations from the mean of these series with increases in the 20th century. PCA algorithms inflate the weights of proxies with the highest deviations. If one proxy series in the group has a relatively high level of deviation from the mean, its weight in the PC1 gets inflated. The MBH algorithm did just this. The PCA procedure would, in effect, sift through a data set and identify series with a 20th-century up-trend, and then load almost all the weight onto these series. In effect, it data-mines for hockey stick trends.

Consider the hypothetical set of proxies shown in Figure 2.10. If PCA based on the mean for the calibration period is used for the flat proxies, no problem arises. However, when the mean for the calibration period is used for the proxy with an uptrend in the calibration period, the majority of deviations from the mean (*b*) will be much greater than deviations calculated from the mean for the entire data set (*c*). The point here is that for the hypothetical set of proxies shown in Figure 2.10, PCA based on the mean for the calibration period will produce a "trend" similar to the one proxy with a slope and will essentially zero out the contribution of all the horizontal proxies to the estimated trend. Yet the preponderance of evidence is that the trend representing the overwhelming majority of data is actually horizontal. In cases such as that shown in Figure 2.10, PCA emphasizes the proxy with the greatest trend but it is not representative at all of the whole data set.

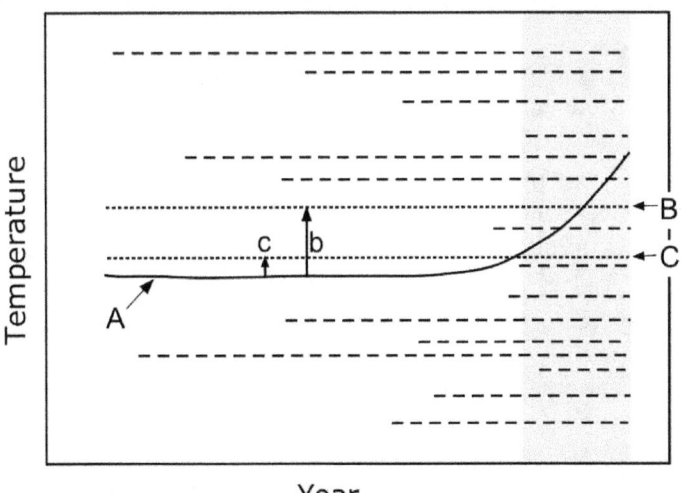

Figure 2.10. Hypothetical set of proxies where all proxies except one are flat and only one rises. The calibration period is shown as the gray rectangle.

Flat proxies are shown as horizontal dashed lines. The one proxy (*A*) with variance is shown as a solid curve. Dotted line (*B*) is the mean for proxy (*A*) over the calibration period, while dotted line (*C*) is the mean for proxy (*A*) over the entire series. Deviations of proxy (*A*) from the two means are shown as (*b*) and (*c*).

Figure 2.11 provides an example of the data-mining effect. It shows 2 of the 90 full-length series in the MBH database. The top panel is a tree ring chronology from a stand of bristlecone pines at Sheep Mountain, California. The bottom panel is a tree ring chronology from Mayberry Slough, Arkansas. In the bottom panel, the mean over the last 80 years is roughly equal to the mean for the previous 500 years, but in the top panel the post-1900 mean is above that for the pre-1900 portion. The MBH algorithm attributes 390 times as much weight to the top series as it does to the bottom series in the first principal component (PC1).

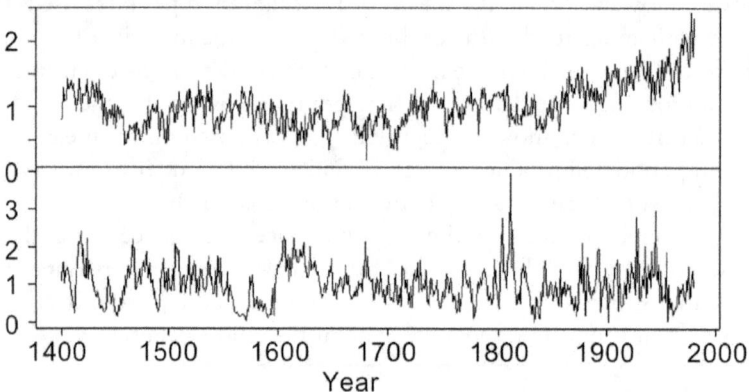

Figure 2.11. Two tree ring temperature anomaly series from the MBH data set. Top: Sheep Mountain, CA. Bottom: Mayberry Slough, Arkansas (*climateaudit.org*).

As it turns out, of 1,082 proxies used by MBH, only a handful exhibit the form shown in the upper panel of Figure 2.11, and all of these probably suffer from the potential CO_2 fertilization problem in the 20th century.

To test the MBH data-mining algorithm, M&M ran an experiment in which they input only trendless random red noise, simulating the data one would obtain from trees in a climate that is only subject to random fluctuations with no warming trend. In 10,000 repetitions, they found that a conventional PC algorithm (using the mean for the entire data set) almost never yielded a hockey stick–shaped PC1, but the MBH algorithm using the mean for only the calibration period yielded a pronounced hockey stick–shaped PC1 more than 99% of the time. The MBH algorithm efficiently looks for those kinds of series and flags them for maximum weighting. It concludes that a hockey stick is the dominant pattern even when pure noise is the input!

M&M extended their study in two ways. First, they showed that the MBH data-mining procedure did not just pull out a random group of proxies—it pulled out an eccentric group of bristlecone pine chronologies. These trees (the Sheep Mountain series in Figure 2.11 is an example) all turned out to exhibit a 20th-century growth spurt that has not been fully explained, but is likely to be at least partly due to CO_2 fertilization and is known not to be a temperature signal since it does not match nearby temperature records. The original authors (and others) have stressed that these series do not constitute proper climate proxies. So, M&M examined the consequences to the MBH results if these 20 bristlecone pine proxies were excluded. The result showed no hockey stick at all. Without these proxies with their rising shapes in the 20th century to mine for, the MBH method generates a result just like that from a conventional PC algorithm, and shows the dominant pattern is not hockey stick– shaped at all. In other words, without the bristlecone pines, even the flawed procedure of MBH would not have had a hockey stick shape.

Since the MBH papers purported to be the ultimate reference for estimation of the Earth's climate for one or two millennia, these results are likely to play a pivotal role in influencing the understanding of climatology. As such, the data and methodologies used by MBH should be readily available for replication and evaluation by others. McKitrick (2005a) provided an informal review of the way that he and McIntyre became interested in the Mann hockey stick, and their efforts to replicate the MBH results in order to understand the basis for their findings. In his discourse, McKitrick (2005a) described the difficulties in obtaining the required information, and the obstructive attitude of the MBH team. It became evident that the MBH team had manipulated the data unwittingly to greatly amplify the weight assigned to the few proxies with hockey stick form. It also became evident that the peer review process did not penetrate into the MBH papers, and probably operates at a rather superficial level in most cases. M&M were thwarted by the journal *Nature* in their attempts to publish their criticisms and *Nature* crassly and cynically bowed to pressure from the *paleoclimatic cabal* and allowed the misleading publications of MBH to stand unchallenged. Montford (2010) provides a detailed chronology of the efforts made by M&M to obtain data from the MBH team and others. Part of the problem here was that M&M were not professors of climatology with extensive publication records, so how could they have the expertise to find errors in the MBH papers? Thus the arguments became a matter of who you were, not the validity of what you had to say.

In a more recent paper by the paleoclimatic cabal, Mann *et al.* (2008) updated their previous work by including additional proxies of various types. The spatial distribution of these was heavily concentrated in the U.S. and Europe (about 85%) with very few in the rest of the world (about 15%). As is usual in papers authored by Mann and co-workers, the paper is difficult to decipher. Oceans, which cover 70% of the Earth, were claimed to be included in some of the studies but it is not clear how ocean temperatures from a thousand years ago were obtained and averaged over all the oceans—if indeed that is what was done. It is difficult to understand how they incorporated ocean data such as it may be from the terse description given in the paper. Of the 1,209 proxies utilized, 59 extended back 1,000 years and 25 extended back 2,000 years from the present. The mean duration of a proxy was about 270 years. Some of their reconstructions utilized all proxies, and some were restricted to a subset of proxies that passed "a screening process for a local surface temperature signal. The screening process required a statistically significant correlation with local instrumental surface temperature data during the calibration interval." The period 1850–1995 was used for calibration.

As Rapp (2010) showed in his Figures 2.5, 2.8, 2.16, and 2.19, when a typical set of proxies from various regions is compared, the differences between proxies are huge compared with the similarities (assuming that similarities exist at all). The proxies used by Mann *et al.* (2008) were no exception. Figure 2.12 shows some of the proxies that they utilized. Evidently, the variations from proxy to proxy outweigh any consistent signal that may underlie these time series. Hence this set of time series represents a dataset with low signal-to-noise ratio, and simply adding up these proxies is bound to produce little more than noise. Nevertheless, Mann *et al.* (2008) remained undaunted. They applied a variety of sophisticated statistical methods in an attempt to unravel a signal from the noise. However, as the saying goes, it is difficult to "convert a sow's ear into a silk purse", or in a more modern vernacular, it may be a case of "garbage in–garbage out." As Burger and Cubasch (2005) showed, almost any desired result can be obtained from interpreting this very noisy data, depending on how it is processed.

In some of the results of Mann *et al.* (2008), they provide estimates of the uncertainties in their results as a standard deviation envelope around the linear curve of temperature vs. time. In their Figure S11, they indicate that when alternate calibration periods are included, the total standard deviation of a temperature estimate for any year is typically about ±0.6°C, which suggests that any variations smaller than 0.6°C are uncertain by more than 100%. It must be concluded that when an uncertainty of at least ±0.6°C is imposed on the estimates by MBH and others, the resultant envelope would be large enough to hide almost any historical temperature profile that one desires.

It is noteworthy that the temperature anomalies shown in Figures 2.8 and 2.9 are essentially all negative prior to about 1950, which proves that the mean that they used to calculate anomalies was based on the calibration period, not the full data set (temperatures were higher on average during the calibration period). This casts doubt on the entire statistical procedure. In this connection, no mention was made of the criticisms of the methodology made by M&M, Wegman and others, nor were any of their concerns addressed. Like the Wizard of Oz when exposed, they seemed to say: "pay no attention to that man behind the screen".

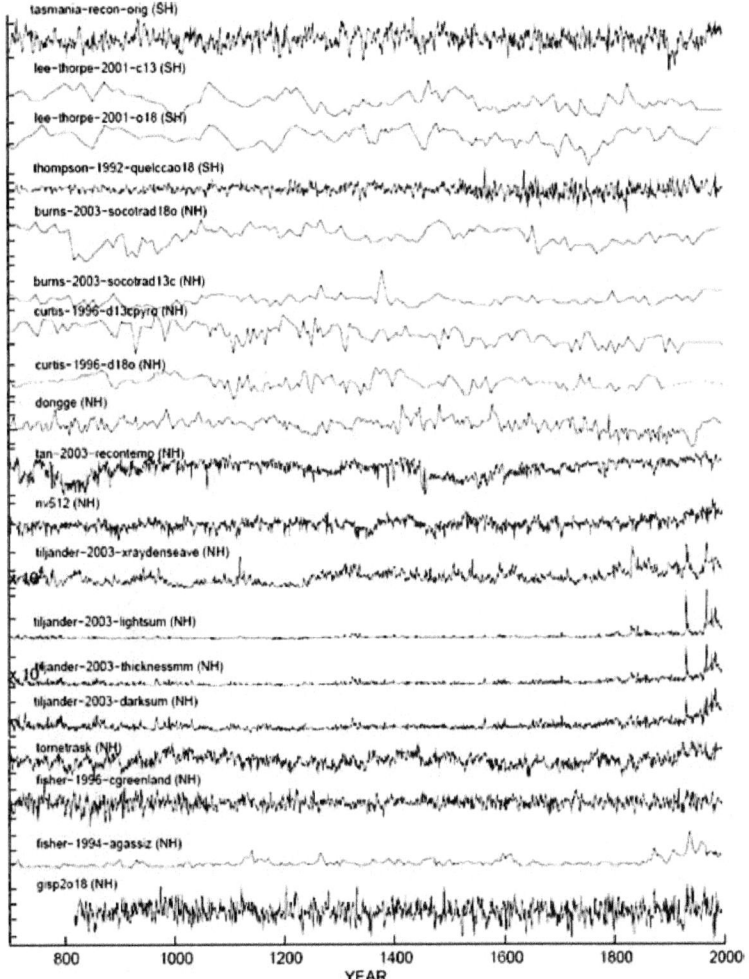

Figure 2.12. Some of the proxies used by Mann *et al.* (2008).

Clearly, the results of MBH and related papers are faulty. Juckes *et al.* (2006) attempted to deal with the criticisms of McIntyre and McKitrick (2003) claiming: "... the deficiencies in the description of the data used and possible irregularities in the data itself. These issues have been largely resolved in [Mann, Bradley, and Hughes (2004)]." However, aside from deficiencies in the data, which Mann, Bradley, and Hughes (2004) did not resolve except to obfuscate the matter, the critical issue of using the wrong mean, resulting in mining for *hockey stick* results was not even mentioned. Juckes *et al.* (2006) is just another in a series of papers by the *paleoclimatic cabal* using flawed statistics in a feeble effort to provide a pseudo-basis for their contention that we are in a state of unprecedented runaway global warming. There is no great point in reproducing their graphs here since they all display *hockey stick*s of one form or another.

It is noteworthy that Richard Muller said:

"... carbon dioxide from burning of fossil fuels will prove to be the greatest pollutant of human history. It is likely to have severe and detrimental effects on global climate. I would love to believe that the results of Mann *et al.* are correct, and that the last few years have been the warmest in a millennium". (*http://muller.lbl.gov/TRessays/32-Global_Warming_Bombshell.htm*)

Hence, he was far from being a confirmed skeptic. Nevertheless, he went on to roundly criticize the methods of Mann *et al.* and concluded that the hockey stick is actually an "artifact of poor mathematics" and said: "A phony hockey stick is more dangerous than a broken one--if we know it is broken".

It is also worthwhile to review the response of *Nature* magazine to McIntyre and McKitrick when they attempted to publish their critique of Mann *et al.* (1998). On January 2004, M&M submitted their critique as a Letter to *Nature*. One Referee provided a favorable review. The other offered some confusion emphasizing the complexity of the details, but said: "In general terms I found the criticisms raised by McIntyre and McKitrick worthy of being taken seriously. They have made an in depth analysis of the MBH reconstructions and they have found several technical errors that are only partially addressed in the reply by Mann *et al.*" *Nature* issued a "favorable revise and resubmit" to which M&M responded in March 2004 with a revised manuscript. *Nature* then asked M&M to reduce the manuscript to 800 words. This was difficult, but was achieved and reduced manuscript was submitted in April 2004. In August 2004, *Nature* declined to publish the article that now (for reasons unexplained) needed to be reduced to 500 words. The main reason given was that the matters involved were "too technical" for a science journal. In other words, in a matter concerning the legitimacy and validity of the most widely accepted model of the Earth's climate over the past millennium, *Nature* decided that they would allow the erroneous publication by Mann *et al.* to stand unchallenged because the issues involved were too complicated. The ironic thing was that the *paleoclimatic cabal* could then claim that the criticisms of M&M could not be taken seriously because they were not published in a peer-reviewed journal.

A team led by Professor Edward J. Wegman performed an independent examination of the *hockey stick* controversy (Wegman, Scott, and Said, 2006). They produced a lengthy report, full of details that echoed the conclusions of M&M.

According to Wegman, Scott, and Said (2006):

"The controversy of Mann's methods lies in that the proxies are centered on the mean of the period 1902–1995, rather than on the whole time period ...Principal component methods are normally structured so that each of the proxy data series are centered on their respective means and appropriately scaled. The first principal component attempts to discover the composite series that explains the maximum amount of variance. The second principal component is another composite series that is uncorrelated with the first and that seeks to explain as much of the remaining variance as possible. The third, fourth, and so on follow in a similar way. In the MBH approach the authors make a simple seemingly innocuous and somewhat obscure calibration assumption. Because the instrumental temperature records are only available for a limited window, they use instrumental temperature data from 1902–1995 to calibrate the proxy data set. This would seem reasonable except for the fact that temperatures were rising during this period, so that centering on this period has the effect of making the mean value for any proxy series exhibiting the same increasing trend to be de-centered low. Because the proxy series exhibiting the rising trend are de-centered, their calculated variance will be larger than their normal variance when calculated based on centered data, and hence they will tend to be selected preferentially as the first principal component. Thus, in effect, any proxy series that exhibits a rising trend in the calibration period will be preferentially added to the first principal component."

Wegman, Scott, and Said (2006) went on to say:

"The centering of the proxy series is a critical factor in using principal components methodology properly. It is not clear that the MBH Team even realized that their methodology was faulty at the time of writing the MBH paper. The net effect of the de-centering is to preferentially choose the so-called *hockey stick* shapes. While this error would have been less critical had the paper been overlooked like many academic papers, the fact that their paper fit some policy agendas has greatly enhanced their paper's visibility. Specifically, global warming and its potentially negative consequences have been central concerns of both governments and individuals. The *hockey stick* reconstruction of temperature graphic dramatically illustrated the global warming issue and was adopted by the IPCC and many governments as the poster graphic. The graphic's prominence together with the fact that it is based on incorrect use of PCA puts Dr. Mann and his co-authors in a difficult face-saving position. We have been to Michael Mann's University of Virginia website and downloaded the materials there. Unfortunately, we did not find adequate material to reproduce the MBH materials."

Wegman, Scott, and Said (2006) performed a calculation similar to that of M&M by comparing the results of an analysis of the North American tree network PC1 using the MBH data with centering based either on the calibration period mean or the mean for the whole time span of the data set. The result is shown in Figure 2.13. In addition to the *hockey stick* shape of the upper panel it is worth noting that the lower panel exhibits considerably more variability. PCA seeks to identify the largest contributor to the variance. The MBH offset of the mean value shifts the main variance from the bulk of the data set to the 20th century data.

The findings of Wegman, Scott, and Said (2006) are quite lengthy and only a very brief summary is given here.

1. In general they found the writing of MBH somewhat obscure and incomplete. (This writer found the same.)
2. In general, they found the criticisms by M&M to be valid and their arguments to be compelling.
3. Use of the temperature profile in the 1902–1995 time span for centering leads to misuse of the principal components analysis. However, the narrative in MBH on the surface sounds entirely reasonable, and could easily be missed by someone who is not extensively trained in statistical methodology.
4. The cryptic nature of some of the MBH narratives requires that outsiders would have to make guesses at the precise nature of the procedures being used.
5. Much of the discussion on the *hockey stick* issue has taken place on competing web blogs. Web blogs are not an appropriate way to conduct science and thus the blogs give credence to the fact that these global warming issues have migrated from the realm of rational scientific discourse. Unfortunately, the factions involved have become highly and passionately polarized.
6. Generally speaking, the paleo-climatology community has not recognized the validity of the M&M papers and has tended to dismiss their results as being developed by biased amateurs. The paleo-climatology community seems to be tightly coupled, and has rallied around the MBH position.
7. The widely quoted assessments that the decade of the 1990s was the hottest decade in a millennium and that 1998 was the hottest year in a millennium cannot be supported by a proper rendition of the MBH analysis …The paucity of data in the more remote past makes the hottest-in-a-millennium claims essentially unverifiable.

8. Use of bristlecone pine proxies are inappropriate because they were probably CO_2-fertilized. It is not surprising therefore that this important proxy in MBH yields a temperature curve that is highly correlated with atmospheric CO_2. There are clearly confounding factors for using tree rings as temperature signals.

9. There are other detailed statistical problems with the MBH treatment that require specialized knowledge to understand.

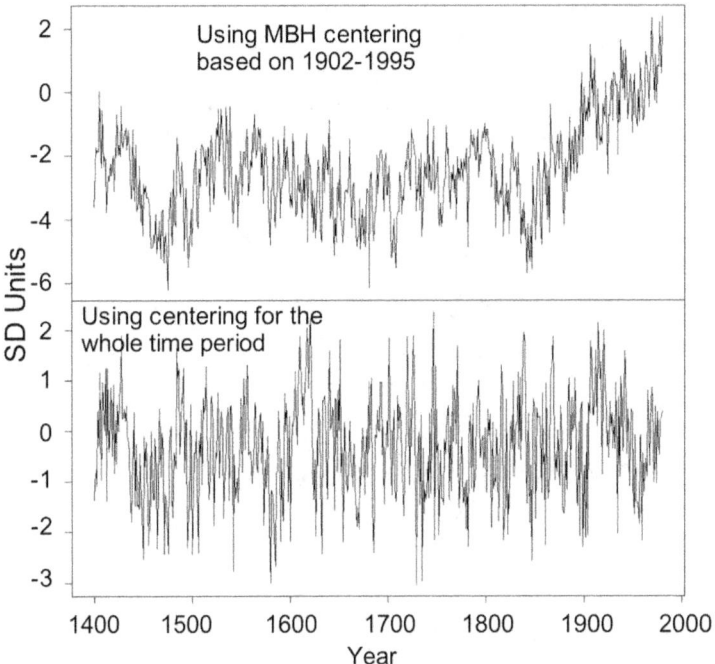

Figure 2.13. Comparison of rework of the North American tree network PC1 using MBH centering vs. using centering across the whole time span of the data set. The *hockey stick* is shown to be an artifice. Adapted from Wegman, Scott, and Said (2006).

Crowley and Lowery (2000) argued that anecdotal reports as well as studies of individual records from MWP suggest that the present warmth of the 20th century is not unusual and therefore cannot be taken as an indication of forced climate change from greenhouse gas emissions. But Crowley and Lowery then asked the question: "Were all of these changes synchronous, with hemispheric amplitudes comparable to or warmer than present?" However, this question seems to imply that present warming is spatially universal and synchronous—which it is not. In fact, it has recently been shown that over the past 200 years, one-third of the land measurement stations showed a decrease in temperature while the Earth was warming. Crowley and Lowery revisited the controversy regarding the existence of the putative MWP by carrying out another proxy analysis, incorporating additional proxy time series not used in previous hemispheric compilations. The 15 proxies used in the study are shown in Figure 2.14.

It is to the credit of Crowley and Lowery that the individual proxies are shown; this is not often the case when assemblages of multiple proxies are analyzed. However, a glance at these fifteen suggests that that represent uncoordinated noise, more than signal. They combined the various proxies to obtain Figure 2.15. However, the process used to combine proxies is not clear to this writer. Furthermore, Crowley and Lowery found that the correlation of proxies with temperature during the calibration period broke down for the middle range of years: 1880-1920 so they only calibrated from 1860 to 1880 and from 1920 to 1965, leaving out a 40-year period in the middle.

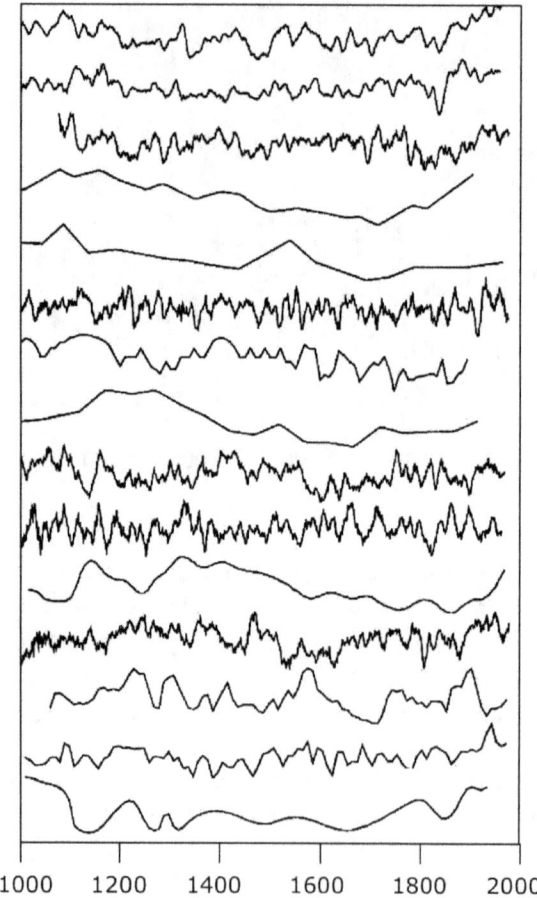

Figure 2.14. Fifteen individual proxies from various locations. Adapted from Crowley and Lowery (2000). Vertical scale is temperature anomaly.

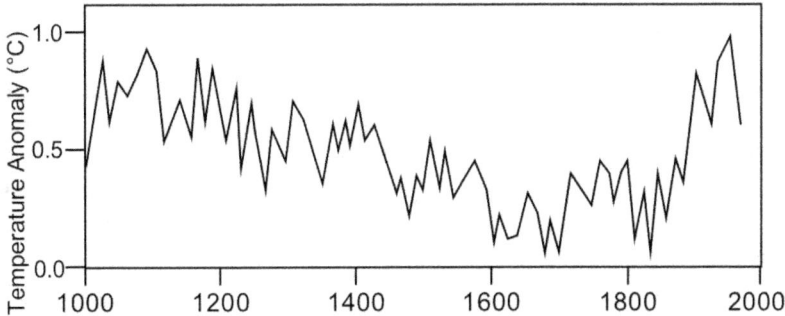

Figure 2.15. Derived NH temperature anomalies. Adapted from Crowley and Lowery (2000).

Based on this result, Crowley and Lowery (2000) reached the rather incredible conclusions:

"Despite clear evidence for medieval warmth greater than present in some individual records, the new hemispheric composite supports the principal conclusion of earlier hemispheric reconstructions and, furthermore, indicates that maximum medieval warmth was restricted to two–three 20–30 year intervals, with composite values during these times being only comparable to the mid-20th century warm time interval. Failure to substantiate hemispheric warmth greater than the present consistently occurs in composites because there are significant offsets in timing of warmth in different regions; ignoring these offsets can lead to serious errors concerning inferences about the magnitude of Medieval warmth and its relevance to interpretation of late 20th century warming."

"Because of uncertainties in the proxy-instrumental temperature calibration, it is still difficult to unequivocally [sic] assert that the late 20th century warming is significantly greater than the peak warmth of the *Medieval Warm Period*. But there is even less justification to assert the opposite—it is not possible to make a robust statement that the *Medieval Warm Period* was warmer than the last two decades."

These conclusions seem far-fetched to this writer because:

1. The huge variation from proxy to proxy suggests that these divergences may not represent true differences in climate from place to place, but rather, noise and error in the proxies themselves.
2. Taking averages of highly divergent individual proxies must tend to average out variations, and add uncertainty and large error bars to the resultant average.
3. The belief that the averages shown in Figure 2.15 can be trusted to the extent that differences between 20-year periods can be affirmed

seems unwarranted by the lack of precision and consistency of the underlying data.

4. The insistence that the MWP must perforce involve uniformly high temperatures at all locations for the entire period is unreasonable. The issue is not whether such uniform warmth occurred, but rather, allowing for spatial and temporal variations, the preponderance of evidence favors relative warmth compared with other eras. Furthermore, even the current warming is far from uniform, spatially and temporally.

McIntyre (2007) examined the results of Crowley and Lowery (2000) in some detail. McIntyre prepared a new figure, similar to Figure 2.15, except that it showed contributions from each of the 15 individual proxies with color-coding. McIntyre pointed out:

"Although Crowley and Lowery (2000) argued [based on Figure 2.14] that there is relatively little synchroneity [sic] between proxies from different regions (and thus no MWP or LIA), the color-coded graphic of their actual data could also be construed as showing a certain amount of coherence between the proxies."

McIntyre went on to say:

"A distinctive 'hockey-stick' shape can be discerned in the 4 lowest records. Indeed, whatever '*hockey stick*iness' exists in Crowley and Lowery (2000) is entirely due to these 4 series, which consist of 2 bristlecone pine series, Briffa's Polar Urals series and Thompson's Dunde series ... The bristlecone pine series are prominent in the MBH99 reconstruction and the Polar Urals series in the Jones *et al.* (1998) series. Both series have problems (as discussed elsewhere in McIntyre, 2007)."

McIntyre (2007) then modified the color-coded version of Figure 2.15 by omitting four suspect proxy series. Instead of deleting the Sargasso Sea and Central Michigan proxies, both of which are claimed by McIntyre to be well linked to temperature, the two bristlecone pine series were excluded (as not being good temperature proxies) and the first century of the Polar Urals series was excluded on quality control grounds. Without the contribution of the bristlecones and Polar Urals, the MWP peak is comparable with the 20th century peak.

Hegerl *et al.* (2007) added to the work of Crowley and Lowery (2000) using "updated records, a modified reconstruction method, and a new calibration technique". The stated goal of the study (as evidenced by the title of the paper) was to detect human influence on the climate.

"The reconstruction consisted of three individual segments. A baseline reconstruction used 12 decadal records and covered the period to 1505. One longer, less densely sampled land temperature reconstruction ... was based on seven records back to A.D. 946, and [the third] consisted of five records back to A.D. 558."

As is usual in such studies, they did not show the actual comparisons between proxies and temperatures during the calibration period so there is little basis to judge their adequacy *a priori*. Nevertheless, Hegerl *et al.* (2007) arrived at a slightly modified hockey stick result. Their hockey stick had a very minor MWP and a ~0.5°C LIA. They used Jones' "trick" of tacking on the recent instrumental record to the proxy results (see Section 3.4.2). However, they show an instrumental record that gained almost 1.5°C since 1900 that is far out of line with other estimates, and provides a very exaggerated view of the rise in temperature in the 20th century. It defies the logical mind to imagine how 12 or 7 or 5 proxy records, each of dubious credibility, centered in Europe, and with no representation from the Southern Hemisphere or the 70% of the Earth covered by oceans, could adequately define the global climate over 1,500 years.

What is missing from the proxy analysis of Hegerl *et al.* (2007) (as well as most of the published proxy analysis that I am aware of) is a presentation of the comparison of each proxy at each location with the temperature as measured at that location during the calibration period (as well as before the calibration period). The variations from proxy to proxy are enormous. It seems likely that these vastly different patterns have little to do with temperature. But if they do properly represent temperature at each location, and the temperature patterns vary by that much, we might ask how many proxies (locations) are needed to approximate a global average temperature? Certainly, use of only 15 proxies appears on the face of it to be grossly inadequate.

Esper, Cook, and Schweingruber (2002) started out by repeating the mantra of the global warming alarmists:

> "... the MBH reconstruction indicates that the 20th century warming is abrupt and truly exceptional. It shows an almost linear temperature decrease from the year 1000 to the late 19th century, followed by a dramatic and unprecedented temperature increase to the present time. The magnitude of warmth indicated in the MBH reconstruction for the MWP, 1000–1300 is uniformly less than that for most of the 20th century."

However, Esper, Cook, and Schweingruber (2002) (ECS) admitted: "the MBH reconstruction has been criticized for its lack of a clear MWP." It was admitted that critics doubt that tree-ring records can preserve long-term, multi-centennial temperature trends. However, as usual in papers written by members of the *paleoclimatic cabal*, no mention is made of the devastating criticism of the MBH reconstruction made by McIntyre (2007). ECS then went on to present a defense of tree-ring reconstructions using centuries-long ring width trends in 1,205 radial tree ring series from 14 high-elevation and middle to high-latitude sites distributed over a large part of the NH extra-tropics. While ECS intended to support Mann, Bradley, and Hughes (1999), the large differences between their results and those of MBH lead this writer to the opposite conclusion. This raises the question whether any reconstruction based on proxies is credible. Furthermore, the anomalies in the result of ECS are mostly negative, suggesting that the mean used for data processing was not the mean for the entire time period, but only for the calibration period.

Nevertheless, based on their result, ECS reached the following conclusions:

1. Multi-centennial temperature variability in long tree ring records can be preserved if the appropriate tree ring data and proper methods of analysis are used.
2. The MWP appears to be more temporally variable than the warming trend of the last century and may have begun in the early 900s.
3. The warmest period covers the interval 950–1045, with the peak occurring around 990.
4. Past comparisons of the MWP with the 20th-century warming back to the year 1000 may not have included all of the MWP and, perhaps, not even its warmest interval.

McIntyre (2007) examined the data in Esper, Cook, and Schweingruber (2002) in considerable detail and wrote at length on their analysis. The issues are intricate and detailed and beyond the scope of the present write-up. McIntyre commented on the difficulty in obtaining the original data: "It's obviously been pulling teeth to get data from Esper. After only two years of trying, I've recently obtained all but one site chronology ... and gobbledy-gook about methodology." Using the 13 site chronologies that he had available, McIntyre plotted the individual proxies as shown in Figure 2.16. This is sometimes referred to as a "spaghetti chart".

McIntyre pointed out that only 2 of the 13 series have strongly elevated closing values. They both entail foxtail pines (interbreeding cousins of bristlecone pines) both from sites very close to Sheep Mountain, California. He cast considerable doubt on the validity of these two proxy sites. McIntyre then went on to present individual plots for each proxy, and perform a simple average. These results show that the proxies vary widely, and cast doubt on the consistency and credibility of the various proxies. While ECS provides us with assurance that "multi-centennial temperature variability in long tree-ring records can be preserved if the appropriate tree-ring data and proper methods of analysis are used," Figure 2.16 suggests otherwise.

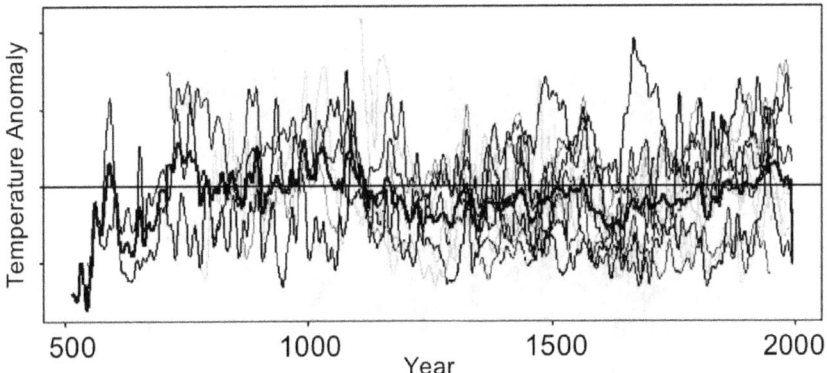

Figure 2.16. "Spaghetti chart" of individual proxies (except Mongolia). Adapted from McIntyre (2007).

McIntyre (2007) presented a simple average of all the proxies as shown in Figure 2.17. The result suggests an MWP and an LIA. Nevertheless, it seems evident that Esper, Cook, and Schweingruber (2002) used statistical data manipulation that unreasonably and illegitimately overemphasized the weighting of the two suspect proxies with high closing values.

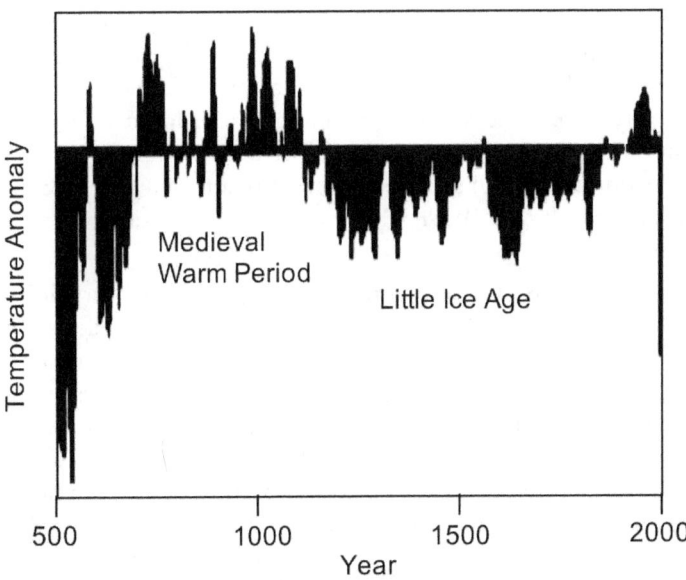

Figure 2.17. Simple average of proxy data from Esper, Cook, and Schweingruber (2002). Adapted from McIntyre (2007).

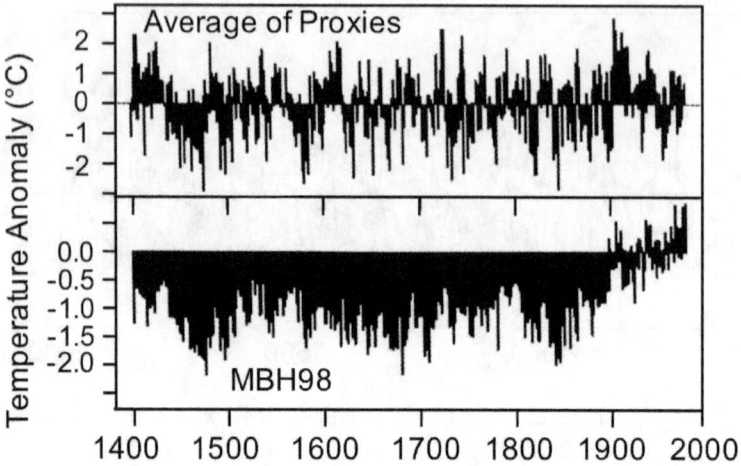

Figure 2.18. Temperature anomalies from MBH. The lower graph is the result of MBH98 while the upper graph is a simple average of their proxies (Montford, 2010).

Some paleoclimatic cabalists have argued that the MBH procedure is supported by other, more recent studies that also lead to a hockey stick. However, amazingly enough, none of these further studies paid any attention at all to the M&M criticisms of the MBH procedure, and they blithely went ahead and used the mean of only the calibration period, making the same mistake as MBH, over and over again. Proof of this assertion is the fact that in all of these reconstructions, the temperature anomaly remains starkly negative at all times prior to the calibration period (as for example, in Figures 2.8, 2.9 and 2.10). Montford (2010) supplied Figure 2.18 attributed to M&M. The lower graph is the result of MBH98 while the upper graph is a simple average of their proxies.

Paleoclimatologists who use PCA to extract a "trend" from the very noisy data seem to lose sight of the fact that in cases where only a few proxies show a trend, PCA will weight these heavily to the exclusion of large numbers of proxies that do not show a trend. While Figure 2.10 is a rather extreme case, it illustrates a point. The object should not be to extract a "trend" but rather to represent the information contained in the entire data set. The average of all proxies in Figure 2.10 will be a line that is horizontal across most of the pre-calibration period and has a very slight rise through the calibration period due to the one proxy that rises. PCA will simply select the one proxy with a trend and ignore the data in all the horizontal proxies. In that respect, PCA is a useless, misleading method.

2.3.4.2 Hiding the Decline

Tree ring proxies are important in attempting to discern historical temperatures over the past millennium or two because they often date back 2,000 years or more. Hence, tree ring proxies are prominent in the MBH and other related reconstructions of global temperatures over one or two millennia. However, as we pointed out in Section 2.1, tree growth is also affected by other factors (water availability, humidity, wind, cloudiness, CO_2 content in the atmosphere, nutrients, etc.) that add noise to the temperature signal. Hence, it is not a simple matter to extract accurate historical temperature data from tree rings (or other proxies, for that matter). There is ample evidence that the climatologists who have developed models for global temperatures over the past millennium or two have had a vested interest in proving that rising temperatures in the 20th century are unique and unprecedented, thus suggesting that natural causes cannot account for this change, and it must be attributed to growth of greenhouse gas concentrations. We can speculate on their motives. One might be a true idealistic desire to save the world from what they believe is an impending catastrophe due to global warming. Another might be the crass fact that funding for climate research will be proportional to the degree of catastrophe that is predicted. Whatever their motives, unfortunately, the behavior of tree ring proxies was not supportive of this belief. Tree ring proxies showed aberrations at various times, but the most serious problem was that tree ring data typically show a downward trend in the late 20th century, while measured global temperatures were rising. The goal of the alarmists was to preserve the hockey stick, which they felt was necessary to show that rising greenhouse gases in the 20th century produced continuously rising temperatures. The "solution" to the problem of this divergence was "the trick" of not showing the down trending proxy data in the late 20th century, and replacing it with measured temperatures that were rising.

> "I've just completed Mike's Nature trick [Michel Mann's publication in Nature where he replaced tree ring proxy data with actual data because the tree ring data went in the 'wrong' direction] of adding the real temperatures to each series for the last 20 years (i.e. from 1981 onwards) and from 1961 for Keith's to **hide the decline**." – excerpt from email by Phil Jones.

It is particularly revealing to note some results of Briffa *et al.* (2001). Figure 2.19 shows eight different reconstructions using various procedures with one preferred reconstruction. Note that all reconstructions decline in the second half of the 20th century while measured temperatures rise. Briffa *et al.* (2001) also compared their results with reconstructions by others as shown in Figure 2.20. The divergence is readily seen and Jones' "trick" produces the hockey stick.

Hegerl *et al.* (2007) also used Jones' "trick" of tacking on the recent instrumental record to the proxy results. Mann, Bradley, and Hughes (1998) and Mann, Bradley, and Hughes (1999) also cleverly substituted the measured temperatures for the modeled temperatures (Jones' "trick") to exaggerate the rise in the 20th century and thus accentuate the *hockey stick*.

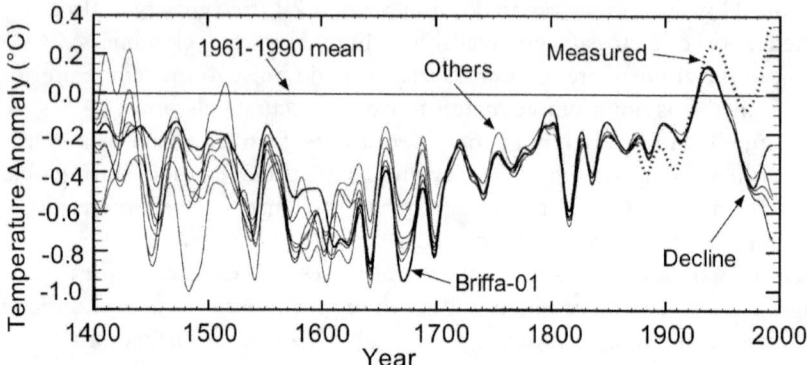

Figure 2.19. Eight reconstructions of historical northern non-tropical summer temperatures using various procedures. The heavy line (Briffa-01) is "preferred". (Adapted from Briffa *et al.*, 2001).

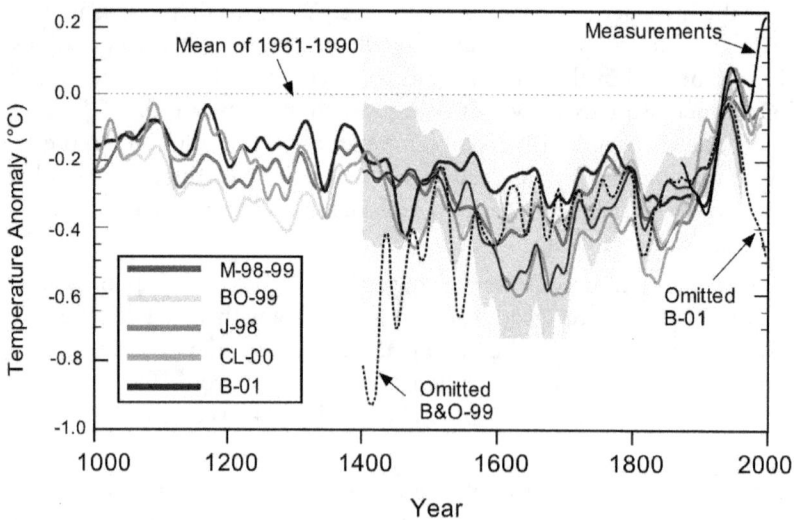

Figure 2.20. Comparison of reconstructions. M-98-99 = Mann *et al.* (1998, 1999). BO-99 = Briffa and Osborne (1999). J-98 = Jones *et al.* (1998). CL-00 = Crowley and Lowery (2000). B-01 = Briffa *et al.* (2001). (Adapted from Briffa *et al.*, 2001).

The ultimate test for reliability of proxies is how well they track temperatures. Of all the many papers on proxies that I have reviewed, very few if any have provided such data in any detail. Briffa *et al.* (1998) is an exception. They compared tree ring proxies with temperatures at many sites in the NH from 1880 to 1990. They said:

"When averaged over large areas of northern America and Eurasia, tree-ring density series display a strong coherence with summer temperature measurements averaged over the same areas, demonstrating the ability of this proxy to portray mean temperature changes over sub-continents and even the whole Northern Hemisphere. During the second half of the twentieth century, the decadal-scale trends in wood density and summer temperatures have increasingly diverged as wood density has progressively fallen. The cause of this increasing insensitivity of wood density to temperature changes is not known ..."

Although Briffa *et al.* (1998) pointed out the discrepancy between tree ring data and temperature after 1950, their assessment that proxies tracked temperatures prior to 1950 may be somewhat optimistic (see Figure 2.21).

Since the publication of this data in 1998, a number of additional papers have appeared dealing in one way or another with tree ring proxies.

Jacoby *et al.* (2000) said:

"Data from annual tree-ring widths are used to reconstruct May–September mean temperatures for the past four centuries. These warm-season temperatures correlate with annual temperatures and indicate unusual warming in the 20th century. However, there is a loss of thermal response in ring widths since about 1970."

Thus they admit to a divergence problem after 1970. However, when one examines their data prior to 1970, the correlation of tree ring data with temperature even prior to 1970 is not impressive to this writer.

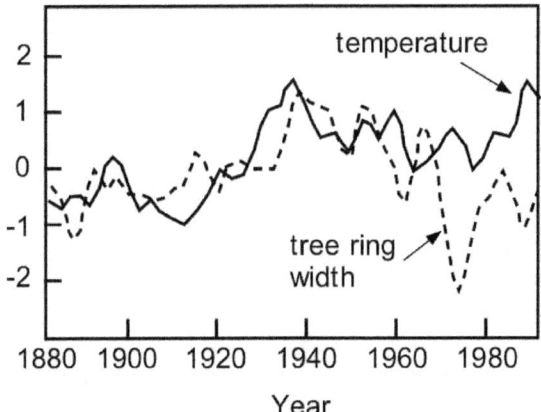

Figure 2.21. Comparison of tree ring width with temperature for NH (Briffa *et al.* 1998).

D'Arrigo *et al.* (2006) came to the defense of tree ring proxies. They began their paper with the usual orthodoxy that "recent warming in the Northern Hemisphere appears to have been unprecedented over the past millennium and that this warming is most likely a result of the anthropogenic release of greenhouse gases into the atmosphere" – which has not much to do with the reliability of tree ring proxies. They mention that D'Arrigo *et al.* (2006a) used "simple averaging of tree-ring records (after accounting for differences in mean and variance over time), followed by linear regression". Simple averaging is a step in the right direction, but how good are the proxies as temperature indicators? As is usual in almost all papers on proxies, the data for the calibration period were not shown. They do allude to the divergence between tree ring proxies and temperature reported by Briffa *et al.* (1998) and others cited in their paper, where they said:

"Theories for the cause (s) of this observed divergence, which may vary from site to site, include decreased temperature sensitivity due to warmer temperatures, drought stress, increased winter snowmelt and ozone effects. This divergence needs to be considered to avoid bias in dendroclimatic reconstructions; however it is not present everywhere. For example, temperature-sensitive elevational treeline sites in Mongolia and the European Alps exhibit dramatic growth increases in recent decades. Greater attention to site selection (e.g. avoidance of drought-prone sites) and careful comparison of adjacent sites with regards to their ecological characteristics can help circumvent this problem. [It has been] demonstrated that the divergence appears to be limited to the recent period (after ~1950) and to trees from some northern locations (at some sites within ~55-70°N), and that there is no evidence for a comparable divergence prior to this time (e.g. during the *Medieval Warm Period*). These observations suggest a unique, anthropogenic cause for the recent divergence and argue very strongly that tree-ring temperature reconstructions for the past millennium should not be called into question based on these recent observations".

One problem with site selection is that if one is attempting to estimate a global average temperature, one needs all the sites one can find. If only a few sites provide reliable data, how can one derive global or even hemispheric temperatures in the past? The claim that "there is no evidence for a comparable divergence prior to this time (e.g. during the *Medieval Warm Period*)" is sheer nonsense because there are no measured temperature data for that period and hence there is no way to ascertain whether such a divergence exists. D'Arrigo *et al.* (2006) closed their paper with further homage to the orthodoxy of "unusual recent anthropogenic warming on a hemispheric to global scale" but their defense of tree ring proxies falls flat.

To hide the divergence between proxies and reality, MBH terminated their calibration phase in 1980 even though more recent data were available. (Unfortunately the proxies went down while measured temperatures went up after 1980.)

Wilmking and Singh (2008) discussed the "divergence effect" between measured temperatures and tree ring proxies in the 2nd half of the 20th century and pointed out that this "seriously questions the validity of tree-ring based climate reconstructions, since it seems to violate the assumption of a stable response of trees to changing climate over time". In their study they claimed to have

"... eliminated the 'divergence effect' in northern Alaska by careful selection of individual trees with consistently significant positive relationships with climate (17% of sample) and successfully attempted a divergence-free climate reconstruction using this subset".

However, they did admit:

"The majority of trees (83%) did not adhere to the uniformitarian principle as usually applied in dendroclimatology. Our results thus support the notion that factors acting on an individual tree basis are the primary causes for the 'divergence effect' (at least in northern Alaska)".

Unfortunately, even the small subset of 17% of trees that are claimed to show good consistency with temperatures over the last century provide somewhat doubtful consistency. The diagram provided by Wilmking and Singh (2008) in their Figure 2 is a tiny little diagram that compresses the excursions between the temperature and tree ring curves. Nevertheless, accepting the claim that 17% of the trees show good correlation with temperature, the question arises as to whether it makes sense to select a subset of trees that happen to fit the temperature curve, and use these for estimating temperatures a thousand years or more ago. Apparently Wilmking and Singh suggest that there occurs a "mixture of trees with stable and non-stable climate growth relationships" and the ones with stable relationships provide a basis for estimating past climates. However, it may be equally likely that all the tree ring records are randomized by other variables than temperature, and by happenstance, about 17% of the records have correlation coefficients with temperature that satisfy the criterion adopted by Wilmking and Singh (which is not impressive to this writer). There is then no great reason to believe that even these 17% of trees would remain as accurate temperature indicators over much longer periods.

2.3.4.3 Sparse Data Set

Aside from all the other problems in reconstruction of millennial temperatures, the data set from which the analyses were conducted was very sparse. When MBH09 extended the time scale of MBH98 back from year

1400 to 1000, they depended on just 13 proxy series. Four were ice cores from a single small ice cap in Peru, and three were derived from southwestern U.S. tree rings. How could one possibly claim to have estimated global temperatures from such a sparse data set?

2.4 The Alarmist Cabals

2.4.1 Paleoclimatic Cabal

There is strong evidence that the climatologists who earn their living by reconstructing paleoclimates over the past few millennia are in frequent communication with one another and are mutually supportive of their various efforts. There is nothing fundamentally wrong with a collegial relationship between scientists in a field and this can, in principle, be a very positive thing. However, in the case of paleoclimatology, the relationship fostered very malicious, insidious, unprofessional behavior by the principals. For a number of years, the various paleoclimatologists published their reconstructions and acted as reviewers for one another's manuscripts. All was well. The hockey stick was widely accepted and became one of the pivotal supporting foundations of global warming alarmism. Then around 2005, the *climateaudit.org* website began reviewing these studies in great detail and found that they were all flawed due to (1) use of a mean for only the calibration period, (2) hiding the decline and using Jones' "trick" of substituting measured temperatures for proxies, as well as (3) various other problems discussed on *climateaudit.org*. This threatened to undermine years of work upon which the paleoclimatologists' reputations were based. Instead of admitting their errors and fixing them, they dug in and became defensive (and indeed paranoid) at first, and then went on the offense against their critics. The most defensive of them all was Michael Mann. Even a cabalist in an email cautioned: "but he would probably go ballistic" in regard to any criticism of his work.

Ball (2007) was critical of how Phil Jones (Head, Climate Research Group, East Anglia University) came up with his estimate of uncertainty in his temperature reconstructions and wrote to Jones asking for an explanation. Ball (2007) claimed that Jones replied in an email:

> "We have 25 or so years invested in the work. Why should I make the data available to you, when your aim is to try and find something wrong with it?"

When a world-leading climatologist is more concerned with protecting his *turf* than finding *truth*, things have taken a very bad turn. Jones also said in an email to Michael E. Mann, professor of climatology, Penn State University:

> "And don't leave stuff lying around on anonymous download sites - you never know who is trawling them. McIntyre and McKitrick have been after the *Climatic Research Unit* ... data for years. If they ever hear

there is a *Freedom of Information Act* now in the United Kingdom, I think I'll delete the file rather than send it to anyone."

These scientists would rather destroy data than allow others to check up on them. In 2009, 2010 and 2011, extensive sets of emails between principal figures in the *paleoclimatological cabal* were made public (by unknown, but clearly illegal means). These emails revealed a deeply imbedded agreement amongst these climatologists to promulgate their orthodoxy that the Earth's climate has hardly wavered over the past 2,000 years, and that CO_2 was the principal cause of unprecedented global warming in the 20th century. The collection of emails is now referred to as *"climategate"*. As Mosher and Fuller (2010) pointed out, they:

> "... ruthlessly suppressed dissent by insuring that contrary papers were never published and that editors who didn't follow their party line were forced out of their position. When *Freedom of Information* requests threatened to reveal their misbehavior, the emails showed them actively conspiring to delete emails to frustrate legitimate requests for information. Worst of all, one scientist threatened to delete climate data rather than turn it over, and that data is still missing."

The defensive posture of the *cabal* seems to have been to disclose nothing, prevent others from delving into their work, totally ignore criticisms, and continue publishing bad science, acting as reviwers for one another's manuscripts. Amazing at it seems, even to this day, I am not aware of any principal paleoclimatologist responding to or even admitting that a criticism was made of their use of the mean for only the calibration period. McIntyre revisited the Wegman Report in May 2011. He pointed out that the combination of criticisms of the MBH hockey stick by the *climateaudit.org* website and the Wegman Report generated a great deal of controversy on the Blogs. The amazing thing is that, as McIntyre put it:

> "Rather than conceding even seemingly indisputable points, Mann and his associates contested every single issue – even the seemingly indisputable and elementary observation that Mann's principal components mined datasets for hockey stick shaped data. To this date [May 23, 2011], neither Mann nor any of his associates has conceded the point."

Dr. Phil Jones, head of the Hadley CRU in 2009, said that the U. S. Department of Energy was funding his data collection – and that officials there agreed that he should not have to release the data. In a 2009 email, he said:

> "Work on the land station data has been funded by the U. S. Dept of Energy, and I have their agreement that the data needn't be passed on. I got this [agreement] in 2007,"

Two months later, Jones said that the data "has to be well hidden. I've discussed this with the main funder (U.S. Dept of Energy) in the past and they are happy about not releasing the original station data." Evidently, the U. S. Dept of Energy is in cahoots with the *cabal* to evade the requirements of the FOIA as well as the basic tenets of science. It should be emphasized that this is not a case of intellectual property produced by some brilliant new concept. It merely represents collecting climate data and storing it in columns. If the U. S. Dept of Energy paid for it, it should be in the public domain.

The *cabal* also went on the offense. Since the principal figures in the *paleoclimatological cabal* were widely published, they tended to be chosen as reviewers for new manuscripts submitted to journals. They were able to act in collusion to prevent contrary papers from being accepted for publication and put pressure on editors who did not cooperate. The cabal refereed one another's papers submitted to journals, communicated improperly in a mutual back-scratch environment subverting the peer review process, pressured journal editors not to publish papers contrary to the orthodoxy, conspired to write rebuttals to any papers that did slip through their barrier to publication of contrary views, and conspired to act in partnership to disparage and ridicule anyone with contrary findings. A particularly egregious episode in the shenanigans of the *cabal* is documented in great detail at the BishopHill website.[1] The challenge to the hockey stick by McIntyre needed to be rebutted in time for inclusion in the 2007 IPCC Report. Two Mann associates, Caspar Amman and Eugene Wahl were chosen to do this. However, they missed the IPCC deadline and their papers were originally rejected by journals (for good reason – they are misguided). However, the cabal managed to circumvent the IPCC deadline and manipulate the journals to their advantage.

In response to the many charges of malfeasance by the *cabal* members that appeared on the blogs, several reviews of the *climategate* activities were carried out by vested interests (e.g. the so-called "Russell Report" – *http://www.cce-review.org/*). These generally provided a whitewash that was only to be expected, considering who were on the review boards.

Wegman, Scott, and Said (2006) have suggested that the field, reconstruction of the temperature history of the Earth, is dominated by a cadre (*cabal*) that is vitally concerned about the potential impacts of global warming, and supports the *hockey stick* result, as well as the procedure used to derive it. Wegman, Scott, and Said (2006) said:

"If there is a tight relationship among the authors, and there are not a large number of individuals engaged in a particular topic area, then one may suspect that the peer review process does not fully vet papers before

[1] *http://bishophill.squarespace.com/blog/2008/8/11/caspar-and-the-jesus-paper.html*

they are published. Indeed, a common practice among associate editors for scholarly journals is to look in the list of references for a submitted paper to see who else is writing in a given area and thus who might legitimately be called on to provide knowledgeable peer review. Of course, if a given discipline area is small and the authors in the area are tightly coupled, then this process is likely to turn up very sympathetic referees. These referees may have co-authored other papers with a given author. They may believe they know that author's other writings well enough that errors can continue to propagate and indeed be reinforced."

It was concluded:

"It is immediately clear that Mann, Rutherford, Jones, Osborn, Briffa, Bradley, and Hughes form a clique, each interacting with all of the others."

McIntyre also discussed the question: "… given the defects of Mann's principal components, how did the methodology pass peer review and then remain unchallenged by specialists in the field?" He stated further:

"The Wegman Report hypothesized that this failure was due to the inter-connectedness of climate scientists through co-authorship and, in particular, by the extent of Mann's network of co-authorship, a level of inter-connectedness that the Wegman Report seemed to think as not existing in their own field. Wegman speculated that members of Mann's closest circle ('clique' in network terminology) reviewed papers of other members of the clique, resulting in non-independent and weak peer review, which, in turn, had resulted in the failure to identify the incorrectness of Mann's principal components in both the original article and subsequently."

It is noteworthy that not only did the *paleoclimatic cabal* ignore the criticisms of its hockey stick and blithely continue to promulgate their incorrect result, they went on the attack to attempt to besmirch the criticizers on a personal level. One mechanism for doing this was to accuse Wegman (and myself) of plagiarism in our writings. This seems to have originated on the *deepclimate.org* website where the scurrilous person hiding behind a cloak of anonymity known as "DC" (or his henchmen) apparently stayed up nights comparing word-for-word the criticisms of MBH with references to identify passages that were copied or slightly modified without attribution. This was further propagated by a report written by one John Mashey. DC and Mashey succeeded in their goal, which was to divert attention from the technical errors in the hockey stick, and focus attention on the issue of trumped-up claims of plagiarism. There have been thousands of nonsensical blog entries about plagiarism in connection with the Wegman Report, whereas the perpetration of false science by the *paleoclimatic cabal* has been ignored by the alarmists. And to this very day, schoolchildren all over the world learn the false science of the hockey stick as if it were fact.

Bouville (2008) wrote a treatise on plagiarism. He said:

"... even though ... copying other people's intellectual contribution is wrong, they do not apply to the copying of words. Copying a few sentences that contain no original idea (e.g. in the introduction) is of marginal importance compared to stealing the ideas of others. The two must be clearly distinguished, and the 'plagiarism' label should not be used for deeds that are very different in nature and importance."

The point is that plagiarism is only a serious malpractice when an innovative intellectual concept is stolen for personal gain. When background material is presented without attribution, that is an inadvertency or an indiscretion, but not a crime. The thrust of the Wegman Report was twofold: (1) the hockey stick was based on bad science, and (2) collusion between members of the paleoclimatic cabal allowed the hockey stick to get repeatedly published despite the errors in the methods used. There was no plagiarism in these elements of the Report. Unfortunately, some of the introductory and background material was not given proper attribution. In the spirit of Bouville's paper; big deal.

It is interesting that both DC and Mashey accused me of plagiarism as well in my book: *Assessing Climate Change*. This book contains 1,348 specific citations to references giving credit to authors for their work. It also includes 411 specific quotations of authors with their own words included in quote signs in my book. It is possible that in a few places, I may have slipped up and used words from a paper and forgot to give attribution. That was not, is not and cannot be plagiarism. It was simply the very small fraction of inadvertent cases where I failed to give credit. Anyone with any sense can immediately see that with 1,348 citations of references and 411 direct quotes with attribution, I could hardly have "intentionally, knowingly and recklessly" failed to reference the authors in a few cases. What did I have to gain? Why would I reference 1,348 but not the other 10 or 20? Ridiculous!

One of the henchmen involved in the attempt cast aspersions on me was a person named Samuel Cohen. Unaccountably, he sent an email out of the blue to me in October 2011:

Dear Dr. Rapp, I am unclear about your affiliation with the University of Southern California. It is listed on your online CV, but no trace can be found on the USC website outside of an old seminar you gave at the behest of that demented crank George Olah. Can you clarify you status as a Research Professor there?

Sincerely, Sam Cohen

(By the way, Olah is a Nobel Laureate).

The real meaning of this was that Mr. Cohen was under the impression that I was on the faculty at USC and he desired to poison that appointment by accusing me of plagiarism. But he was (and remains) confused on my relationship to USC. As it turns out, I did have a one-year interim appointment in 2010 but that had lapsed by the time Mr. Cohen went on the attack. After I left USC, Mr. Cohen sent a scurrilous complaint to USC accusing me of plagiarism but it had no effect; I was already long gone. As he said in an email:

"I assume you have seen the detailed complaint made to USC?"

I heard from Mr. Cohen again on 2/22/2012. He said:

Dear Dr. Rapp,

How's your good buddy Edward Wegman doing lately? Heard anything about the bullshit plagiarism charges that were brought against him? I guess we will now see how the NIH and Department of Defense handle sanctions. Could be fun to watch. They certainly will act faster than GMU. Talk to Ed lately? Sam

I tried to explain to him the difference between academic sloppiness in not making an attribution on well known and widely accepted introductory material vs. stealing someone's intellectual innovation for personal gain. I quoted Bouville (2008) who wrote a treatise on plagiarism. The point is that plagiarism is only a serious malpractice when an intellectual concept is stolen for personal gain. When background material is presented without attribution, that is an inadvertency or an indiscretion, but not a crime. The thrust of the Wegman Report was two fold: (1) the hockey stick was based on bad science, and (2) collusion between members of the paleoclimatic cabal allowed the hockey stick to get repeatedly published despite the errors in the methods used. There was no plagiarism in these elements of the Report. Unfortunately, some of the introductory and background material was not given proper attribution. To Mr. Cohen, however, there is no distinction between failing to give attribution on some quotes vs. stealing ideas for personal gain. In fact, if you used one word from another publication that would be plagiarism in his view. I asked Mr. Cohen what he ever accomplished in his life and he said:

"Besides turning Wegman in to the NIH for plagiarism for the CSDA article, a few other things such as becoming a full professor at a top-tier research university, that sort of stuff."

He seems proud to proclaim that he "turned in" Ed Wegman. Evidently, he tried to bestow the same favor on me. When I asked him if he had no shame (ala McCarthy trials) he responded:

"I feel no shame whatsoever. I did what I consider right, and my actions are supported by the vast majority of the academic community. My actions have been lauded by many, many people. The McCarthy allusion is false."

The orthodoxy seems to have enlisted Mr. Cohen as their hatchet man using accusations of plagiarism as their weapon. And while these charges prevail, the real message that Wegman was technically correct that the hockey stick is bogus gets lost. Meanwhile Cohen admitted:

"I don't know Wegman and I have no real understanding of his work"

Cohen claims to be " an expert in academic and research misconduct."

Despite his admitted lack of knowledge, Mr. Cohen went on to proclaim:

"The so-called hockey stick has been validated many times, most recently by the group at UC Berkeley (http://berkeleyearth.org/). They were experts, they were skeptical, they found that the paleoclimate data supported global warming, they were honest in both their approach and their conclusions (as so many denier are not), so I trust they did the work correctly and I trust their conclusions."

The Berkeley group never dealt with long-term climate change. They never dealt with the hockey stick. They only dealt with measured temperatures over the past couple of hundred years. One third of their sites reported cooling, not warming. The *climateaudit.org* site has clearly demonstrated that the hockey stick is *phonus balonus*. Mr. Cohen also said:

"The hockey stick model is a fact. No one has ever demonstrated that it is an inaccurate representation of global temperatures, as far as I know. If you have citations to the primary literature demonstrating this, please send them to me."

Now it matters little what Sam Cohen thinks about the hockey stick but this exchange of emails provides some insight into the sinister thinking and actions of the *deepclimate.org* mentality. These people are stupid and dangerous. What is most sinister about Sammy is that he seems to have taken delight in disparaging professional scientists who commited minor indiscretions and wringing his hands in joy, relishes his attempts to destroy their reputations and their careers. As he said: "It could be fun to watch" [the downfall of Ed Wegman]. This man is a cruel, vicious, nasty hangman. The *deepclimate.org* crowd has enlisted him to damage the reputations of those who disagree with them, while never refuting the arguments of those who disagree with them. Sammy is certain the the hockey stick yet he admits he does not know squat about it. What makes Sammy run?

We have previously reviewed the response of *Nature* magazine to McIntyre and McKitrick when they attempted to publish their critique of Mann *et al.* (1998). As we said, "in a matter concerning the legitimacy and validity of the most widely accepted model of the Earth's climate over the past millennium, *Nature* decided that they would allow the erroneous publication by Mann *et al.* to stand unchallenged because the issues involved 'were too complicated'." There is at least the appearance (and more likely the reality) that the *paleoclimatic cabal* was in complete control of the situation.

Kevin Trenberth, Senior Scientist as the National Center for Atmospheric Research has emerged as a defender of the *cabal* (*http://www.cgd.ucar.edu/cas/Trenberth/Presentations/ClimategateS.pdf*).

He did admit to "lack of openness in sharing data and violations of the Freedom of Information Act" but he pointed out that five investigations failed to find any of the alleged misconduct. Unfortunately, these five investigations were conducted by friends of the *cabal*. He also asserted that "scientists would not make up stuff that could be disproven by others!" but the nature of paleoclimatic data is that it is not susceptible to proof, disproof, verification or validation, and hence is a very safe field to work in. He cited an excerpt from a Phil Jones email:

> "I can't see either of these papers being in the next IPCC report. Kevin and I will keep them out somehow – even if we have to redefine what the peer-review literature is!"

and implied that this was Jones' invention and he (Trenberth) had nothing to do with this. Whether this is true or not, this excerpt reveals the intellectual environment of the climate cabal. However, I have to agree with one slide in Trenberth's presentation that says the Internet is "An open sewer of untreated, unfiltered information and the American public is incapable of deciphering between facts, fiction and opinion".

As McLean (2007a) observed:

> "The peer-review process was established for the benefit of editors who did not have good knowledge across all the fields that their journals addressed. It provided a 'sanity check' to avoid the risk of publishing papers which were so outlandish that the journal would be ridiculed and lose its reputation. In principle this notion seems entirely reasonable, but it neglects certain aspects of human nature, especially the tendency for reviewers to defend their own (earlier) papers, and indirectly their reputations, against challengers. Peer review also ignores the strong tendency for papers that disagree with a popular hypothesis, one the reviewer understands and perhaps supports, to receive a closer and often hostile scrutiny. Reviewers are selected from practitioners in the field, but many scientific fields are so small that the reviewers will know the authors. The reviewers may even have worked with the authors in the past or wish to work with them in future, so the objectivity of any review is likely to be tainted by this association. Some journals now request that authors suggest appropriate reviewers, but this is a sure way to identify reviewers who will be favorable to certain propositions ... In 2002 the editor-in-chief of the journal *Science* announced that there was no longer any doubt that human activity was changing climate, so what are the realistic chances of this journal publishing a paper that suggests otherwise? The popular notion is that reviewers should be skilled in the

relevant field, but a scientific field like climate change is so broad, and encompasses so many subdisciplines, that it really requires the use of expert reviewers from many different fields. That this is seldom undertaken explains why so many initially influential climate papers were later found to be fundamentally flawed. In theory, reviewers should be able to understand and replicate the processing used by the author(s). In practice, climate science has numerous examples where authors of highly influential papers have refused to reveal their complete set of data or the processing methods that they used. Even worse, the journals in question not only allowed this to happen, but have subsequently defended the lack of disclosure when other researchers attempted to replicate the work."

Dr Willie Soon, is an astrophysicist at the Solar, Stellar and Planetary Sciences Division of the Harvard-Smithsonian Centre for Astrophysics. He and Dr. Baliunas published an article in 2003 that is quite a good study of regional proxy evidence for significant MWP and LIA (Soon and Baliunas, 2003a,b).

Soon and Baliunas (2003a,b) argued that the procedure (used, for example, by Mann, Bradley, and Hughes, 1998) of mathematical decomposition techniques to the reconstruction of a global 1,000-year temperature history is limited by both the inhomogeneous spatio-temporal sampling gaps in proxy records and the very short length of surface thermometer record available for calibration–verification purposes. The classification of proxies using local proxy data is complementary to the mathematical decomposition process but avoids some of their difficulties (albeit at the expense of quantitative results). The different sensitivities of proxies to climate variables and the potential time dependence of the proxy–climate correlation require careful calibration and verification on a location-by-location basis; the emphasis on local results by Soon and Baliunas (2003a,b) avoided the difficulty of inter-comparing disparate proxies but did not generate a synoptic global view. Thus, Soon and Baliunas (2003a,b) gave up on quantitative synthesis of many proxies into global or NH average temperatures, because even for the same location, different proxies may yield different climate expressions simply because of their different sensitivities to local climatic variables. Soon and Baliunas (2003a,b) suggested that a compact mathematical representation of individual proxy variations (e.g., Mann, Bradley, and Hughes, 1998), without full understanding of proxy–climate calibration relations, may yield overconfident results.

Soon and Baliunas (2003a,b) provided a very long table listing the various proxies used in the study. For each proxy, they provided the spatial extent, latitude, and longitude (where applicable), type of proxy, reference, and qualitative evaluations of whether MWP and LIA trends were discernible. These included 14 worldwide proxies, and >100 proxies that are regional or local. The results were provided in several figures accompanied by a lengthy discussion (about 25 pages) of detailed information regarding specific proxy results that led to these figures. These results indicated that:

- The proxy data suggest that the LIA existed as a distinguishable climatic anomaly in almost all regions of the world that were assessed. Only two records, did not exhibit any persistent or unusual climatic change over this period.
- The MWP is a distinguishable climatic anomaly with only two unambiguous negative results.
- Most of the proxy records do not suggest the 20th century to be the warmest or the most extreme in its local representations. There are only three unambiguous findings favoring the 20th century as the warmest of the last 1,000 years. An interesting feature is that the warmest or most extreme climatic anomalies in the proxy indicators often occurred in the early to mid-20th century, rather than throughout the century.

The reaction of the *paleoclimatic cabal* was quick and forceful. In a series of emails that were not revealed until they were hacked in November 2011, *cabal* members strategized to repair the damage from the Soon and Baliunas paper. In one email from Malcolm Hughes to a dozen members of the *cabal*, he cautioned that "an appeal to the National Academy of Sciences (NAS) could be counterproductive – remember the poor treatment of high-res paleo in the NAS report requested by the White House the other year." [Note that he is apparently referring to the Wegman Report that was sanctioned by the NAS]. Michael Mann in his usual arrogant stance referred to "two awful papers written by those clowns" and yet, what could be more awful than Mann's publications? Mann also referred to their paper as "an assault on the science of climate change" – which is exactly what his papers constitute. The outcome of this cabalistic exchange was the publication:

Mann, M. E., Ammann, C. M., Bradley, R. S., Briffa, K. R., Crowley, T. J., Hughes, M. K., Jones, P. D., Oppenheimer, M., Osborn, T.J., Overpeck, J. T., Rutherford, S., Trenberth, K. E., Wigley, T. M. L. (2003) "On past temperatures and anomalous late 20th century warmth" *EOS*, **84**, July 8, 2003.

that has little technical content and represents mainly an affirmation of the faith of the *cabal* in its orthodoxy. Many websites lit up with the news: *"Leading Climate Scientists Reaffirm View that Late 20th Century Warming Was Unusual and Resulted From Human Activity"*

(e.g.: *http://www.agu.org/news/press/pr_archives/2003/prrl0319.html*).

This report was authored by the founding members of the *cabal*: Michael Mann, Caspar Ammann, Kevin Trenberth, Raymond Bradley, Keith Briffa, Philip Jones, Tim Osborn, Tom Crowley, Malcolm Hughes, Michael Oppenheimer, Jonathan Overpeck, Scott Rutherford, and Tom Wigley. The report was not made public but was available only to journalists. Dozens of websites blared this headline but few details were revealed. The point was: 13 *cabalists* can hardly be wrong – or can they?

In late 2011, additional emails between *cabalists* were hacked. Some of these exposed some chinks in the armor. Self doubt began to creep in. Tim Osborne was quoted as saying: "Also, we set all post-1960 values to missing in the MXD data set (due to decline), and the method will infill these, estimating them from the real temperatures – another way of 'correcting' for the decline, though may be not defensible!" Richard Alley (who is not really a *cabalist*, but a fellow traveler in alarmism) was quoted as saying:

"Unless the 'divergence problem' can be confidently ascribed to some cause that was not active a millennium ago, then the comparison between tree rings from a millennium ago and instrumental records from the last decades does not seem to be justified, and the confidence level in the anomalous nature of the recent warmth is lowered. I think the best way to sum up all of this is: Where does all this lead us? It is very likely that the NH mean temperature has shown much larger past variability than caught by previous reconstructions. We cannot from these reconstructions conclude that the previous 50-year period has been unique in the context of the last 500-1000 years. *Of course we all know that the IPCC reports differently.*" [Emphasis added].

Another hacked email was reported to have said:

"I am afraid the Mike [Mann] and Phil [Jones] are too personally invested in things now (i.e. the 2003 GRL paper that is probably the worst paper Phil has ever been involved in - Bradley hates it as well), but I am willing to offer to include them if they can contribute without just defending their past work - this is the key to having anyone involved. Be honest. Lay it all out on the table and don't start by assuming that ANY reconstruction is better than any other" (email refers to Mann and Jones, 2003).

Jonathan Overpeck was quoted as saying:

"... what Mike Mann continually fails to understand, and no amount of references will solve, is that there is practically no reliable tropical data for most of the time period, and without knowing the tropical sensitivity, we have no way of knowing how cold (or warm) the globe actually got.... Unsatisfying, perhaps, since people will want to know whether 1200 AD was warmer than today, but if the data doesn't exist, the question can't yet be answered. A good topic for needed future work."

Tim Osborne was quoted as saying: "Also we have applied a completely artificial adjustment to the data after 1960, so they look closer to observed temperatures than the tree-ring data actually were."

It seems to me to be unfortunate that since publishing that excellent article in 2003, Dr. Soon has published some articles and made a number of public presentations expounding a very extreme skeptical point of view. He has been quoted as saying:

> "Most of the weather and climate variations we observed are essentially related to the sun and the changing seasons – not by CO_2 radiative forcing and feedback. The climate system is constantly readjusting naturally in a large way – more than we would ever see from CO_2. The CO_2 kick [impact of CO_2 emissions] is extremely small compared to what is happening in a natural way. Within the framework of a proper study of the sun-climate connection, you don't need CO_2 to explain anything".

I personally think he is completely misguided in this viewpoint. But his views are his right and privilege. It is unfortunate that organizations allied with the cabal went on campaign to besmirch Soon's reputation and the Internet is full of accusations and attacks. In fact, if you dial "willie soon climate change" into *Google*, you obtain mainly derogatory claims (apparently planted by *Greenpeace*) saying for example: "Climate skeptic Willie Soon received $1M from oil companies, papers show..." (e.g. *http://www.examiner.com/seminole-county-environmental-news-in-orlando/harvard-astrophysicist-dismisses-agw-theory-challenges-peers-to-take-back-climate-science#ixzz1fDDWbbz1*). What these websites don't reveal is that *cabal* members have received many tens of millions of dollars to fund their alarmist research.

2.4.2 The Climate Alarmism Cabal

Initially, the *cabal* consisted of paleoclimatologists. However, as time progressed, the *paleoclimatic cabal* was joined by other climatologists not necessarily involved in reconstruction of past climates, who had vested interests in climate alarmism, and viewed attacks on the *paleoclimatic cabal* as destructive to their cause. So the *paleoclimatic cabal* became a subset of the *climate alarmism cabal* dedicated to propagating the orthodoxy of alarmism. A new set of emails within *climategate* appeared in 2011. An exchange of emails between members of the *climate alarmism cabal* was revealed. Members appear to include (amongst others): Tom Wigley, Jonathan Overpeck, Caspar M Ammann, Raymond Bradley, Keith Briffa, Tom Crowley, Malcolm Hughes, Phil Jones, Tim Osborn, Kevin Trenberth, Ben Santer, Steve Schneider, Malcolm Hughes, Michael E. Mann, Andrew Dessler and Michael Oppenheimer. See: *http://junkscience.com/2011/11/27/climategate-2-0-mann-suggests-harvard-take-action-against-soon-baliunas/*. The goals of the *climate alarmism cabal* seem to be to prevent contrary papers from getting published, to harass editors that pass contrary papers, to immediately combat any contrary papers or influential blog entries with counter papers and blog entries, and unfortunately in some cases it appears that attacks of a more personal nature might be considered. They have pompously and arrogantly claimed that their interpretations are *climate science* while work by other climatologists reaching different conclusions is something other than *climate science*. We see evidence of this in many publications and press releases. In particular, in regard to the effect of clouds, Dessler (2011) said: "In recent papers, Lindzen and Choi (2011), and Spencer and Braswell (2011) have argued that ... clouds are the cause of, and not a feedback on, changes in surface temperature. If this claim is correct, then significant revisions to *climate science* may be required". In other words, he regards "*climate science*" as that which the orthodoxy subscribes to. It is not his interpretation of climate science – it is *CLIMATE SCIENCE*!

Another bizarre aspect of Dessler's publication was discussed by Pielke, Sr. (*http://pielkeclimatesci.wordpress.com/2011/09/06/comments-on-the-dessler-2011-grl-paper-cloud-variations-and-the-earths-energy-budget/*). He said: "Dessler's paper was received 11 August 2011 and accepted 29 August 2011. This is some type of record ... and indicates that the paper was fast-tracked. This is certainly unusual..." – to say the least. He went on to say:

> "It is not clear whether the Editor of *GRL* included Roy Spencer as one of the referees, [and if they did not] they were derelict in their responsibilities. Dessler's paper should have been submitted to *Remote Sensing* as a Comment [on Spencer's paper]. Then Roy Spencer would submit a Reply."

We are now witnessing a phenomenon in climatology publications that is occurring repeatedly. The climatology orthodoxy seems to have united into an informal association dedicated to (1) prevent contrary analyses and interpretations from being published, and (2) to quickly respond to those few contrarian publications that slip through their net with vitriolic attacks on the paper on orthodoxy blogs, and in the literature via rapid rebuttal publications such as that of Dessler (2011). It seems evident that many editors are in cahoots with the orthodoxy; certainly the editor of *GRL* is, and the editor of *Remote Sensing* who let Spencer and Bradwell's paper through the net, suddenly resigned for unclear reasons.

One topic that gets *cabal* members upset is the claim by some contrarians that persistent El Niños since 1976 were dominant in causing warming in the NH in the latter part of the 20th century. If this were true, it would suggest that the role of CO_2 in climate change may be far less than the orthodoxy believes. Thus, when the article by McLean *et al.* (2009) appeared in the literature suggesting an important role for El Niños as a dominant cause of warming in the NH in the latter part of the 20th century, it produced great animosity and consternation amongst the members of the *climate alarmism cabal*. McLean *et al.* (2009) concluded that the El Niño index

"... is a dominant and consistent influence on mean global temperature. Shifts in temperature are consistent with shifts in the [El Niño index] that occur about 7 months earlier. The relationship weakens or breaks down at times of volcanic eruption in the tropics... Since the mid-1990s, little volcanic activity has been observed in the tropics and global average temperatures have risen and fallen in close accord with the [El Niño index] of 7 months earlier. Finally, this study has shown that natural climate forcing associated with ENSO is a major contributor to variability and perhaps recent trends in global temperature, a relationship that is not included in current global climate models."

According to their estimates, changes in the El Niño index could account for about 70% of the variance in the global tropospheric temperature over the past 50 years. This paper was reviewed and accepted by three independent referees. One referee commented in part: "I found the paper to be well-organized, well-written, and clear on the importance of the research... The findings are likely to be of interest to a wide variety of readers." A second referee commented in part: "This very clear and well-written manuscript is an analysis of the relationship between MSU-derived and radiosonde-based tropospheric temperature variability and the Southern Oscillation, as modified by major tropical volcanic eruptions."

After McLean *et al.* (2009) was published, a flurry of emails was exchanged between *cabal* members, strategizing on how to carry out damage control for their orthodoxy by preparing a rebuttal. Soon afterward, a group of cabalists (Grant Foster, James Annan, Phil Jones, Michael Mann, Jim Renwick, Jim Salinger, Gavin Schmidt and Kevin Trenberth) decided to prepare a rebuttal, and to insure speedy publication, they pressured the editor of the *Journal of Geophysical Research* and suggested the following persons as possible reviewers for their submitted critique: Ben Santer, Dave Thompson, Dave Easterling, Tom Peterson, Neville Nicholls, and David Parker (with Tom Wigley, Tom Karl and Mike Wallace also mentioned). All of these were professionally associated in some way to the Foster *et al.* group and are thought to be members of the *climate alarmism cabal*. Phil Jones commented: "All of them know the sorts of things to say - about our comment and the 'awful original', without any prompting." (They all subscribe to the same orthodoxy). McLean *et al.* describe the whole sordid story at

(*scienceandpublicpolicy.org/originals/censorship_at_agu.html*).

In their rush to rebut the original McLean article, the *climate alarmism cabal* posted their rebuttal on a website, in violation of JGR rules. The results of McLean *et al.* (2009) would seem to be a major stumbling block for alarmists who attribute most of the warming of the 20th century to greenhouse gases. It is therefore not surprising that the alarmists struck back with members of the cabal publishing Foster *et al.* (2010), that claimed that the results of McLean *et al.* (2009) "are seriously in error" and concluded "In fact, the general rise in temperatures over the 2nd half of the 20th century is very likely predominantly due to anthropogenic emissions of greenhouse gases". Foster *et al.* (2010) fell back on climate models that attribute only 15-30% of temperature variation in the 20th century to variability of the El Niño index. Foster *et al.* (2010) constituted a rather vicious criticism of McLean *et al.* (2009), but JGR refused to publish McLean's response. Evidently, the JGR is acting in collusion with the alarmist cabal, and probably regrets that McLean *et al.* (2009) "slipped through". McLean (2010) provides all the details.

McLean *et al.* attempted to rebut the criticism by Foster *et al.* (2010), but the *Journal of Geophysical Research* (JGR) refused to publish it. Their rebuttal, which will be referred to as "McLean2010" is available at: *icecap.us/images/uploads/McLeanetalSPPIpaper2Z-March24.pdf* and *http://scienceandpublicpolicy.org/originals/censorship_at_agu.html*.

There are several important aspects of this episode that require further elaboration. These include (1) technical aspects, (2) attitudes and collusion amongst *cabal* alarmists, and (3) collusion of the JGR with *cabal* alarmists.

In regard to technical aspects, the issue revolves about methods used for filtering in statistical processing of data. Foster *et al.* (2010) appear to have made some valid criticisms of specific details, but these do not negate the strong correlation of the El Niño index with climate change. Perhaps the contribution of the El Niño index is less than the 70% claimed by McLean *et al.* (2009), but clearly the El Niño index is an important factor in climate change. It seems doubtful that climatology has sufficient data and analytical insight to pin down its quantitative share in influencing climate change. A number of authors, even members of the alarmist *cabal* have admitted that climate models do not adequately account for El Niño effects. McLean (2010) presented excerpts from the *climategate* emails that clearly show that the alarmist *cabal* regarded McLean *et al.* (2009) as a threat to their orthodoxy, and they colluded together to disparage McLean *et al.* (2009). The *cabal* regards itself as a police force to eradicate any contrary evidence or analysis that would refute their emphasis on greenhouse gases.

McLean *et al.* submitted a response to the published comment by Foster *et al.* but the JGR sent their response to the Foster cabal for review – like asking the fox to guard the henhouse. Needless to say, the McLean response was rejected and never published by JGR, although it appears at *http://scienceandpublicpolicy.org/*. One does not need to be an expert on statistics of large data sets to see that persistent El Niños since about 1976 have contributed significantly to warming in the NH (see Figures 4.11 to 4.13). The science of climatology is not capable of assigning accurate estimates of the percentage contribution of El Niños to total warming. Skeptics suggest perhaps 70%; alarmists suggest less than 30%.

A rather parallel situation occurred in regard to the paper by Douglass *et al.* (2007) that examined measured tropospheric temperature trends and compared them with "Climate of the 20th Century" model simulations. They concluded that observed temperature trends were in significant disagreement with model predictions in most of the tropical troposphere. These conclusions contrasted strongly with those of recent publications by *cabalists*. It has been claimed that a major problem for climate models is the disparity between the temperature trends observed at the Earth's surface and the much smaller trends observed in the lower troposphere that is just the opposite of what global climate models (GCM) predict. (Figure 4.24 shows that the forcing due to doubling CO_2 from the pre-industrial value is much higher at the tropopause than at the Earth's surface). Douglass *et al.* (2007) compared tropical temperature trends with climate model predictions for temperatures in the so-called "characteristic emission layer" (CEL) (2-6 km altitude) where the role of water vapor is most important. Over the period from 1979 through 2004, the models predicted a rising temperature trend of roughly 0.2 to 0.3°C per decade, whereas satellite temperature measurements indicate essentially no increase below 10 km altitude, and a negative trend above 10 km. This was cited as evidence of the inadequacy of current climate models.

It has come to pass that a few determined skeptics (Douglass, Lindzen, McLean, Spencer, McIntyre, …) continue to publish contrarian papers (in those rare cases where the *cabal* does not succeed in censoring publication), and immediately thereafter, a flurry of emails is exchanged between *cabal* members (Mann, Jones, Schmidt, Trenberth, …) castigating the skeptics, and strategizing to achieve damage control to protect their orthodoxy that rising CO_2 is essentially the sole cause of global warming. The most pugnacious and aggressive of these is Michael Mann. It is ironic that his own research, responsible for the *hockey stick*, is far less believable than the work of those he would criticize.

After publication of Douglass *et al.* (2007), the cabal came forth with Santer *et al.* (2008) as a rebuttal. This paper begins with the sentence: "There is now compelling scientific evidence that human activities have influenced global climate over the past century" which aside from the fact that the statement is not true, reveals the orthodoxy to which the authors subscribe religiously. The details of the statistical processing of large data sets are complex. The issue is whether tropical tropospheric temperatures have risen more than surface temperatures as climate models would predict for the effect of greenhouse gases on climate. Douglass *et al.* (2007) concluded that models and data disagreed to "a statistically significant extent". Santer *et al.* (2008) claimed to achieve a "partial resolution of the long-standing 'differential warming' problem" although they also said:

"We may never completely reconcile the divergent observational estimates of temperature changes in the tropical troposphere. We lack the unimpeachable observational records necessary for this task. The large structural uncertainties in observations hamper our ability to determine how well models simulate the tropospheric temperature changes that actually occurred over the satellite era. A truly definitive answer to this question may be difficult to obtain."

Yet, this did not prevent Santer *et al.* from producing a so-called "Fact Sheet" that said "We've gone a long way towards such a reconciliation" [between climate models and tropical tropospheric temperatures]. (*https://publicaffairs.llnl.gov/news/news.../NR-08-10-05-factsheet.pdf*).

In 2009, McIntyre *(http://climateaudit.org/2009/01/27/submited-article-on-tropical-troposphere-trends/ and http://climateaudit.org/2009/04/14/tropical-troposphere-march-2009/)* pointed out that when the data used by Santer *et al.* (2008) that ended in 1999 is extended through 2008, the discrepancy reported by Douglass remains, and "the claim by Santer *et al.* (2008) to have achieved a 'partial resolution' of the discrepancy between observations and the model ensemble mean trend is unwarranted". McIntyre also noted the difficulty in obtaining data from Santer *et al.*, and indicated that the *International Journal of Climatology* (IJC) was stalling in responding to him. It appears that this article will never pass through the cabal's lock on the IJC, and McIntyre had to be content with merely archiving his article *(http://arxiv.org/abs/0908.2196)*. Yet, alarmists continue to refer to Santer *et al.* (2008) as evidence that climate models have been adequately tested.

Douglass and Christy

(http://www.americanthinker.com/2009/12/a_climatology_conspiracy.html) presented evidence for their claim that Ben Santer, Phil Jones, Timothy Osborn, Tom Wigley, and 13 other *climate alarmism cabal* members apparently conspired to compromise the peer review process, with the willing cooperation of the editor of the *International Journal of Climatology* (IJC), Glenn McGregor. This evidence involved dozens of e-mails over nearly a year, suggesting "(a) unusual cooperation between authors and editor, (b) misstatement of known facts, (c) character assassination, (d) avoidance of traditional scientific give-and-take, (e) using confidential information, (f) misrepresentation (or misunderstanding) of the scientific question posed by Douglass *et al.* (2007), (g) withholding data, and more." Douglass and Christy provide the entire sordid story; there is no need to reproduce the details here.

An example of the need by the *climate alarmism cabal* to respond to challenges by contrarians is the paper by Santer *et al.* (2011). This paper was concerned that measurements indicated that tropospheric temperatures had not risen since 1998 despite continued growth in CO_2 concentration. The paper had seventeen authors in an expression of support by the *cabal* although it is difficult to figure out what contributions (if any) were made by the various authors. The listing of these authors seems to be more a political than a scientific statement. The goal was to produce an analysis that concludes that temporary periods with no temperature gain may be viewed as noise imposed on an underlying upward trend due to rising greenhouse gas concentrations. The logic of the paper is quite shaky however as we discuss in Section 4.5.

Lindzen and Choi (2009) examined data on the outgoing radiation budget from the Earth Radiation Budget Experiment (ERBE) in the tropics in an attempt to determine whether observations of the Earth's radiation imbalance can be used to infer feedbacks and climate sensitivity. From this, they concluded that the climate sensitivity is considerably less than the values predicted by climate models. Later, Lindzen and Choi admitted: "This work was subject to significant criticism by Trenberth *et al.* (2009), much of which was appropriate". As a result, they wrote a revised paper (Lindzen and Choi, 2011) that was "an expansion of the earlier paper in which the various criticisms are addressed and corrected...." As might be expected, Lindzen and Choi (2011) found that feedbacks were primarily negative, resulting in relatively low climate sensitivity. This is contrary to the alarmist position that feedbacks are positive leading to higher climate sensitivity (and therefore produce a greater increase in global temperature as greenhouse gas concentrations increase). The manuscript by Lindzen and Choi (2011) was rejected by the *Proceedings of the National Academy of Sciences* (PNAS). The revelation of the reviewers and their comments led to a very extensive series of blog entries at *climateaudit.org*. In the course of these blog entries, we find (along with the usual trivia) several nuggets of information worth mentioning. Lindzen is a member of the NAS and it is very rare that a paper submitted by a member would be rejected (96% are accepted). In a highly unusual move, the PNAS rejected Lindzen's suggestion for reviewers, and instead chose reviewers who were obviously antagonistic to Lindzen's viewpoint. The reviews of this paper were incredibly detailed and penetrating. It seems likely that papers expressing the alarmist agenda glide through the review process with little friction and no depth of review. One blog contributor was a reviewer for the paper by Wahl and Ammann (2007). His review was discarded by the *Journal of Climate* because it was not in conformity with the alarmist agenda. It appears that most of the papers in climatology are based on inadequate data: lacking in spatial and temporal coverage. The sophisticated data processing used to cover this up, whether filtering, smoothing, use of principal components, or otherwise, hides the fact that the foundations are typically very weak. Had other landmark papers in climatology that are repeatedly referred to in biblical tones been given the same kind of penetrating review as Lindzen's manuscript, they would also have been rejected. Indeed, most of the literature in climatology would have to be cleared out. Finally, the Lindzen and Choi paper was published in the *Asia-Pacific Journal of Atmospheric Science*.

The entire set of pirated emails in *climategate* provides strong evidence that there is indeed a *climate alarmism cabal*, including both paleoclimatologists who seek to show that the Earth's climate has been relatively constant for thousands of years prior to the 20th century, climate modelers who seek to use climate models to infer that most of the 20th century warming was due to greenhouse gas buildup, and the members of this *cabal* are dedicated to their preconceived beliefs, conspire with one another to prevent publication of dissenting views, conspire with one another to oppose and rebut dissenting papers that slip through their net of referees for major journals, and make frequent alarmist press releases in a losing effort to win over the public. It seems possible that the motivation for all this is to create a climate of fear so that governments will exponentially increase funding for climate research; in that respect they have been very successful.

2.5 The Climate Debate Revisited

In Section 1 we noted that the first two points of controversy in the climate debate are:

Aspect	Alarmists	Extreme Skeptics	Moderate Skeptics
Variability of climate over past few thousand years	Minor	Significant	Significant
Temperature rise over past century compared to natural fluctuations	Significant difference in 20th century	Within past variations	Probably within past variations

In the preceding sections, we have shown that variability of climate over past few thousand years was significant but the data are not good enough to determine how the temperature rise over past century compares to the range of natural fluctuations.

Scientists abhor a vacuum. They can't seem to shrug their shoulders and admit that we just don't know the answer to a vexing problem. They demand explanations, however speculative they may be. Thus, we have theories that have gelled into beliefs on how life started on Earth, how life begins from inanimate matter, how the universe began, and how the climate of the Earth varied over past millennia. In this regard, climate scientists seem to be at the far end of the spectrum; they tend to reach firm conclusions regardless of how flimsy the evidence is to back them up.

There is evidence that the Earth has been primarily in a warming trend during much of the 20th century, although there was a definite hiatus in this rise from 1945 to 1978 and the warming has neither been continuous nor universal. The 20th century also saw a steady rise in CO_2 concentrations in the atmosphere, due presumably to the burning of fossil fuels, land clearing, and cement production. Many scientists (and others) have legitimately become concerned that the greenhouse effect due to this CO_2 increase may be responsible for some or most of this observed rise in temperature, and if left unchecked, could possibly lead to disastrous consequences in the future. In principle, if a sufficiently good global climate model can be produced, the effect of CO_2 concentration on Earth temperature can be calculated. Unfortunately, there are so many variables and unknowns in the Earth system that such estimates can only be made as rough approximations. As in any detective story, if direct evidence is not available, one falls back on circumstantial evidence.

One central issue in this regard is a comparison of the observed temperature rise in the 20th century with estimated variations of temperature in the past millennium or so. If past temperature fluctuations were small compared with the temperature rise in the 20th century, it would suggest that the temperature rise in the 20th century is unique, unprecedented, and likely to be due to factors unique to the 20th century (e.g., greenhouse gases). On the other hand, if past fluctuations prior to industrialization were as large as, or greater than those observed in the 20th century, it might suggest that the temperature rise observed recently might just be another fluctuation such as has occurred in the past. While this argument is not ironclad in either direction, it does provide some valuable insights. Accordingly, the quest for better space–time resolution of historical temperatures has become an important part of the effort to understand the causes of climate change.

There have been a number of studies of historical temperatures conducted in the past, either based on anecdotal records, models of solar variability and climate responses to variable solar intensity, or more likely, based on proxies for past temperature such as tree rings, ice cores, etc. Although numerous papers have pointed out the confounding factors inherent in proxies, "in the land of the blind, a one-eyed man is king" (In modern terms, we may say: "It is the only game in town."). Therefore, despite the problems inherent in the use of proxies, many studies of proxies have abounded in the literature. Some of these fragmentary glimpses of the past have evoked a picture of significant variations in the past climate, with a notable warm period during medieval times, and a relatively cold period called the LIA from about 1600 to about 1850, depending on the criteria used for selection.

The first major global, synoptic, encompassing study of historical global average temperatures from proxies was the "MBH" study reported in 1998 (Mann, Bradley, and Hughes, 1998). This was an audacious effort, encompassing over 1,000 proxies, which provided an unprecedented breadth to the study of historical temperatures. To aid in processing all these data, a sophisticated statistical data-processing methodology (PCA) was utilized. This was particularly remarkable because it was primarily the product of a Ph.D. dissertation by Michael Mann at the University of Massachusetts. This initial paper was followed by several more that extended the analysis further back into the past. The end result of these studies was a historical temperature profile that had the so-called *hockey stick* shape with a relatively flat profile for a thousand years or so, followed by a sudden sharp rise in the 20th century. These papers were compact, full of jargon, and difficult to follow. Sufficient data for others to make independent checks were typically difficult to obtain. Nevertheless, they were impressive papers and gave the impression of being ironclad. As a result of this work, Michael Mann was catapulted from a newly graduated Ph.D. to a position of fame and renown and almost instantly became recognized as a world leader in paleoclimatology.

An assemblage of scientists (and others), hungry for evidence of human-induced global warming, seized on the MBH results as a landmark. The *hockey stick* figure was reproduced and disseminated widely, being offered up as strong evidence of CO_2 induced global warming in the 20th century. The *hockey stick* was adopted by the Intergovernmental Panel on Climate Change of the U.N. (IPCC), Al Gore, and in general, most of the paleoclimatology science community. It appears in many government reports and schoolbooks. The claim was made that the warming in the 20th century was unprecedented, that the 1990s was the hottest decade on record, and 1998 was the hottest year in at least the past millennium or two.

A few years later, McIntyre and McKitrick (M&M) rained on the *hockey stick* picnic. In a series of papers and informal reports, they clearly showed that

1. MBH made an innocent-looking mistake in the principal components analysis by standardizing with a mean based only on the calibration period, instead of a mean based on the entire time period covered by the data. As it turns out, this led to a chain of events that placed undue emphasis on a few highly suspect proxies that produced the *hockey stick* result.
2. Use of certain tree-ring data by MBH was unjustified because much of the observed growth in the 20th century was due to CO_2 fertilization and other factors, rather than a rise in temperature.
3. When a proper recalculation of the MBH data is performed, the result shows that although there was indeed a significant temperature increase in the 20th century, there were comparable high temperatures earlier in the millennium.

Thus, the bases for the claim that the 20th century exhibited an unprecedented temperature rise, and that the 1990s and 1998 were the hottest in a millennium or two, were undermined. In addition to this, there is another factor not usually discussed by the critics. The correlation of the proxies with measured temperature during the calibration period is usually poor, and the extrapolation backward in time for periods much longer than the calibration period is a matter of faith. The use of all proxies democratically mixes in many poor ones with the few good ones, and produces mainly noise.

The responses to the findings of M&M are interesting. Instead of issuing a *mea culpa* and going on from there, Mann dug in his heels and protected turf from truth. He issued a response to the M&M charges that is a masterpiece of evasion and obfuscation, not even mentioning M&M. Most of the paleo-climatology community, which by and large adopted the *hockey stick* as its motif, cooperated by controlling which papers get published in the journals. In general, the U.N., Al Gore, and the climatological alarmists have simply ignored M&M and continued to vouchsafe the *hockey stick*, pretending that the criticism of M&M did not exist (think of The Emperor's New Clothes). For those who are determined to raise the alarm to the world on the dangers of global warming, the *hockey stick* is too valuable as a public message to allow truth to interfere.

Anon. (N) presented a very detailed and generally objective review of surface temperature reconstructions for the past two millennia. Their conclusions are summarized below:

- "The instrumentally measured warming of about 0.6°C during the 20th century is also reflected in various proxy measurements."
- "Large-scale surface temperature reconstructions yield a generally consistent picture of temperature trends during the preceding millennium, including relatively warm conditions centered near 1000 (identified by some as the *Medieval Warm Period*) and a relatively cold period (or *Little Ice Age*) centered near 1700. The existence of a *Little Ice Age* from roughly 1500 to 1850 is supported by a wide variety of evidence including ice cores, tree rings, borehole temperatures, glacier length records, and historical documents. Evidence for regional warmth during medieval times can be found in a diverse but more limited set of records including ice cores, tree rings, marine sediments, and historical sources from Europe and Asia, but the exact timing and duration of warm periods may have varied from region to region, and the magnitude and geographic extent of the warmth are uncertain."
- "It can be said with a high level of confidence that global mean surface temperature was higher during the last few decades of the 20th century than during any comparable period during the preceding

four centuries. This statement is justified by the consistency of the evidence from a wide variety of geographically diverse proxies."[2]

- "Less confidence can be placed in large-scale surface temperature reconstructions for the period from 900 to 1600. Presently available proxy evidence indicates that temperatures at many—but not all—individual locations were higher during the past 25 years than during any period of comparable length since 900. The uncertainties associated with reconstructing hemispheric mean or global mean temperatures from these data increase substantially backward in time through this period and are not yet fully quantified."[3]
- "Very little confidence can be assigned to statements concerning the hemispheric mean or global mean surface temperature prior to about 900 because of sparse data coverage and because the uncertainties associated with proxy data and the methods used to analyze and combine them are larger than during more recent time periods."[4]

---

[2] This statement reflects a generally prevailing implicit view that "the preceding four centuries" were normal, while the relatively higher temperatures at the end of the 20th century are comparatively abnormal. However, the preceding four centuries extend across the LIA, and therefore one might state the proposition differently: Temperatures during the preceding four centuries were colder than they were at the end of the 20th century.

[3] This author cannot find any substantial evidence that temperatures were (as claimed) generally higher in the past 25 years than they were in 900.

[4] This author has very little confidence in estimates of temperature prior to 1600.

3 GLOBAL SCARES, SUBJECTIVE SCIENCE AND CLIMATOLOGISTS

3.1 Global Scares

Global-warming alarmism can be viewed from the broader perspective of global scares as just another in a series of panics that rise and fall; global warming is still in the rise period. Booker and North (2007) documented the details of the rise and fall of various global scares ranging from a wide variety of food scares, to mad cow disease, to dioxins, to the millennium bug, to lead and asbestos, to passive smoking, and finally to global warming. (They failed to cover excessive quarantine of returning astronauts or depletion of the ozone layer by SSTs.) In the introduction to the book *Scared to Death*, Booker and North defined the unifying characteristics of global scares:

> "Each was based on what appeared at the time to be scientific evidence that was widely accepted. Each has inspired obsessive coverage by the media. Each has then provoked a massive response from politicians and officials, imposing new laws that inflicted enormous economic and social damage. But eventually the scientific reasoning on which the panic was based has been found to be fundamentally flawed. Either the scare originated in some genuine threat that had then become wildly exaggerated, or the danger was found never to have existed at all.

By now, however, the damage has been done. The costs have amounted in some cases to billions, even hundreds of billions, of pounds, imposing an enormous hidden drain on the economy. Yet almost all of this money has been spent, it turns out, to no purpose.

What does it say about the psychology of our time that such an extraordinary thing can happen, not just once but again and again? When we examine the pattern behind these scares we find further elements that each has in common.

- The source of the supposed danger must be something universal, to which almost anyone in the population might be exposed, such as eggs or beef, asbestos or climate change.
- The nature of the danger it poses must be novel, a threat that has never appeared in this form before.
- While the scientific basis for the scare must seem plausible, the threat must also contain a powerful element of uncertainty. It must in some way be ill-defined, maximizing the opportunity for alarmist speculation as to the damage it might cause.
- Society's response to the threat must be disproportionate. It is this more than anything which defines a true 'scare'; that, even where the threat is not wholly imaginary, the response to it is eventually seen to have been out of all proportion to its reality."

In Ionescu's play *Rhinoceros*, written for the theater of the absurd, he explores the pressures on people to conform to trends and adopt expanding belief systems. As more and more people turn into rhinos, the pressure to conform by doing likewise becomes intense. Today, we witness just such a pressure on politicians, scientists, and the public to jump on the global-warming bandwagon. Even George W. Bush, who in his tenure as President of the United States had opposed every single effort to legislate even the most mild and moderate steps to improve or protect the environment, began to weaken on global warming in 2007. Former Vice-President Gore led a national campaign to raise consciousness about the dangers of global warming, based heavily on the *hockey stick* model. His efforts netted him the Nobel Peace Prize. The United Nations, through its Inter-governmental Panel on Climate Change (IPCC) has similarly taken an alarmist position, also dependent on the *hockey stick*. The Union of Concerned Scientists, and a number of U.S. governmental agencies have taken similar positions. In addition, quite a large number of scientists have also become very concerned regarding the potential impacts of global warming.

Matt Ridley wrote an essay on the history of apocalyptic predictions.[5] He said:

"Predictions of global famine and the end of oil in the 1970s proved just as wrong as end-of-the-world forecasts from millennialist priests. Yet there is no sign that experts are becoming more cautious about apocalyptic promises. If anything, the rhetoric has ramped up in recent years. Echoing the Mayan calendar folk, the Bulletin of the Atomic

[5] Ridley, Matt (2012) "Apocalypse Not" *http://www.rationaloptimist.com/blog/apocalypse-not.aspx*

Scientists moved its Doomsday Clock one minute closer to midnight at the start of 2012, commenting: 'The global community may be near a point of no return in efforts to prevent catastrophe from changes in Earth's atmosphere.'

Over the five decades since the success of Rachel Carson's Silent Spring in 1962 and the four decades since the success of the Club of Rome's The Limits to Growth in 1972, prophecies of doom on a colossal scale have become routine. Indeed, we seem to crave ever-more-frightening predictions-we are now, in writer Gary Alexander's word, apocaholic. The past half century has brought us warnings of population explosions, global famines, plagues, water wars, oil exhaustion, mineral shortages, falling sperm counts, thinning ozone, acidifying rain, nuclear winters, Y2K bugs, mad cow epidemics, killer bees, sex-change fish, cell-phone-induced brain-cancer epidemics, and climate catastrophes."

He discussed these specific fears in some detail. His final conclusions were:

"Over the past half century, none of our threatened eco-pocalypses have played out as predicted. Some came partly true; some were averted by action; some were wholly chimerical. This raises a question that many find discomforting: With a track record like this, why should people accept the cataclysmic claims now being made about climate change? After all, 2012 marks the apocalyptic deadline of not just the Mayans but also a prominent figure in our own time: Rajendra Pachauri, head of the IPCC, who said in 2007 that 'if there's no action before 2012, that's too late … This is the defining moment.'

So, should we worry or not about the warming climate? It is far too binary a question. The lesson of failed past predictions of ecological apocalypse is not that nothing was happening but that the middle-ground possibilities were too frequently excluded from consideration. In the climate debate, we hear a lot from those who think disaster is inexorable if not inevitable, and a lot from those who think it is all a hoax. We hardly ever allow the moderate 'lukewarmers' a voice: those who suspect that the net positive feedbacks from water vapor in the atmosphere are low, so that we face only 1 to 2 degrees Celsius of warming this century; that the Greenland ice sheet may melt but no faster than its current rate of less than 1 percent per century; that net increases in rainfall (and carbon dioxide concentration) may improve agricultural productivity; that ecosystems have survived sudden temperature lurches before; and that adaptation to gradual change may be both cheaper and less ecologically damaging than a rapid and brutal decision to give up fossil fuels cold turkey.

We've already seen some evidence that humans can forestall warming-related catastrophes....

Just as policy can make the climate crisis worse – mandating biofuels has not only encouraged rain forest destruction, releasing carbon, but driven millions into poverty and hunger-technology can make it better.... Humanity is a fast-moving target. We will combat our ecological threats in the future by innovating to meet them as they arise, not through the mass fear stoked by worst-case scenarios."

3.2 Dealing With Subjective Science

3.2.1 Nature of Subjective Science – Emergence of Consensus

There are many phenomena in nature that, for one reason or another, are not susceptible to verification by independent testing. These typically include events that either occurred a long time ago or that occurred at distant sites not accessible to us, or both. Examples include the expansion of the Universe after the Big Bang, the variations in climate of the Earth in the past or in the future, the origin of life on Earth, putative existence of life elsewhere in the Universe, the evolution of species on Earth, continental drift, and other such topics. There is no way to go back into the past or travel great distances to directly verify hypotheses. Although the remnants of the past may be discernible to some extent in proxies that exist in the present, these tend to have significant limitations. For such phenomena that occurred long ago and/or in distant locations, scientists create hypotheses that would "explain" how these processes might have occurred in conformity with the known laws of science. If these hypotheses provide a reasonable explanation of phenomena and are in conformity with scientific laws, they acquire the elevated status of a theory. Such a theory is typically not unique and represents one viewpoint—often a preconceived viewpoint. It provides one possible explanation for events that cannot be verified by any known means. Conjecture for things improvable is a safe venture—no one can ever prove you wrong. It is far more dangerous to predict tomorrow's weather than it is to predict the climate 100 years from now—tomorrow's weather is subject to practical test. I call this kind of science "subjective science". It is not amenable to detailed verification such as the laws of motion. While some subjective science has strong foundations (e.g. evolution, continental drift) the foundation of almost every subjective aspect of climate change are weak.

Scientists do not seem to be able to shrug their shoulders and admit that we just don't know the answers to some questions. What happens is that one of the unprovable hypotheses in a subjective science gains popularity amongst scientists and is regarded by the majority as the most credible. When a significant number agree, a consensus evolves. The consensus acts like a gigantic gravitational field, drawing in more and more scientists. Eventually, the consensus gels, and ultimately hardens into a belief system — an orthodoxy. The foundations are often weak, and not understood by the public. The emergence of the consensus as the essence of reality in science has replaced scientific skepticism, and as Lindzen (2008) has noted: "simulation and programs have replaced theory and observation, where Government largely determines the nature of scientific activity." As Lindzen (2008) has emphasized, "the bulk of the educated public is unable to follow scientific arguments; 'knowing' that all scientists agree relieves them of any need to do so." Taking issue with the consensus "serves as a warning to scientists that the topic at issue is a bit of a minefield that they would do well to avoid." It should also be noted that many climatological publications are so full of jargon and so obscure that they are unreadable except to a very few narrow specialists. So, not only the general public, but even much of the science community is unable to digest these abstruse treatises.

The consensus acquires legitimacy in proportion to the number and prominence of the scientists who subscribe to it. As the consensus becomes firmly imbedded in culture, it acquires the respect usually accorded to fact. However, as Crichton (2003) said:

> "Let's be clear: the work of science has nothing whatever to do with consensus. Consensus is the business of politics. Science, on the contrary, requires only one investigator who happens to be right, which means that he or she has results that are verifiable by reference to the real world. In science consensus is irrelevant. What is relevant is reproducible results. The greatest scientists in history are great precisely because they broke with the consensus. There is no such thing as consensus science. If it's consensus, it isn't science. If it's science, it isn't consensus. Period."

Curry and Webster[6] wrote a review of the notion of consensus in science, with particular regard to climate change. They review a number of philosophical questions reagrding the role of consensus in science. They concluded:

> "Arguing from consensus to enforce conclusions does not work with the extended peer community. What is needed are serious attempts to engage the extended peer community with the modes of expert reasoning used to reach those conclusions."

There seems to be quite a bit of confusion in the world of climate science as to what the consensus is consenting to. For example, Curry and Webster focused on a IPCC conclusion that warming in the 20th century was primarily caused by anthropogenic generation of greenhouse gases. The report by Curry and Webster seems to follow two paths. One is the "consensus findings" of the IPCC regarding the role of greenhouse gases on warming over the past

---

[6] J. A. Curry and P. J. Webster (2012) "Climate change: no consensus on consensus", CAB Reviews (in press).

century, and the other is a review of philsosophical views of consensus in science. While the role of greenhouse gases in warming over the past 100 years is somewhat relevant to the issue, the real issue we face is: For various future scenarios of world energy consumption by sector and fuel, what are the expected environmental consequences? Assuming that the more the future resembles business as usual, the environmental impacts are greater, what is the technical feasibility and cost of shifting from future scenarios with greater environmental impact to scenarios with lesser environmental impact? Is there a consensus that we know the answers to these questions? I think not.

Ward (2008) also discussed the scientific method, for which he emphasized: "No hypothesis is considered proven until it has undergone rigorous scientific review and testing, and other scientists must be able to replicate the tests or experiments and achieve the same results." Unfortunately, his notion of scientific rigor seems to be that it is supported by a consensus. His report emphasizes the necessity of repeatability in the scientific method; for example, it says: "A scientist tells how he or she arrived at a conclusion in enough detail so that another investigator can follow the same trail, examine the same data, and get the same answer." Yes, repeatability is *necessary*, but not *sufficient*. If the method of processing the data is arbitrary or possibly improper, successive investigators can reproduce the results but they will be just as arbitrary or inaccurate as they were before they were replicated. Ward (2008) mixes up two requirements of the scientific method. There is a huge difference between, on the one hand, applying a hypothesis to a large array of independent phenomena vs. on the other hand, applying a hypothesis to a single case (with adjustable parameters) in such a manner that those who follow can duplicate the results by repeating the same steps. The problem with climate change is that very little in this field of endeavor is "proven" and most of the "accumulated advances in understanding of climate change" over the past 20 years range from arbitrary (climate model results) to improper (the so-called *hockey stick* result).

3.2.2 Consensus to Orthodoxy

Orthodoxy is a belief system. Thus, in regard to global warming, we have believers (alarmists) and non-believers (skeptics). Otherwise intelligent people often discuss whether or not they "believe in global warming."

But like any religion, scientific orthodoxies cannot tolerate disagreement with the orthodoxy. Therefore the alarmists have politicized the science of climatology to enforce their views. Lindzen (2008) described the politicalization of climate science:

> "All such organizations, whether professional societies, research laboratories, advisory bodies (such as the national academies), government departments and agencies (including NASA, NOAA, EPA, NSF, etc.), and even universities are hierarchical structures where

positions and policies are determined by small executive councils or even single individuals. This greatly facilitates any conscious effort to politicize science via influence in such bodies where a handful of individuals (often not even scientists) speak on behalf of organizations that include thousands of scientists, and even enforce specific scientific positions and agendas.

The temptation to politicize science is overwhelming and longstanding. Public trust in science has always been high, and political organizations have long sought to improve their own credibility by associating their goals with 'science'—even if this involves misrepresenting the science."

Occasionally, a counter-argument is published as, for example, the *New York Times* article[7] of Freeman Dyson's opposition to the orthodoxy on global warming, but this only serves to confuse the public a bit.

The world of science seems to have lost its foundation of skepticism. Instead of doubt and dialectic opposition, science has adopted orthodoxy and consensus. Scientists, like the public at large, seem unable to shrug their shoulders and simply admit that we just don't know the answers. The fierce competition for funding in an environment dominated by orthodoxy pressures scientists to bias their viewpoints. We note a significant rise in the number of news releases and papers by scientists with phrases such as "there might be ...," or "it is possible that ..." What science cannot seem to do these days is accept that:

"Sometimes there is no alternative to uncertainty except to await the arrival of more and better data" (Wunsch, 1999).

It seems likely that scientific (or economic) progress in climatology will be impeded by the fact that data and models are routinely biased to adhere to a belief system. The IPCC has led the way with a plethora of conclusions and predictions regarding the role of CO_2 emissions on the Earth's climate and the potential impact on humanity. These represent mainly political, not scientific conclusions. The majority of recognized climatologists have aligned like weather vanes to the prevailing wind, making it all but impossible to get contrary views published in journals. As a result, there has arisen a blogopolis in which contrary views are available on websites but not in the literature. While many of these blogs are populated by moronic entries, a few are full of detailed analysis and data that you cannot find in the journals.

---

[7] *http://www.nytimes.com/2009/03/29/magazine/29Dyson-t.html?_r=1&scp=1&sq=freeman%20dyson&st=cse*

3.2.3 Counting Adherents to the Orthodoxy

To further the ends of the alarmists, some academicians have carried out counting studies where they sum up the number of scientists who subscribe to the alarmist persuasion. Oreskes (2004) built her argument in favor of anthropogenic global warming based on a list of who supports the hypothesis, rather than the scientific basis for it. She emphasized that the IPCC, the National Academy of Sciences, the American Meteorological Society, the American Geophysical Union, and the American Association for the Advancement of Science (AAAS) "all have issued statements in recent years concluding that the evidence for human modification of climate is compelling." In a study of 928 abstracts published in refereed scientific journals between 1993 and 2003, she did not "find one paper that disagreed with the consensus position." However, at the end, she provided an escape clause:

"The scientific consensus might, of course, be wrong. If the history of science teaches anything, it is humility, and no one can be faulted for failing to act on what is not known ... Many details about climate interactions are not well understood, and there are ample grounds for continued research to provide a better basis for understanding climate dynamics."

Three years later, she concluded that global warming due to greenhouse gases "is an established scientific fact" (Oreskes, 2007a). In 2007, she gave a presentation of 109 slides to the American Meteorological Society (2007b). That a social scientist would have 109 slides' worth of information to convey about climate change speaks to the role of consensus as a force in science. Her first slide quoted no less an expert on climatology than Arnold Schwarzenegger. The Terminator said "the debate is over" (pun intended). Her second slide proclaimed: "There is a scientific consensus over the reality of anthropogenic global warming." She then proceeded to quote various authorities to show that a consensus exists. Unfortunately for her, she ventured briefly into the science of climate change where she got in well over her head. Using the "*hockey stick*" result, she concluded that carbon dioxide as the cause of warming in the 20th century is "not just a correlation—it's a confirmation of a prediction—the scientific method." It is not clear which scientific method she refers to – certainly none that I am aware of. Of course, the real issue is not whether anthropogenic generated CO_2 contributed to warming in the past ~ 100 years; the real issue is what will be the consequence of further emissions in the 21st century? That is a totally differnt and more difficult question. Again I refer to Miyogi: "Answer only matter if ask right question".

Oreskes, Conway, and Shindell (2008) wrote a 70-page treatise on CO_2 as the putative cause of global warming in which the entire argument is based on a comparison of the number and credentials of those who *believe in it* vs. the number and credentials of those who *disbelieve* it. It is worth noting that the issue here has degenerated down to what opposing camps *believe* rather than a question of what the data tell us. The discussion was highly biased. For example, the authors refer disparagingly to "challenges to climate science" as if those of the alarmist persuasion are climate scientists whereas the challengers are something other than climate scientists. These authors are social scientists and don't appear to have a clue regarding the science of global warming. As it turns out, the majority of paleoclimatologists do accept the thesis widely promulgated by Al Gore, the U.N., NOAA, the National Academy of Sciences, and other predominant professional organizations, that CO_2 emissions were the cause of global warming in the 20th century, and this warming will increase in the 21st century in proportion to further emissions, causing great misery for humanity. This is the orthodox institutional viewpoint that is taught to schoolchildren and widely promulgated by academia. As institutions and organizations continue to dominate over individuals in these matters, a consensus builds up on each topic.[8]

However, the degree of consensus has been exaggerated (Schulte, 2008). It is also noteworthy that a number of prominent European and Russian climatologists and scientists have aligned themselves with the skeptics.

The French journal *La Meteorologie* criticized a French scientist for taking "an opposing stance" to "*the prevailing ideology.*" The French scientist said "The term 'consensus', a term often employed by the IPCC, implies only that the 'good' keepers of the true faith are in the majority, meaning that mere weight of numbers (not necessarily synonymous with better quality) may control and dismiss discordant voices from publications and papers." Kondratyev *et al.* (2003) describe the IPCC Report as "less than scientific, speculative opinions" and describe climate models as being "forced to fit the observational data" through adjustment of parameters. They claim that reductions of greenhouse gas emissions based on such models "are senseless." Bischof (2000) said:

---

[8] An anecdote illustrates the point. At a recent NSF workshop *Reversing Global Warming: Chemical Recycling and Utilization of CO₂* I presented a talk showing why the *hockey stick* representation of past temperatures was incorrect. A representative of the NSF raised a question at the end of my talk. She asked: "Why should I believe you when the National Academy of Sciences says otherwise?" She was relying on the institution over the individual. Ignoring the data that I presented, she fell back on reliance on the consensus. The issue was no longer whether my data and analysis were accurate, but rather, whether more prestigious organizations took a contrary position. Ayn Rand must be turning over in her grave! While I was giving my talk, one attendee of the alarmist persuasion stomped out the meeting hall audibly cursing.

"The reader should understand that in science, as in other sectors of public life, the outcome of a study is often guided, if not determined, by an *a priori* idea, a tenet. In the case of global warming, this belief was that, if enormous amounts of greenhouse gases are released into the atmosphere, a temperature rise must occur. This prior assumption has guided scientific thinking and triggered a true deluge of investigations all desperately trying to prove just that."

3.3 Hyperbole on Impacts of Global Warming

One of the favorite tactics used by alarmists is to focus on a short-term upward trend and predict doomsday from extrapolation of this trend into the future. Usually, the trend turns out be an upward lobe of an oscillation. The warming induced by the great El Niño of 1997–1998 provided grist for this mill and the alarmists were in their glory in the aftermath of that year. Michaels and Balling, 2009 (pp. xii and 127) provide a good example of selective presentation of short-term data to create the impression of a climatic trend.

Alarmists tend to see long-term danger in all short-term variations. In 1974, *Time Magazine* ran an article[9] entitled "Another Ice Age?" in which it was stated:

"... when meteorologists take an average of temperatures around the globe they find that the atmosphere has been growing gradually cooler for the past three decades. The trend shows no indication of reversing. Climatological Cassandras are becoming increasingly apprehensive, for the weather aberrations they are studying may be the harbinger of another ice age."

The article goes on to state: "the atmosphere has been gradually cooling for three decades". It cites expanding pack ice and snow cover, and changing polar winds. Ironically, it claims that humans may be involved in the cooling trend via dust released into the atmosphere from farming and fuel burning. Twenty years later, alarmists completely reversed their worries.

On April 26, 2007, James E. Hansen (perhaps the most well-known global-warming alarmist) gave testimony on the dangers of global warming to the Select Committee of Energy Independence and Global Warming of the U.S. House of Representatives. Hansen provided the case for alarmists in considerable detail. Only a few quotations are given here.

According to Hansen, the greatest near-term danger is sea level rise. He said that "sea level is already rising at a rate of 3.5 cm per decade and the rate is accelerating" due primarily to "ice sheet disintegration." He said: "there is increasing realization that sea level rise this century may be measured in

---

[9] *http://www.time.com/time/magazine/article/0,9171,944914,00.html*

meters if we follow business-as-usual fossil fuel emissions," and that "adaptation to a continually rising sea level is not possible." Hansen concluded "increasingly rapid changes on West Antarctica and Greenland ... are truly alarming."

One of the major slow feedback processes that Hansen identified is "the effect of warming on emissions of long-lived greenhouse gases," caused by the "melting of tundra in North America and Eurasia," which "is observed to be causing increased ebullition of methane from methane hydrates."

Hansen said: "continued business-as-usual greenhouse gas emissions threaten many ecosystems," and that "very little additional [climate] forcing is needed ... to cause the extermination of a large fraction of plant and animal species."

He also said: "Earth's history shows that climate is remarkably sensitive to global forcings" and that "positive feedbacks predominate," causing "the entire planet to be whipsawed between climatic states."

Summarizing, Hansen said: "The dangerous level of CO_2 is at most 450 ppm, and it is probably less ... Ignoring the climate problem at this time, for even another decade, would serve to lock in future catastrophic climatic change and impacts that will unfold during the remainder of this century and beyond." The Earth "is close to dangerous climate change, to tipping points of the system with the potential for irreversible deleterious effects ... The planet is on the verge of dramatic climate change." We "are forced to find a way to limit atmospheric CO_2 more stringently than has generally been assumed ... We cannot shrink from our moral responsibilities ... to preserve the planet for future generations."

Idso and Idso (2007) reviewed this testimony and provided a skeptical commentary and critique.

Alarmists have found it rewarding to engage in a contest of "Can you top this?" by issuing a constant barrage of press releases about what supposedly "may", "might", or "could" happen in the future as a result of putative global warming. If you go to *Google* and punch in "global warming" you get thousands of responses predicting disaster from global warming. These include claims such as:

(1) the role of obese people in contributing to global warming by requiring extra resources; (2) "climate change may be century's greatest health threat"; (3) "pets may be the latest victims"; (4) "climate change may halve South Africa"; (5) "increased incidence of tropical diseases, food shortages, natural disasters and heat waves threaten global ..."; (6) "climate change may drive refugees to Australia"; (7) "how climate change may be threatening national parks"; and (8) "climate change will overload humanitarian system"—with many more like this.

In technical press releases, there is a strong bias to portray data in the worst possible light. Temperatures have meandered since the hot year of 1998 induced by a huge El Niño. As it turns out, 2008 was an unusually cold year in which much of the warming of the previous 20 years (or more) was mitigated. But in early 2009, the U.S. National Weather Service (NWS) published a news release that said: "2008 was the 39th warmest year since 1895." Of course, 1895 was a very cold baseline year. Since 1930, 39 out of 79 years were warmer than 2008—hardly a basis for alarm. Yet the NWS made it seem as if 2008 was an exceptionally hot year! Similarly, the NOAA website reports the 2009 temperature as "the nth warmest year on record out of 130 years" where n is typically in the range 6 to 9. But the 6 to 9 years that were warmer than 2009 all occurred since the hot year of 1998 induced by a huge El Niño; hence 2009 is stacking up as a relatively cold year compared with the previous decade!

Most recently, Hansen *et al.* (2008) now claim:

> "If humanity wishes to preserve a planet similar to that on which civilization developed and to which life on Earth is adapted, paleoclimate evidence and ongoing climate change suggest that CO_2 will need to be reduced from its current 385 ppm to at most 350 ppm."

In order to reduce the CO_2 content of the atmosphere to below 350 ppm in the 21st century, such draconian measures would be needed that it would imply the end of a modern technological world.

3.4 Spreading the Gospel

Ward (2008) reported on a series of workshops dealing with communication of science results to the public, with particular emphasis on climate change in which it was claimed that "the nation's top climate scientists and leading science and environmental journalists [met] together to discuss media coverage and communication of climate change science." The principal funder of the workshops project and its report was the Paleoclimate Program, Division of Atmospheric Sciences, National Science Foundation. Some financial support was also provided by grants from the U.S. Environmental Protection Agency's Office of Air Programs, and limited in-kind support was provided by the National Centers for Coastal and Ocean Science (NCCOS), the scientific research arm of the National Oceanic and Atmospheric Administration's National Ocean Service (NOAA/NOS) in the U.S. Department of Commerce; and by the National Aeronautics and Space Administration (NASA). David Verardo, Ph.D., of the National Science Foundation, was the primary person who enabled this science/journalism project. While this report is portrayed as an effort to promote better communication on climate change, its real intent seems to be to promote the alarmist viewpoint. For example, it begins:

"Climate scientists were frustrated by what they saw as a failure of the general public to understand and appreciate the seriousness of the climate change issue [i.e., the grave dangers according to the alarmist viewpoint]. Many scientists said they were frustrated that the accumulated advances in understanding of climate change over more than two decades of research had not led to a better-informed public [i.e., the use of climate models to predict large temperature increase in the 21st century] ... The workshops focused in particular on what scientists call 'anthropogenic climate change'—that caused by human activities and not part of a natural cycle."

Like ("Fair and Balanced") Fox News on TV, this report suggests:

"The preponderance of scientific evidence had since accumulated to a point where responsible reporters should give the scientific consensus on anthropogenic climate change much greater weight than dissenting claims challenging the mainstream scientific conclusions. The journalistic tenet of accuracy now demands that *the established science be given total or near total prevalence* in coverage of certain aspects of climate change science." (*emphasis added*)

In other words, only the prevailing orthodoxy should be reported by the media. This attempt to muzzle the opposition is outrageously anti-scientific and anti-American. The report goes on to say:

"Many participating reporters said they were having trouble convincing their editors of the virtues of reporting in an accurate and fair [*i.e., alarmist*], rather than quantitatively balanced fashion [*i.e., roughly equal time to both sides*]. Their reporting on new scientific findings often met with an editor's insistence that they also report the perspectives of climate science contrarians who lack comparable scientific expertise and standing, as if covering a political campaign or a public policy dispute" [*i.e., contrarians are thereby characterized as uneducated and occupying positions of less knowledge and importance compared with alarmists.*]

This viewpoint was echoed by Curry *et al.* (2006) who said:

"Boykoff and Boykoff (2004) demonstrated that superficial balance in coverage of global warming by the U.S. "prestige press" (e.g., *New York Times, Washington Post, Los Angeles Times, Wall Street Journal*) can actually be a form of informational bias. Boykoff and Boykoff state that by giving equal time to opposing views, the major newspapers are significantly downplaying scientific understanding of the role humans play in global warming. Pitting what 'some scientists have found' against what 'skeptics contend' implies a roughly even division within the scientific community. In the media debate on global warming and hurricanes, greenhouse-warming deniers (which, in addition to scientists, includes lawyers and others with at best minimal scientific credentials) are set side by side with scientists who have actually done the work and published papers on the subject".

Note that Curry *et al.* (2006) said that Boykoff and Boykoff "demonstrated" their claim. Curry *et al.* describe skeptics as "greenhouse-warming deniers" and are said to include some with "at best minimal scientific credentials". But intelligent skeptics do not deny the greenhouse effect. They merely doubt that we can quantitatively affirm feedback effects based on present knowledge. Furthermore, there are a great many strong supporters of climate alarmism who are clearly lacking in "scientific credentials". Naomi Oreskes, who has written articles and books and made presentations in favor of alarmism is a prime example. In addition, the fact that a climate scientist may have done work and published papers means little in most cases. Most scientists are narrow specialists and do not have a synoptic view of the field. Many are immersed in their narrow slice of the pie. In fact, as Curry *et al.* (2006) emphasized in their paper, several of the logical fallacies that they attribute to deniers of the relation between SST and hurricanes, are the very same fallacies utilized over and over again by alarmists: *scientific credentials, appeal to authority, appeal to motive, begging the question, hasty generalization, and fallacy of the single cause.*

Now there is an entire website dedicated to improving communicating climate change and climate science to the public (that is, the alarmist view of climate change).[10]

As climate change has evolved into a big business with plenty of funding, more and more peripheral (typically non-scientific) organizations have sought a piece of the action, introducing economic, social science and communication aspects. A common theme of some studies (such as that of Ward, 2008) is to seek better ways of communicating *the* climate science (of the orthodoxy) to the public. For example, the Carsey Institute[11] posed the following three questions in a survey under the heading: "What do you personally *believe*". Hence they framed the issue as a *belief* system.

[On] the issue of global warming or climate change, how much do you feel you understand about this issue—would you say

- a great deal,

- a moderate amount,

- only a little, or

- nothing at all?

Which of the following two statements do you think is more accurate?

---

[10] http://talkingclimate.org/guides/communicating-climate-science/
[11] Carsey Institute Issue Brief No. 26: "Climate Change Partisanship, Understanding, and Public Opinion" http://www.carseyinstitute.unh.edu

• Most scientists agree that climate change is happening now, caused mainly by human activities.

• There is little agreement among scientists whether climate change is happening now, caused mainly by human activities."

Which of the following three statements do you personally *believe*?

• Climate change is happening now, caused mainly by human activities.

• Climate change is happening now, but caused mainly by natural forces.

• Climate change is not happening now."

But these are all silly questions. Climate change has always occurred and is always occurring. Regardless of the *beliefs* of the public, human activity *is* contributing to climate change. The real question is how additional emissions of greenhouse gases will affect the climate in the future quantitatively, and what will be the impact on humanity? These questions are rarely posed, and indeed, there are no clear answers.

George Mason University and Yale University also conducted a similar study[12] asking: "Do you think that global warming is happening?" and similar questions. They categorized the public into six groups: alarmed, concerned, cautious, disengaged, doubtful and dismissive in accordance with their "*beliefs*". They managed to fill up 57 pages with detailed breakdowns of *belief* systems.

Even the Catholic Church has gotten the message. They urge their brethren to: "Contact your members of Congress and urge greater U.S. leadership to address climate change, especially its disproportionate impact on poor and vulnerable people here and abroad".[13]

3.5 Climatologists

Weiler *et al.* (2011) provide a very interesting insight into the personalities of climate scientists. Personality types of interdisciplinary, Ph.D. climate change researchers were collected based on a Jungian type personality assessment (described below). Each person was characterized by four personality traits as shown in Table 3.1. Climate researchers were compared with the general public as shown in Table 3.2. One thing stands out. There is a huge statistical inversion between climate scientists vs. the public in that climate scientists greatly lean toward intuition whereas the public heavily leans toward sensing. This implies the climate scientists "focus on theories" and

---

[12] George Mason University, "Climate change in the American mind Americans' Global Warming Beliefs and Attitudes in May 2011"
environment.yale.edu/climate/files/ClimateBeliefsMay2011.pdf;
environment.yale.edu/climate/files/SixAmericasMay2011.pdf
[13] United States Conference of Catholic Bishops, "Global Climate Change", February 2011.

"follow hunches to reach conclusions" whereas the public tends to "focus on experience" and "build carefully and logically towards conclusions". The strange thing is that one would expect that the very nature of the scientific method requires that scientists should focus on sensing, rather than intuition. In addition, there is also a much stronger tendency of climate scientists to prefer judging to perceiving, and there is a somewhat greater tendency of climate scientists to prefer thinking to feeling. Thus climate scientists tend to "prefer to make decisions quickly, come to closure and move on". This is clearly evident in the many papers in climatology that utilize a penny's worth of data to draw a dollar's worth of conclusions. In fact one might say in a Churchillian sense: Rarely have so many drawn so many conclusions from so few reliable data.

Table 3.1. Personality Traits (Weiler *et al.*, 2011).

Extraversion	Intraversion
Think out loud in discussions, talk more than listen	Process information internally, listen more than talk
Share ideas immediately	Share ideas after careful reflection
Sensing	**Intuition**
Focus on experience	Focus on theories
Build carefully and logically towards conclusions	Follow hunches to reach conclusions
Want details	Want big picture, become bored or impatient with details
Anchored in the present, relate to the past	Oriented towards the future
Prefer step-by-step information or instructions	Talk in general terms
Ask "what" and "how" questions	Ask "why" questions
Look for facts	Look for patterns and possibilities
Prefer practical, plain language to symbols, metaphors, theories or abstractions	Use metaphors, analogies and other symbolic language
Thinking	**Feeling**
Present information using cause-and-effect reasoning	Use personal situations, stories and examples to communicate
Analytical	Empathetic
Need to know "why"	Connect with people
Judging	**Perceiving**
Prefer to make decisions quickly, come to closure and move on	Prefer to stay open to new information and last-minute options
Uncomfortable with free-flowing discussions	Feel confined by detailed plans and final decisions
Prefer focused discussion and options	Prefer open discussion to explore linkages between topics

Table 3.2. Comparison of Personality Traits of Climate Scientists with those of the general public (Weiler *et al.*, 2011).

Personality Trait	Climate Scientists vs. Public
Extraversion/Intraversion	Climate scientists similar to general public (roughly 50% extravert and 50% intravert)
Sensing/Intuition	Climate scientists were far more likely to use intuition (82%) over sensing (18%) than the general public that preferred sensing (73%) vs. intuition (27%)
Thinking/Feeling	Climate scientists were somewhat more likely to use thinking (49%) over feeling (51%) than the general public that preferred feeling (60%) vs. thinking (40%)
Judging/Perceiving	Climate scientists were far more likely to use judging (73%) over perceiving (27%) than the general public but was more even with judging (54%) vs. intuition (46%)

Martin (1979) wrote an interesting report in which he described the biases that inevitably creep into scientific research and reporting. According to Martin, scientists "do not disinterestedly look at the available evidence, do not make a balanced analysis, and do not present results in a neutral manner." Instead, he suggested "from the beginning [they] support or favor a particular conclusion, and in a number of ways organize their scientific work so as to selectively support this conclusion." He labeled this as "pushing the argument". Martin argued that pushing scientific arguments is inevitable, and therefore pushing should not reflect unfavorably upon the competence or integrity of the scientist. He said:

"The important thing is not to eliminate pushing, which is impossible anyway, but to recognize that it exists ...Neither does the existence of pushing automatically imply that a scientist's results or conclusions are unjustifiable or wrong. While a scientist's argument may be judged on the basis of current understanding to be pushed, it may eventually be vindicated. Or it may not."

He went on to say:

"A scientist in developing an argument to support an hypothesis draws evidence from a number of sources. In presenting evidence one must always be selective — all the evidence and arguments cannot be presented. Often different authorities support different viewpoints, present different 'facts', and offer different interpretations of evidence. Depending on the field, a scientist may draw sound support for many points of view and find some support for nearly any view. Therefore it is easy for a scientist, knowingly or unknowingly, to push an argument by selective choice and use of available evidence."

Most of Martin's treatise was framed in terms of the debate during the early 1970s as to whether emissions from high-flying supersonic transports (SSTs) would destroy the ozone layer and thereby endanger the Earth's population by exposure to excessive radiation. For purposes of discussion, he presented detailed summaries and reviews of two prime scientific papers in the field with contrasting approaches. One paper was said to contain "the built-in assumption that the burden of proof lies with those who claim that SSTs are safe: that all that he must demonstrate is that there is at least some small possibility of danger." By contrast, the other paper used "the [implied] assumption that the burden of proof lies with those who claim that SSTs are dangerous to ozone: that all [they needed to] demonstrate was that the likelihood of significant danger was small."

Above all, Martin emphasized that scientists are human beings, motivated by various forces and factors in their lives. According to Martin:

"People tend to selectively observe and interpret information in a way that supports their preconceived ideas. Because of this, the personal commitments of individual scientists can help to explain the link between the scientists' presuppositions and their pushing of arguments ...In a scientist, this process might operate as follows. The scientist starts with an original idea or hypothesis, perhaps arrived at as a creative solution to a certain problem. In testing or validating the idea, the scientist will tend to notice and use supporting evidence and arguments. Data that seems mainly supportive will be studied, analyzed and applied so that every possible advantage can be drawn from it. Seemingly irrelevant or inconclusive items will be filtered from advantageous components, or interpreted in a way that promotes the argument. Evidence that seems mainly to contradict or challenge the argument at hand may be ignored completely or explained away or reinterpreted and twisted into support for the argument.

Some of the ways in which a person may deal with a challenging item of information are (1) flat denial of the item; (2) skepticism about the source of the item; (3) ascription of a motive to the source of the item; (4) isolation of the item from the context of one's attitude; (5) minimization of the importance of the item; (6) interpretation of the item to suit one's purpose; (7) misunderstanding of the item; and (8) thinking away or just forgetting the item."

According to Martin, one may often detect a deep-rooted personal commitment or bias of a scientist by examining a series of published papers and detecting a constancy of attitude that repeats itself from year to year. He claims:

"The idea that scientists are often strongly committed or biased is quite compatible with the fact that scientists are human beings ...That is, they are subject to motivations and failings similar to those of other people. They may strive for money, power and prestige; they may work for the satisfaction of a job well done or for revenge or to relieve boredom; they may make terrible blunders as well as have brilliant insights. It is sometimes said or suggested that scientists, at least when it comes to their work, live on a higher moral plane than other mortals. Don't believe it!"

In examining the literature on SST emissions and their impact on the ozone layer, Martin concluded:

"From my point of view, the authors do not disinterestedly look at the available evidence, do not make a balanced analysis, and do not present results in a neutral manner. Rather, it appears to me that the authors from the beginning support or favor a particular conclusion, and in a number of ways organize their scientific work so as to selectively support this conclusion."

These claims made by Martin (1979) are backed up by lengthy and detailed discussions and analyses that seem quite credible to this writer.

Michaels and Balling (2009) devoted a chapter to "Pervasive Bias and Climate Extremism". They began with an analogy. If one starts with a particular weather prediction for a locality, the next update of that prediction might forecast an increase or a decrease in the predicted temperature. There is an equal probability for either outcome. However, in regard to the climate-CO_2 connection and the predicted severity of future global warming, the overwhelming majority of new published papers find that the climate-CO_2 connection is strengthened and the predicted severity of future global warming is increased. Only rarely do papers get published with opposite conclusions. Alarmists might argue that the original hypothesis of CO_2-induced global warming becomes solidified as more data and better analyses accumulate. On the other hand, we have already demonstrated that *cabals* rule the review process for major journals, favoring those of the alarmist persuasion. We have also noted that the majority of published climatologists favor the alarmist position and this influences their perspectives. The alarmist position is also favorable to receiving funding for research. Michaels and Balling (2009) discussed the so-called "file-drawer problem" in which

"Negative results are generally considered not noteworthy.... Scientific journals are skewed by a prejudice for the publication of statistically significant, 'positive' results and prejudiced against findings of no relationship between hypothesized variables.... For any given research area, one cannot tell how many studies have been conducted but never reported."

They also quote Stephen Jay Gould who said publication bias results from "prejudices arising from hope, cultural expectation or ... a particular theory dictate that only certain kinds of data will be viewed as worthy of publication , or even documentation at all".

In the 30 years that have passed since Martin wrote this report, several major changes have taken place in the way that scientific information is distributed. With the advent of the Internet, the monopoly of scientific journals has been weakened. Other cultural changes have taken place. Of some relevance is the fact that scientists are now far more prone to issue press releases on their work prior to publication, and these tend to find their way onto many websites. Other scientists, disagreeing with the orthodoxy of the consensus, have difficulty getting published in the journals. A number of so-called web blogs dealing with climate change have emerged over the past several years, and these have become foci for discussions and commentary. Some blogs are rabidly one-sided and present forums for either alarmists or naysayers to agree with one another. Any moron can voice his or her opinion. Two blogs that stand out above the others are *climateaudit.org* which has become a universal watchdog for reviewing statistical analysis of large data sets, and *judithcurry.com* which provides an even-handed forum for both sides. The *judithcurry.com* blog has emerged in 2010-2012 as by far the best source of new ideas in climate science analysis with many stimulating new posts by Judith Curry. Unfortunately, the responses on these blogs have become so numerous (typically many hundreds) that the wheat often gets lost in the chaff. It is particularly disappointing to observe that a limited number of adherents clog up the responses to Judith Curry's stimulating posts with mostly irrelevant, trivial or nonsensical entries. Most of these responses are contributed with supercilious attitudes under psuedonyms. *pielkeclimatesci.wordpress.com/* is also a very informative website.

Ward (2008) extolled the peer review system and

"...warned about the growing number of unvetted publications being distributed through an expanding number of electronic and online outlets ... Publishing online ... has become an increasingly popular way to circumvent more rigorous peer review altogether ... the public and the media need to be attuned to these trends and distinguish them from highly respected professional peer-reviewed journals."

However, in many cases, peer review has become subject to political correctness, and assures that only one viewpoint will be heard. Michaels and Balling (2009) provided a number of examples of publication bias by the journals *Science* and *Nature*. Perhaps the most egregious example of politicization of science is the journal *Nature* that has become an alarmist propaganda medium. For example, the April 30, 2009 issue of *Nature* includes three articles that are essentially alarmist propaganda (Meinshausen *et al.*, 2009; Allen *et al.*, 2009; Schmidt and Archer, 2009). There is absolutely no doubt in these articles that CO_2 emissions were the prime cause of global warming in the 20th century. The only issue discussed is how rapidly CO_2 emissions must be reduced to save the world from disaster. The approach taken by these authors is statistical. The wide swath of modeled estimates of future CO_2 emissions and future temperatures are treated as votes, and the winning candidate is the result with the greatest preponderance of ballots. The conclusion is that, to save humanity, future CO_2 emissions must be draconically reduced, a recipe guaranteed to produce much more financial hardship in the world than putative global warming.[14]

Furthermore, most of the peer-reviewed articles in climatology are narrow, highly detailed, and represent new measurements or models. Most of the material that "circumvents peer review" is typically interpretive, synoptic, or in the nature of a review. Relatively little of it presents fundamental new measurements or calculations. One very important role for non peer–reviewed reports is the activity of the climateaudit.org blog that checks out many of the publications passed by peer reviews, attempts to reproduce the results (usually they cannot), and reviews these papers to put them into perspective—a task ignored by the peer review system. The *climateaudit.org* blog run by Steve McIntyre performs a valuable role in checking out the details of many published papers relevant to global warming—a task not done well (and usually not at all) by peer reviews. As it turns out, McIntyre has uncovered many errors and biases in these papers, most notably the MBH "*hockey stick*" papers.

A significant characteristic of climatology is that in general, data are very sparse and noisy, and of inadequate duration. Chaos seems to reign over the data and unless very good long-term data are available, the signal-to-noise ratio tends to aproach zero. Neverthless a significant characteristic of climatologists is that they seem willing to draw incredibly firm conclusions from poor data.

A case in point is the vital issue of cloud feedback. When the Earth warms due to increases in greenhouse gas concentrations, does this warming produce deterministic changes in cloud cover that produce a feedback, and is the feedback positive (amplifying the greenhouse warming) or negative (opposing the greenhouse warming)? McIntyre discusses the debate on this issue at some length (*http://climateaudit.org/2011/09/06/the-stone-in-trenberths-shoe*). The effect of a change in greenhouse gas concentration is expressed as an equivalent forcing at the top of the atmosphere in W/m^2. Similarly, the feedback from cloud cover changes is also expressed as an equivalent forcing ΔR_{cloud} also in W/m^2. The debate between Spencer and Braswell (2010, 2011)

---

[14] Perhaps unrelated to the politicalization of *Nature*, it is noteworthy that every journal except *Nature* has gladly and willingly given me permission to reproduce figures in my books, whereas *Nature* would have charged me as much as $700 for the right to reproduce a single figure.

and Dessler (2010) centers about whether ΔR_{cloud} is positive or negative. The raw data provided by Dessler (2010) are shown in Figure 3.1.

Unfortunately, as Dessler admits, this global cloud feedback data was taken "in response to short term climate fluctuations" where "the primary source of climate variations [was] the El Niño–Southern Oscillation (ENSO)". The data suffer from two lacks: (i) the data are short-term, and (ii) the data are not based on greenhouse gas warming. While Dessler derived the straight-line fit shown in the figure, indicating a weak positive feedback, it seems evident that cloud cover was not driven by Earth surface temperature at all, and varied chaotically during this short period due to unknown factors. Had additional data been taken, it is possible that the slope of the linear fit might turn negative.

As McIntyre pointed out, an argument could be made that there is a time lag of several months between a change in temperature and a change in cloud cover, so he replotted the data with a four-month time lag and obtained Figure 3.2. While McIntyre now obtained a negative feedback, the scatter in the data are even greater than before. Only a climatologist or a statistician could believe there is any significant information contained in this plot.

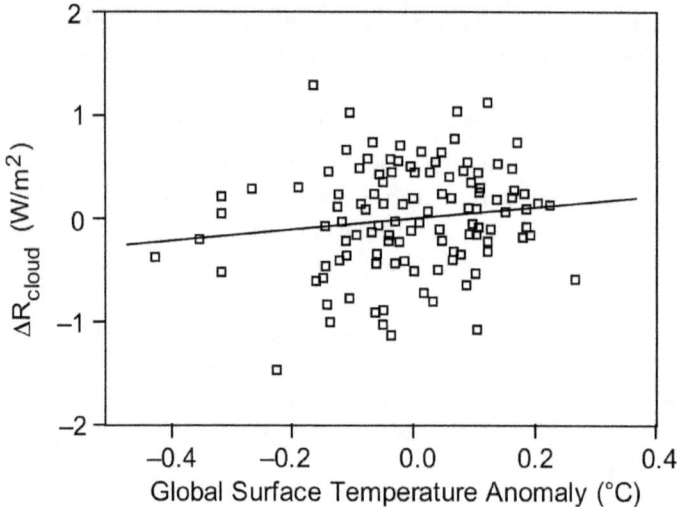

Figure 3.1. Measured cloud feedback vs. Earth surface temperature (Dessler, 2010).

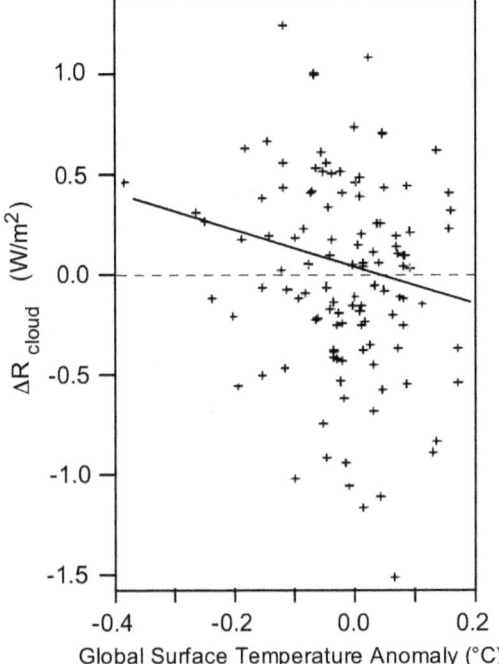

Figure 3.2. Cloud feedback plot with four-month time lag. *(http://climateaudit.org/2011/09/06/the-stone-in-trenberths-shoe/)*

Judith Curry posted[15] an interesting report on false positives in scientific research based on a paper[16] published in an on-line journal. Simmons *et al.* found that "flexibility in data collection, analysis, and reporting dramatically increases actual false-positive rates. In many cases, a researcher is more likely to falsely find evidence that an effect exists than to correctly find evidence that it does not".

Over the past few years, the *http://www.judithcurry.com* website has emerged as the centerpoint of dicussion of a myriad of topics relevant to climate change and climatologists with quite a number of insightful articles attended by hundreds of bogger entries. (Unfortunately, most of the blogger entries are not up to the high level of the original postings by Curry). Nevertheless, there is much to learn from this website. An article appeared August 3, 2012 written by Stephen Mosher on "post-normal science" (PNS). PNS deals with situations in which:

    1. Facts are uncertain

---

[15] *http://judithcurry.com/2012/01/12/false-positives*

[16] Joseph P. Simmons, Leif D. Nelson, and Uri Simonsohn, "False-Positive Psychology: Undisclosed Flexibility in Data Collection and Analysis Allows Presenting Anything as Significant" *dionysus.psych.wisc.edu/lit/articles/Simmons]2011a.pdf*

2. Values are in conflict

3. Stakes are high

4. Immediate action is claimed by one side to be required (this according to Mosher is the defining characteristic of PNS)

In the case of climate change, the stakes are high because of the extremity of the immediate action, the cost of that action, the impact of that action on our lives, the political difficulty in obtaining worldwide agreement to this action, and the technical difficulty in implementing it even if all the other problems could be overcome.

In normal science, "Because facts are uncertain, [both sides] listen to various conflicting theories. They try to put those theories to a test. They face a shared uncertainty and in good faith accept the questions and doubts of others interested in the same field... Because the field of personal values is never in play, personal attacks are minimized. Personal pride may be at stake, but values rarely are. In normal science, ... we can view the behavior of those doing science as puzzle solving. The details of a paradigm are filled out slowly and deliberately." For example, in regard to continental drift, the facts are uncertain (yet far more confirmed than in climate change), and the values are not in conflict, the stakes are not high, and there is no need for immediate action. In regard to evolution of species, the facts are not absolutely certain (yet far more confirmed than in climate change), and the values are in conflict with certain religious yahoos. Yet the stakes are not high (they only pertain to textbook content in some redneck states), and there is no need for immediate action (at a large scale).

According to Mosher: "In all PNS situations it is almost always the case the one side sees the need for action, given the truth of their theory, while the doubters must of necessity see no need for immediate action. They must see no need for immediate action because their values are at risk and because the stakes are high. Another way to put this is as follows. When you are in a PNS situation, all sides must deny it. Those demanding immediate action, deny it by claiming more certainty than is present; those refusing immediate action, do so by increasing demands for certainty. This leads to a centralization and valorization of the topic of uncertainty.... That is decidedly not normal science".

According to one blog entry in response to Mosher, anthropogenic global warming (AGW) proponents are justified in their alarm for the planet, but their view is that the public is always too selfish and parochial to appreciate the dire need to act, The IPCC and many climate scientists view the public as a bloc that needs to manipulated into compliance. Thus the IPCC manipulated the evidence to appear one-sided. It's this arrogant *noblesse oblige*, that serves as justification for their tribal mentality. Somewhere, the mission to convince the public ended up distorting the ability of the science to self-correct (paraphrased from blog entry).

3.6 The Lunatic Fringe

Perhaps the most absurd aspect of the climate alarmist movement is the putative relationship between obesity and global warming. If you enter "obesity and global warming" into Google, you obtain 1,100,000 responses. Typical responses in the queue are: (1) Is Obesity Causing Global Warming? A new study has suggested that obesity is affecting the planet ...by raising carbon emissions ...; (2) Do Obese People Aggravate Global Warming?— ABC News; (3) Scoop: Burning the Fat: Obesity and Global Warming, a study in the latest issue of the *International Journal of Epidemiology* by Phil Edwards and Ian Roberts plays out a grim scene: a world of overweight ...; (4) thinner is better to curb global warming, study says—CNN.com; and there are thousands more like this. Some claim the effect is through excessive use of resources, while others blame it on increased flatulence.

The website *http://www.numberwatch.co.uk/warmlist.htm* lists a huge number of ailments attributed by alarmists to global warming with links to the appropriate websites.

3.7 The Climate Debate Revisited

We have seen that it is in the very nature of global scares that (1) the supposed danger must be universal, (2) the danger must be novel, (3) the science must be plausible but contain elements that are very uncertain, and (4) society's response to the threat must be dispropotionate. The alarmist view of various aspects of the climate debate as embodied in Table 1.1 may now be compared to these criteria for global scares. See Table 3.3.

Table 3.3. Comparison of alarmist view of climate change with criteria for global scares.

Aspect	Danger is universal	Danger is novel	Plausible but uncertain	Response disproport-ionate
Variability of climate over past few thousand years			√	
Temperature rise over past century compared to natural fluctuations			√	
Rising CO_2 was the cause of rising temperatures in 20th century			√	
CO_2 concentration will rise further in 21st century in business-as-usual			√	
Climate models provide reasonable estimates of future warming in business-as-usual scenario			√	
Impacts of future global warming are known to be disastrous	√	√		
Only remedy is immediate draconian reduction in fossil fuel usage	√	√		√
Draconian reduction in fossil fuel usage can be accommodated by renewable energy supply	√	√	√	√

While this does not prove that climate alarmism is incorrect, the good fit of climate alarmism to the criteria for a global scare is highly suggestive.

4 EARTH TEMPERATURES OVER THE PAST ~ TWO HUNDRED YEARS

4.1 Introduction

Surface temperatures on land are available dating back about 100 years in some cases, and as far back as ~200 years for a limited number of sites. Meteorological scientists have methodically examined these data and attempted to derive the best overall data sets that the data allow. Studies of temperature change over land areas are routinely made by several groups based on measurements of the meteorological station network. It is beneficial to estimate global temperature change from both the meteorological station data alone, and a combined analysis with ocean data, because the land and ocean data have their own measurement characteristics and uncertainties.

Hansen *et al.* (1999, 2001) limited their studies primarily to the period since 1880, because of the poor spatial coverage of stations prior to that time and the reduced possibility of checking records against those of nearby neighbors. Analyses for the earlier years need to be carried out on a station-by-station basis with an attempt to discern the method and reliability of measurements at each station. Data collected and recorded by thousands of individuals with equipment and procedures subject to change over time inevitably contain many errors and inconsistencies, some of which will be impossible to identify and correct. However, Hansen *et al.* (1999, 2001) believe that the influence of errors is not dominant, because many of the errors in recording temperature are believed to be random in nature, and would therefore cancel out when a large sample is taken. Regional and global temperatures were estimated by combining local temperature records. Homogeneity adjustments were made to local time series of temperature with the aim of removing non-climatic variations in the temperature record. The non-climatic factors include changes of the environment of the station, the instrument or its location, observing practices, and the method used to calculate the mean temperature. The Climate Research Unit at the University of East Anglia in the UK has also maintained a very detailed database of measured temperatures around the world (see: *http://www.cru.uea.ac.uk/*).

The U.S. Historical Climatology Network (USHCN)[17] and the National Oceanic and Atmospheric Administration (NOAA) Climate Reference Network (USCRN)[18] are widely used temperature measurement networks.

Highly touted and heralded in advance, the long promised "Berkeley Earth Surface Temperature Analysis" (BEST) finally appeared (in preliminary PR form) in October 2011. Five relevant papers appeared on the Internet: Brillinger *et al.* (2011), Wickham *et al.* (2011), Muller *et al.* (2011a), Muller *et al.* (2011b), and Rohde *et al.* (2011). The stated objectives of the program were to:

1) Merge existing surface station temperature data sets into a new comprehensive raw data set with a common format that could be used for weather and climate research.

2) Review existing temperature processing algorithms for averaging, homogenization, and error analysis to understand both their advantages and their limitations.

3) Develop new approaches and alternative statistical methods that may be able to effectively remove some of the limitations present in existing algorithms.

4) Create and publish a new global surface temperature record and associated uncertainty analysis.

5) Provide an open platform for further analysis by publishing our complete data and software code as well as tools to aid both professional and amateur exploration of the data.

It should be noted at the outset that the work reported in October 2011 pertains only to land temperatures, and since 70% of the globe is covered by water, this represents a minority of the Earth. As Judith Curry (a co-author) said:

> "In concluding, I will remind everyone that the REAL problem with the surface temperature data set lies with the ocean data. I hope that the Berkeley group will be able to extend their efforts to include ocean data." *(http://judithcurry.com/2011/10/20/berkeley-surface-temperatures-released/#more-5425)*

---

[17] *cdiac.ornl.gov/epubs/ndp/ushcn/ushcn.html*
[18] *www.ncdc.noaa.gov › NESDIS › NCDC*

In late July 2012, the BEST team released an update to their previous findings (Rohde, *et al.* 2012). This is discussed further toward the end of Section 4.4.

4.2 Quality and Reliability of Land Temperature Measurement Networks

There are a number of factors that must be taken into account in judging the reliability of global or hemispheric temperature averages derived from network of ground measurement stations. (e.g., Pielke *et al.*, 2007a, c; Hoyt, 2006; Ball, 2007; Davey and Pielke, 2005). The issues of concern include the following:

1) Is there sufficient data spatially and temporally distributed to provide a basis for a global average or hemispheric average temperature over any extended time period?

2) How many stations are poorly sited to make good measurements?

3) How many sites have changed over the years, with changes in the surroundings, changes in the instrumentation or recording plan, or actual movement of the site?

4) How many sites were affected by the so-called "urban heat island" effect due to heat stored in urban construction?

5) If there is a preponderance of measurement sites at middle and upper-middle Northern Hemisphere latitudes, is proper account taken for the phenomenon of "arctic amplification"?

6) Have the various teams that have developed estimates of historical Earth as far back as the 19th and even 18th centuries properly corrected for the above effects?

Pielke *et al.* (2007c) concluded: "The use of temperature data from poorly sited stations can lead to a false sense of confidence in the robustness of multi-decadal surface air temperature trend assessments." They suggested that there are problems in the existing temperature databases due to: (1) time-of-observation bias because, at many sites, the observing time has changed during the station's history, (2) changes in instrumentation at stations, (3) station moves or relocations, and (4) bias caused by station urbanization. In addition, three primary issues were identified relevant to

"...land use/land clearing (LULC) and changes in LULC related to placement of climate stations. First, a station may be initially placed in what might be considered a poor LULC environment (e.g., near a highway or other man-made environment that could influence the observed temperature based on day of week, holiday, etc.). Second, a station may have been initially located at what might be considered a good LULC environment only to have that environment change over time. And third, possibly due to one of the above situations, a station may be moved from one LULC environment to another."

Donald Rapp

Pielke *et al.* (2007a) discussed unresolved issues in using surface temperature trends as a metric for assessing global and regional climate change. The issues include warm bias in night-time minimum temperatures, poor siting of the instrumentation, effect of winds, effects of surface atmospheric water vapor content on temperature trends, uncertainties in the homogenization of surface temperature data, and influence of land-use/land-clearing (LULC) change on surface temperature trends.

The observed average surface temperature at any site over land is computed by averaging observed daily maximum and minimum temperatures. While the daily maximum may be accurate, Pielke *et al.* (2007a) claim that the nightly minimum temperature (typically about 1.5 m to 2 m above the land) is subject to variation, depending on surface characteristics (heat capacity) and wind speed. They pointed out:

"As the boundary layer cools at night under light winds, the greatest decrease in temperature occurs near the surface. Unlike the daytime boundary layer where convective turbulence tends to reduce vertical gradients, in the nocturnal boundary layer the cooling suppresses turbulence and enhances vertical gradients. Thus, the vertical variation in temperature in light winds can be huge with temperature changes of 6°C or more often occurring within 25 vertical meters of the surface. This is why great care must be taken to avoid contaminating the climate record with measurements from sites that have changed even a meter or two in their height of observation."

Davey and Pielke (2005) conducted a study of temperature measurement stations in Eastern Colorado.

"It is important to know the site of stations relative to various structures and surfaces. Generally, near-surface air temperature observations should be representative of the free-air conditions over as much of the vicinity as possible, at a height approximately 1.5 m above the ground. The site should be level, without locally significant topographical variations or steep slopes or hollows, and should offer free exposure to both sunshine and wind (not too close to trees, buildings, or other obstructions). It thus becomes critical to conclusively determine how much of any potential regional change in observed air temperatures might be due to land-use changes at the site itself. These changes may include local-scale urban development around the site, changes in local vegetation characteristics, etc."

Local-scale exposure characteristics are therefore important in evaluating station data. Prior to the 1980s, a site sketch was available, illustrating the location of the weather station instrumentation and any nearby obstructions. Currently, however, only vague documentation regarding site exposure characteristics is typically available. This is particularly true for sites' terrain and surface features. Davey and Pielke (2005) visited 57 sites in Eastern Colorado, with emphasis on the 10 sites in the U.S. Historical Climate Network (USHCN). Typical findings were (1) sensors close to buildings, (2) sensors over patchworks of different land coverings, (3) vegetation around sensor locations, or urbanized sensor locations. Evidently, the temperature measurement network needs refurbishment.

The temperature measurements that we require are for the purpose of defining the global climate. If a significant portion of measurement sites are located near local "hot spots" in the form of urban heating islands (UHIs) the data can be very misleading, particularly because urbanization has increased with time, and that would produce an artificial apparent increase in global temperature. On the other hand, if urbanization becomes so widespread with the growing world population, that urban heat islands become the norm—rather than the exception—then inclusion of UHI sites may become appropriate in defining the climate. However, in that case, a significant part of global warming would be induced by urban and industrial activity. It is possible that we are approaching that point in some regions of the Earth. In general, the effect of UHIs is to slightly decrease the daily maximum temperature (due to a variety of factors: reflection from buildings, shading, etc.) and to significantly increase the nightly minimum temperatures as stored solar energy (in concrete and structures) is released overnight. Thus, urbanization is expected to reduce the diurnal temperature range (DTR) as well as to increase the average temperature. For example, in the urban heat island of Houston, Texas between 1990 and 2000, the relative increase in air temperature had an average magnitude of 1.25°C at night but was largely absent during the day. The urban heat island had an area of about 1,000 km² (Streutker, 2003). Other studies report temperature increases in city centers of several degrees, with contours of decreasing temperature toward the suburbs.

Hoyt (2006) discussed the differences between urban sites and rural sites. It is claimed that those who support the IPCC viewpoint consider a town with a population of less than 10,000 people to be rural and not to require any adjustment for urbanization. But according to Hoyt (2006), quoting a 1973 paper by Oke, the temperature increase of an urban center compared with its surroundings is approximated as 0.73 log (population), so that even a town of 1,000 people may have an urban temperature rise of 2.2°C.

Corrections for urbanization are especially difficult in China, where rapid growth has changed the demographics at a very rapid rate. From 1978 to 2000, China's gross domestic product grew at an average annual rate of 9.5%, compared with 2.5% for developed countries and 5% for developing countries; the number of small towns soared from 2,176 to 20,312, the number of cities increased from 190 to 663; and the proportion of urban population rose from 18% to 39%.

Using rural–urban temperature differences to estimate the impacts of urbanization on climate in China may be inappropriate for several reasons.

1. Most Chinese stations are located in or near cities, with only a few in mountainous or remote regions or on small islands.
2. Although China is comparable in size with the U.S., it has considerably fewer meteorological stations, and each city generally has only one station. For example, each of China's two biggest cities, Beijing and Shanghai, has only one station available in the Chinese network.
3. It is impossible to find a corresponding rural station for most of the urban ones, especially in eastern and southern China.
4. China's rapid urbanization in the past two decades could transfer a station from rural into urban in a very short period.
5. Chinese cities have a much higher density of population and urban buildings than do cities in most developed countries (Zhou *et al.*, 2004).

Pielke, Sr.[19] reported on a new paper[20] that modeled summer climate effects of urban sprawl in the rapidly expanding sunbelt in Arizona. "Basically, the more concrete you have in terms of buildings and roads, the hotter it gets during the day. And, because concrete and asphalt absorbs and then releases heat, it cools off less at night." Thus day temperatures rise and night temperatures rise even more as urban spawl expands. Further expansion in Arizona could raise peak temperature several degrees.

According to McLean (2007a):

"In the early days thermometers could only show the temperature at the moment of reading and so the data recorded from that time was for just one reading each day. Later the thermometers were able to record the minimum and maximum temperatures, and so the daily readings were those extremes in the 24 hour period. Only in the last 20 or 30 years have instruments been available that record the temperature at regular intervals throughout the 24 hours, thus allowing a true time-based daily average to be calculated. The so-called 'average' temperatures both published and frequently plotted through time are initially based on only a single daily value, then later on the mathematical average of the minimum and maximum temperatures. Although time-based averages are now available for some regions, they are not generally used because the better

---

[19] *http://pielkeclimatesci.wordpress.com/*
[20] Georgescu, M., M. Moustaoui, A. Mahalov, J. Dudhia "Summer-time climate impacts of projected megapolitan expansion in Arizona" Nature Climate Change *http://www.nature.com/nclimate/journal/vaop/ncurrent/full/nclimate1656.html*

instrumentation is not uniformly installed throughout the world and the historical data is at best a mathematical average of two values. The problem is that these averages are easily distorted by brief periods of high or low temperatures relative to the rest of the day, such as a brief period with less cloud cover or a short period of cold wind or rain."

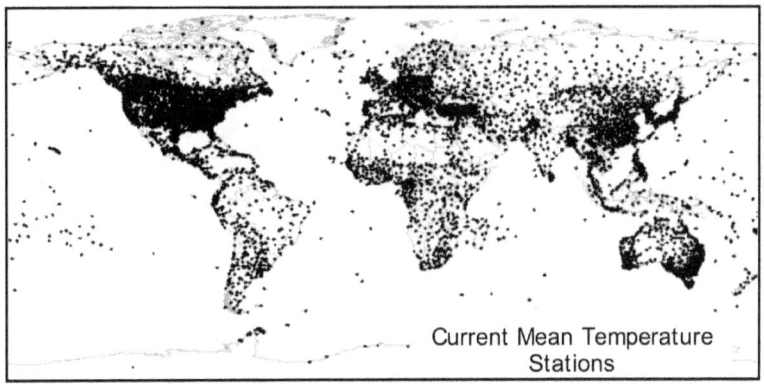

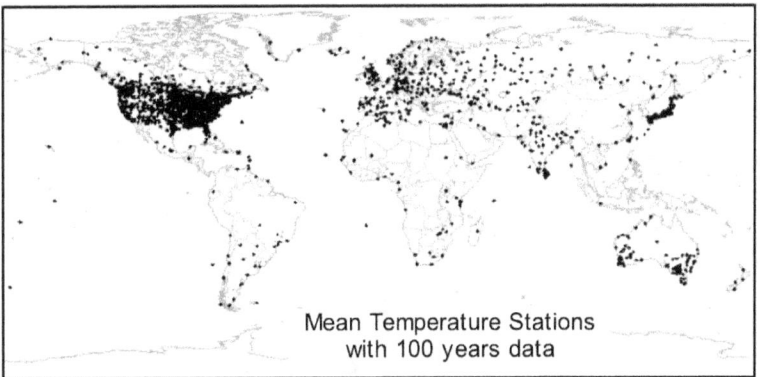

Figure 4.1. Locations of GHCN mean temperature station locations. Upper: All stations. Lower: Stations with at least 100 years of data. Adapted from Peterson and Vose (1997).

In a 1999 report entitled *Adequacy of Climate Observing Systems*, the National Research Council said:

"Climate researchers have used existing, operational networks because they have been the best, and sometimes only, source of data available. They have succeeded in establishing basic trends of several aspects of climate on regional and global scales. Deficiencies in the accuracy, quality, and continuity of the records, however, still place serious limitations on the confidence that can be placed in the research results."

Figure 4.1 is a plot of the mean temperature stations around the world. The upper plot shows all stations and the lower plot shows stations with 100 years (or more) of data.

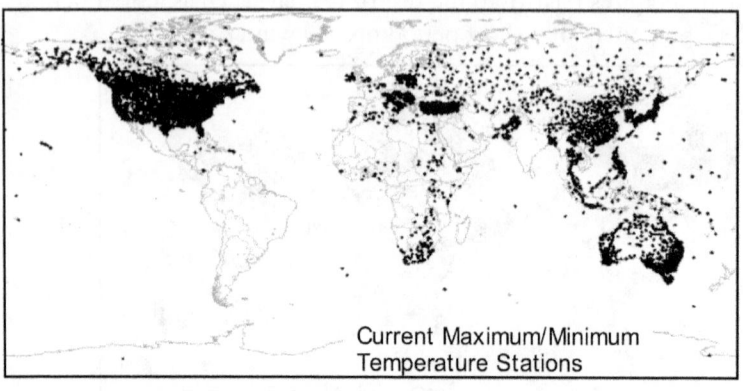

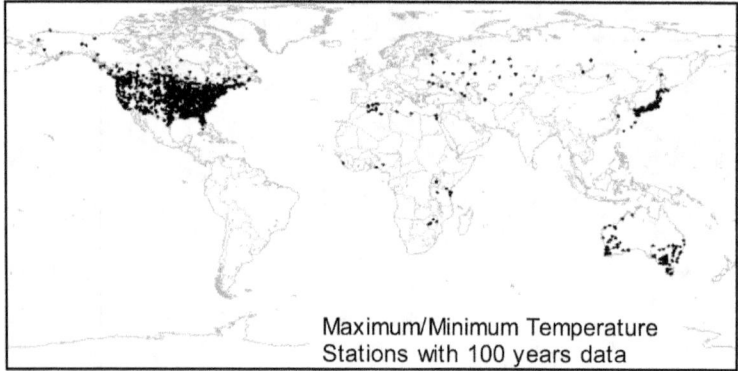

Figure 4.2. Locations of GHCN maximum/minimum temperature station locations. Upper: All stations. Lower: Stations with at least 100 years of data. Adapted from Peterson and Vose (1997).

The U.S. and Western Europe play a dominant role. In most of the world, we don't have adequate station coverage for temperature and even less for precipitation. Seventy percent of the world is ocean, for which the database is uncertain. The Arctic and the Antarctic have very skimpy data. African and South American stations are sparse. Figure 4.2 shows the same information for stations that measure maximum and minimum temperature. Figure 4.3 shows the longevity of existing stations, the rise in the number of stations in the 20th century, and the decrease in the number of stations after 1970.

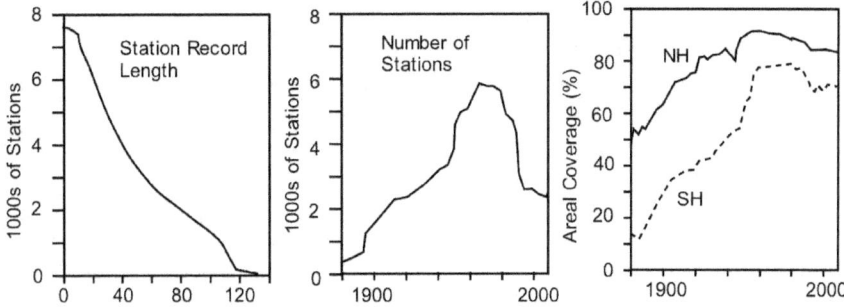

Figure 4.3. World temperature measurement stations. Left: Number of stations vs. record length. Middle: Number of stations vs. year. Right: Areal coverage by hemisphere. Adapted from Bell (2007).

In a quotation available in various places on the Internet, Kevin Trenberth, a leading climatologist, said:

> "It's very clear we do not have a climate observing system.... This may come as a shock to many people who assume that we do know adequately what's going on with the climate but we don't."

McKitrick and Michaels (2007) summarized the situation:

> "The number of reliable monitoring sites around the world has fallen dramatically since the mid-1970s. The Global Historical Climatology Network reached a peak of 6,000 unique contributing sites in the late 1960s, but the number fell to fewer than 3,000 as of the late 1990s, with the most dramatic drop in the early 1990s when the number of stations fell by nearly half in four years."

Fall *et al.* (2011) surveyed 82.5% of the *U.S. Historical Climatology Network* stations and provided a classification based on exposure conditions of each surveyed station, using a rating system employed by the *National Oceanic and Atmospheric Administration* to develop the *U.S. Climate Reference Network*. This study:

> "... examined temperature differences among different levels of siting quality without controlling for other factors such as instrument type. Temperature trend estimates vary according to site classification, with poor siting leading to an overestimate of minimum temperature trends and an underestimate of maximum temperature trends, resulting in particular in a substantial difference in estimates of the diurnal temperature range trends. The opposite-signed differences of maximum and minimum temperature trends are similar in magnitude, so that the overall mean temperature trends are nearly identical across site classifications.... the most poorly-sited stations are warmer ... than are other stations, and a major portion of this bias is associated with the siting classification rather than the geographical distribution of stations.

According to the best-sited stations, the diurnal temperature range in the lower [48] states has no century-scale trend."

The classification scheme and percent of sites that fit each class are given below:

CRN 1: (1.2%) A clear flat surface with sensors located at least 100 meters from artificial heating and vegetation ground cover <10 centimeters high (error < 1°C).

CRN 2: (6.7%) Same as CRN 1 with surrounding vegetation within 25 centimeters and artificial heating sources within 30 meters (error < 1°C).

CRN 3: (21.5%) same as CRN 2, except no artificial heating sources within 10 meters (error ≥ 1°C).

CRN 4: (64.4%) artificial heating sources within 10 meters (error ≥ 2°C).

CRN 5: (6.2%) sensor located next to/above an artificial heating source (error ≥ 5°C).

Based on the classification scheme of Fall *et al.* (2011), Muller *et al.* (2011) concluded that the state of the U.S. station network is very poor with 70% of the stations having projected errors > 2°C and 90% having projected errors > 1°C. Muller *et al.* (2011) presented a map of the sites, color coded by quality (see Figure 4.4).

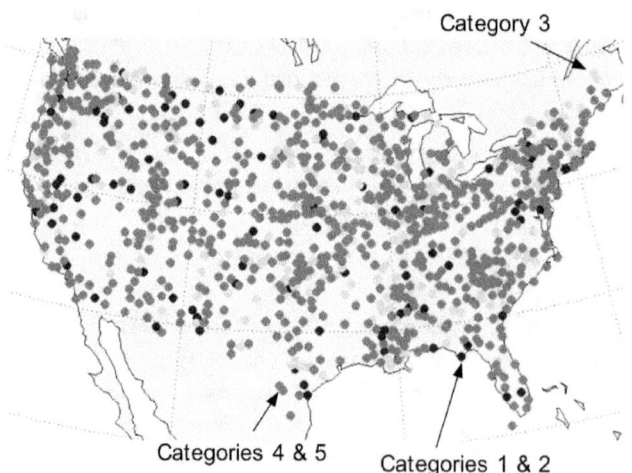

Figure 4.4. Distribution of temperature stations in the U. S. Stations are ranked 1 and 2 (errors < 1°C), ranked 3 (errors > 1°C), and ranked 4 and 5 (errors > 2°C or 5°C).

But the point that was made by Ball (2007) (brushing aside issues regarding whether it makes sense to even talk about a single global temperature or how one deals with the oceans), is how can we know what the global temperature rise was "since the late 19th century" when (i) there hardly were any stations in the late 19th century, (ii) station sparsity was meager until about 1950 and still inadequate after that date, (iii) the number of stations with record lengths of 100 years is minimal, (iv) areal coverage was low until about 1950? The short answer is: we can't know.

The decrease in the number of stations in the late 20th century is alarming. If this is due to "weeding out" poorly equipped and situated stations, that might improve the quality of the residual database but it exacerbates the sparseness. Furthermore, it raises questions regarding the quality of the thousands of stations that were phased out late in the 20th century.

A more upbeat view of the network for temperature measurement on Earth was given by Brohan *et al.* (2006). A new upgrade to the temperature measurement network (TMN) was described. The database was derived from a collection of homogenized, quality-controlled, monthly-average temperatures for 4,350 stations. "Station normals" (monthly averages over the normal period 1961–1990) were generated from station data for this period where possible. However, "normals" were derived from station data "wherever possible" and various approximations were used in many cases. Using these "normals", anomalies were calculated as deviations from the means of the "normals". The discrete set of station anomaly data were then converted to a temperature distribution across a global grid by interpolation, typically at 5° × 5° resolution.

The entire situation regarding the reliability of the ground temperature network was jolted by the appearance of a "pre-print draft discussion paper" at the end of July 2012 by Watts *et al.*[21] This paper provides many references relevant to previous analysis of the reliability of surface land temperature measurement networks. Whereas previous studies (e.g. Menne *et al.*, 2010, Rhode *et al.*, 2012) took into account heat sources and sinks in the view-shed only in terms of their distance, they utilized an improved siting classification system (based on Leroy, 2010) that included a method for including the surface area of heat sinks and heat sources within the view-shed of thermometer. This resulted in a dramatic and statistically significant improvement in the binning of stations quality ratings. These factors, combined with station siting issues, have led to a spurious doubling of U.S. mean temperature trends in the 30 year data period covered by the study from 1979 – 2008.

The new rating system for classification of local sites is given below:

---

[21] Watts, A., E. Jones, S. McIntyre and J. R. Christy (2012) "An area and distance weighted analysis of the impacts of station exposure on the U.S. Historical Climatology Network temperatures and temperature trends": *http://wattsupwiththat.com/*

Class 1 (compliant)

Flat, horizontal land, surrounded by an open space, slope less than 1/3 (19°); Ground covered with natural and low vegetation (< 10 cm) representative of the region; Measurement point situated:

- More than 100 m from heat sources or reflective surfaces (buildings, concrete surfaces, car parks, etc.)
- More than 100 m from an expanse of water (unless significant of the region)
- away from all projected shade when the Sun is higher than 5°.

Class 2 (compliant)

Flat, horizontal land, surrounded by an open space, slope inclination less than 1/3 (19°); Ground covered with natural and low vegetation (<10 cm) representative of the region; Measurement point situated:

- More than 30 m from artificial heat sources or reflective surfaces (buildings, concrete surfaces, car parks, etc.)
- More than 30 m from an expanse of water (unless significant of the region)
- Away from all projected shade when the Sun is higher than 7°.

Class 3 (non-compliant, additional estimated uncertainty added by siting up to 1°C)

Ground covered with natural and low vegetation (<25 cm) representative of the region; Measurement point situated:

- More than 10 m from artificial heat sources and reflective surfaces (buildings, concrete surfaces, car parks, etc.)
- More than 10 m from an expanse of water (unless significant of the region)
- Away from all projected shade when the Sun is higher than 7°.

Class 4 (non-compliant, additional estimated uncertainty added by siting up to 2°C)

Close, artificial heat sources and reflective surfaces (buildings, concrete surfaces, car parks, etc.) or expanse of water (unless significant of the region, occupying:

- Less than 50% of the surface within a circular area of 10 m around the screen
- Less than 30% of the surface within a circular area of 3 m around the screen
- Away from all projected shade when the Sun is higher than 20°.

Class 5 (non-compliant, additional estimated uncertainty added by siting up to 5°C)

Sites do not meet the requirements of Class 4. i.e. more than 50% of the surface within a circular area of 10 m around the screen. More than 30% of the surface within a circular area of 3 m around the screen.

The number of sites across the U.S. in each class was as follows:

Class	Number	%
Class 1	48	6
Class 2	112	14
Class 3	247	32
Class 4	277	36
Class 5	95	12
All	779	100

The results of Watts *et al.* (2012) for the time period 1979 to 2008 indicated that there was a universal tendency for Class 3/4/5 sites to report higher temperatures than nearby Class 1/2 sites with 3/4/5 sites showing a temperature gain from 1979-2008 that was about 60% higher than that seen by Class 1/2 sites. The adjustments made by NOAA in their USHCNv2 compilation of Earth temperatures are actually worse than simply using raw data. The results for Class 1/2 sites indicate a temperature rise of 0.15°C per decade whereas after adjustments, the NOAA compilation indicates 0.31°/decade, or double the value obtained from the best sites with airports included, and about 2.5 times the value obtained from the best sites with airports not included.

The analysis demonstrates clearly that siting quality has a significant effect on reported temperature changes. Watts *et al.* demonstrated that "not only does the NOAA USCHNv2 adjustment process fail to adjust poorly sited stations downward to match the well sited stations, but actually adjusts the well sited stations upward to match the poorly sited stations".

In addition to this, it was demonstrated that "urban sites warm more rapidly than semi-urban sites, which in turn warm more rapidly than rural sites. Since a disproportionate percentage of stations are urban (10%) and semi-urban (25%) when compared with the actual topography of the U.S., this further exaggerates reported trends in [temperature change]".

They concluded:

"Taken *in toto*, these factors identified in this study have led to a spurious doubling of U.S. T_{mean} trends from 1979 – 2008".

It is very likely that similar problems exist in worldwide measurement stations.

However, in a more recent posting on *http://wattsupwiththat.com/* it was noted that some problems in this analysis were discovered and it is currently being reworked. Therefore, the conclusions drawn in the preprint might be revised under further scrutiny. Nevertheless, problems with the measurement site network remain as a concern.

Aside from all the other problems involved in monitoring the Earth's near surface temperature (as discussed by Watts *et al.*, 2012) there is a question of what height to place the sensor above the ground. The current standard for monitoring stations is 1.5 m above the surface. This is supposed to be representative of a boundary layer of air. Presumably, air currents mix the boundary layer during the day, but at night, it is relatively stable and striated. Most monitoring stations measure maximum and minimum temperatures diurnally. Presumably these occur during the day and night, respectively. It has been observed that over the past century or so that minimum temperatures have warmed nearly three times more than maximum temperatures and there is no satisfactory explanation for this effect. Climate models are not consistent with this observation.

McNider *et al.* (2012) developed an explanation based their model that indicated that slight increases in incoming long-wave radiation from greenhouse gases at night (as greenhouse gas concentrations rise) might destabilize the boundary layer of air near the surface. This will introduce some turbulence that will cause mixing of air from above with the boundary layer. At night, the surface cools by radiation to space, which in turn cools the boundary layer to temperatures below that of the air above it. If the air above the boundary layer mixes with the boundary layer at night, it will warm the boundary layer. In their model, they subjected a nocturnal boundary layer to an added increment of downward radiation (4.8 W/m^2) and compared the boundary layer with and without this forcing. They found that the 1.5 m height air temperature warmed substantially due to destabilization caused by additional downward IR irradiance. Most of the warming at 1.5 m height was due to the warm air mixed from aloft.

McNider *et al.* (2012) concluded:

"Thus, it may be better for current climate models, when they test replication of past climates and to project future global warming, to only use maximum temperatures rather than the current metric of using the mean daily temperature, which contains the [amplified] minimum temperature".

4.3 Quality and Reliability of Ocean Temperature Measurements

About 70% of the Earth's surface is covered by ocean. Ocean temperatures are important for two main reasons. One reason is that sea surface temperatures play a 70% role in estimating the global average temperature. Another important reason is that the heat capacity of the oceans is about a thousand times the heat capacity of the atmosphere. Therefore, if the Earth is storing heat due to an imbalance between incoming radiation and outgoing radiation, it must be mainly stored in the oceans. In this connection, deep sea measurements are relevant.How heat is added to the oceans is a complex, intricate matter.

From the point of view of estimating changes in the global average temperature, sea surface temperatures (SST) are of principal concern. Sea surface temperatures are commonly considered to be those measured in the upper 20 m of the oceans.

In this connection, Soon and Baliunas (2003a,b) pointed out:

"The sea surface temperature (SST) record also is complicated by a change in procedure. Prior to the 1940s, SST was determined by measuring the temperature of a dunked bucket of seawater with a thermometer. After the 1940s, SST was taken by measuring the temperature of the seawater at the intake to the engine cooling system. Large adjustments had to be made to the older data to make it compatible with the new data. These adjustments affected average SST by 0.10 to 0.45°C; the upper end of this range is three quarters of the observed change in global average surface temperature for the 20th century. Problems with defining a global-scale mean climatology for SST still exist ... as late as 1961–1990."

Houghton (2004) echoed this:

"In the case of ships, the standard method of observation used to be to insert a thermometer into a bucket of water taken from the sea. Small changes of temperature have been shown to occur during this process; the size of the changes varies between day and night and is also dependent on several other factors including the material from which the bucket is made—over the years wooden, canvas and metal buckets have been variously employed. Nowadays, a large proportion of the observations are made by measuring the temperature of the water entering the engine cooling system."

Nevertheless, Brohan *et al.* (2006) concluded that uncertainties in sea surface temperatures are lower than for land surface.

Marine data used by Brohan *et al.* (2006) consist of a gridded data set made from *in situ* ship and buoy observations from the new International Comprehensive Ocean– Atmosphere data set. For each grid box, mean temperature anomalies, measurement and sampling error estimates, and bias error estimates are available. Blending a sea surface temperature (SST) data set with land air temperature data makes an implicit assumption that SST anomalies are a good surrogate for marine air temperature anomalies. It was claimed that this is the case, and that marine SST measurements provide more useful data and smaller sampling errors than marine air temperature measurements would. So it was claimed "blending SST anomalies with land air temperature anomalies is a sensible choice." Over the period from 1850 to

1940, the predominant SST measurement process changed from taking samples in wooden buckets, to taking samples in canvas buckets, to using engine room cooling water inlet temperatures. A bias correction was applied to remove the effect of these changes on the SSTs, but this adds some uncertainty. Brohan *et al.* (2006) said:

> "As with the land data, the uncertainty estimates cannot be definitive: where there are known sources of uncertainty, estimates of the size of those uncertainties have been made. There may be additional sources of uncertainty as yet unquantified."

A surprising aspect of uncertainties in temperature measurements is that according to Brohan *et al.* (2006), the uncertainty in sea surface temperatures is fairly flat at ±0.1°C from 1850 to 2000. By contrast, the uncertainty in land temperatures is claimed to be range from about ±0.4°C around 1850 to about ±0.15°C in 2000. Since the oceans occupy about 70% of the global land area, the overall uncertainty would be expected to depend more on the ocean data. According to Brohan *et al.* (2006):

> "A notable feature of the global time series is that the uncertainties are not always larger for earlier periods than later periods. The uncertainties are smaller in the 1850s than in the 1920s, at least for the smoothed series, despite the much larger number of observations in the 1920s."

These claims are hard to believe.

Figure 4.4a shows the smoothed monthly temperatures for the two hemispheres and the globe as derived by Brohan *et al.* (2006). The widths of the swaths are measures of the uncertainties associated with each set of measurements. A comparison of the smoothed mean temperature anomalies for the Northern Hemisphere and Southern Hemisphere shows the difference in uncertainties between the two hemispheres. According to Brohan *et al.* (2006) , the difference in the uncertainty ranges for the two series stems from the very different land/sea ratio of the two hemispheres. The NH has more land, and so a larger station sampling and measurement error, but it has more observations and so a smaller coverage uncertainty. The bias uncertainties are also larger in the NH both because it has more land, and because the SST bias uncertainties are largest in the NH western boundary current regions.

Brohan *et al.* (2006) calculated the global temperature as the mean of the NH and SH series (to stop the better-sampled NH from dominating the average). Figure 4.4b shows the resultant smoothed monthly temperatures based on land measurements, sea measurements, and overall.

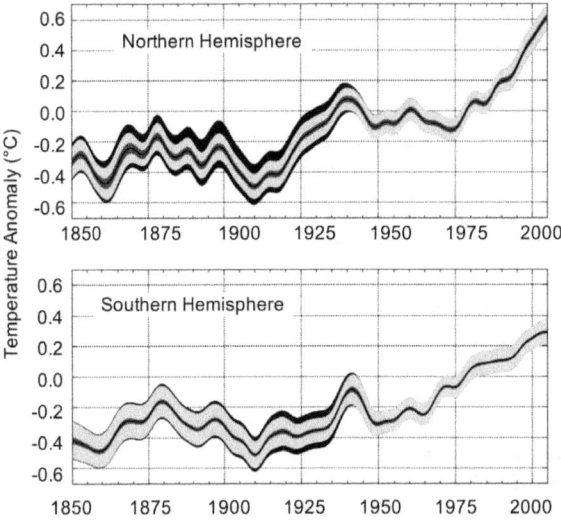

Figure 4.4a. Smoothed monthly sea surface temperatures for the two hemispheres. Adapted from Brohan *et al.* (2006).

There are several aspects of Brohan *et al.* (2006) that are confusing. First, consider the point that it is claimed that sea temperature measurements are more precise than land temperature measurements. On the one hand, this would seem to be intuitively correct because the sea measurements are not afflicted with site aberrations due to urbanization, shade or sun, exposure to winds, changes in land use, etc. However, as we pointed out previously, doubts have been raised about the accuracy of sea temperature measurements due to the changes in measurement procedures over the years. Next, consider that despite the huge preponderance of sites in the NH compared to the SH, and the fact that the number of SH sites probably falls off much more rapidly as one goes back in time, Brohan *et al.* (2006) claim that the measurements of temperature in the SH are more precise. This, of course is based on the putative accuracy of sea temperatures and the preponderance of ocean in the SH. Nevertheless, this seems counter-intuitive and worthy of further examination. But the strangest thing about Figures 4.4a and 4.4b is that they do not seem to link to one another very well. For example, the upper panel in Figure 4.4b is the global temperature anomaly and it should be an arithmetic mean of the SH and NH curves in Figure 4.4a. However, it is not. For example, at the far right of these graphs, the values are global = 0.62, NH = 0.62 and SH = 0.28. By averaging the NH and SH, one should get 0.45 for the global, not 0.62. Something is amiss here. Another problem is that if we take 30% of the land curve and 70% of the sea curve in Figure 4.4b, we do not end up with the global curve (upper panel in Figure 4.4b). Once again, at the far right of the graphs, we find 0.7 × 0.35 + 0.3 × 0.67 = 0.45 whereas the global curve indicates 0.62. It is also difficult to reconcile the width of the uncertainty in the upper panel.

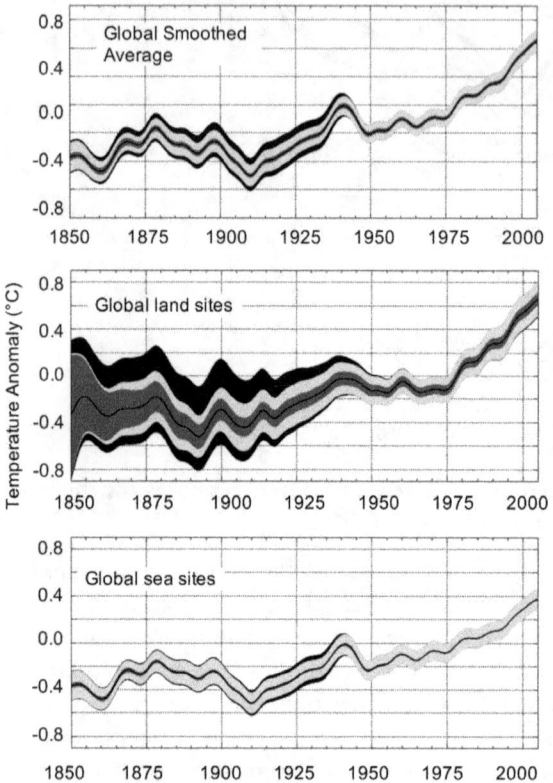

Figure 4.4b. Global smoothed monthly sea surface temperatures. Upper panel: all sites. Middle panel: land sites. Lower panel: sea sites. Adapted from Brohan *et al.* (2006).

Prior to the advent of the ARGO network for monitoring subsurface sea temperatures about 2004, "the historical ocean subsurface dataset was quite sparse over much of the World Ocean" (Harrison and Carson, 2006). Harrison and Carson (2006) pointed out that various "global-scale studies typically involved removing a climatological ocean temperature field, followed by different amounts of interpolation, extrapolation, and averaging, or have otherwise depended on some analyzed field" and they implied that such manipulations might involve some artifaciality. They thought it "worthwhile to look at subsurface thermal variability from a perspective that minimizes manipulations of the data themselves and also focuses on those depths and areas where statistically significant trends can be identified with minimum smoothing or averaging".

Harrison and Carson (2006) studied ocean temperatures down to 500 m depth over the 51-year period: 1950-2001. They found

"... highly structured patterns of 51-yr trends of alternating sign at 100-, 300-, and 500-m depth.... Each of the ocean basins exhibits both warming and cooling trends over this 51-yr analysis period.... The upper ocean evidently is replete with variability in space and time, and multidecadal variability is quite energetic.... These results suggest that trends based on records of one or two decades in length are unlikely to represent accurately longer-term trends. Further, the magnitude of the 20-yr trend variability is great enough to call into question how well even the statistically significant 51-yr trends identified here represent longer-term trends".

Over the period 2004-2007, the Argo system has been in operating to monitor ocean temperatures. "Argo consists of a large collection of small, drifting oceanic robotic probes deployed worldwide. The probes float as deep as 2 km. Once every 10 days, the probes surface, measuring conductivity and temperature profiles to the surface. From these, salinity and density can be calculated. The data are transmitted to scientists on shore via satellite. The initial project goal was to deploy 3,000 probes, completed in November 2007. The Argo program was designed to operate on the same 10-day duty cycle to match the existing satellite measurements of the ocean's sea surface. These satellites, called Topex/Poseidon and Jason 1, measure changes in the surface topography of the ocean. With such measurements, information about temperature, mass redistribution, or surface currents can be inferred. The Argo floats measure subsurface changes in temperature and salinity, hence the float measurements are complementary to the altimetry.

The number of floats is continually changing as floats are lost or expire, while others are deployed. Nominally, some 750 floats are deployed each year to sustain the system. The floats have a nominal 300-km spacing, although the exact separations depend on the randomness of the float drift.

The Argo temperature and salinity measurements are yielding valuable information about the large-scale water properties and currents of the ocean, including the variability of these properties over time scales from seasonal to decadal. However the short duration of this system does not allow any trends to be confirmed.

4.4 Measured Global Average Temperature

A number of institutions have estimated global average temperatures over the past century and in some cases dating back two centuries. Some of these combined land measurements with ocean measurements, and some are restricted to land. In addition, estimates exist for the U. S. lower 48 states, for various latitude ranges and for the NH and the SH.

4.4.1 U. S. Temperatures

NASA's Goddard Institute for Space Studies (GISS) has analyzed U.S. and world temperature patterns since about 1880 in great detail. The GISS data were updated a number of times, most recently by Hansen *et al.*, (2010). Their estimate for the U.S. mean temperature history is shown in Figure 4.5. The results for the U.S. show some warming early in the 20th century, followed by a cooling trend after about 1940, and a subsequent return to warming after around 1978. The effects of the Krakatoa, El Chichon and Pinatubo volcanic eruptions are evident. Most recently, the effects of two major El Niños are also evident. As of 2010, U. S. average temperatures are not much different than they were in the 1930s.

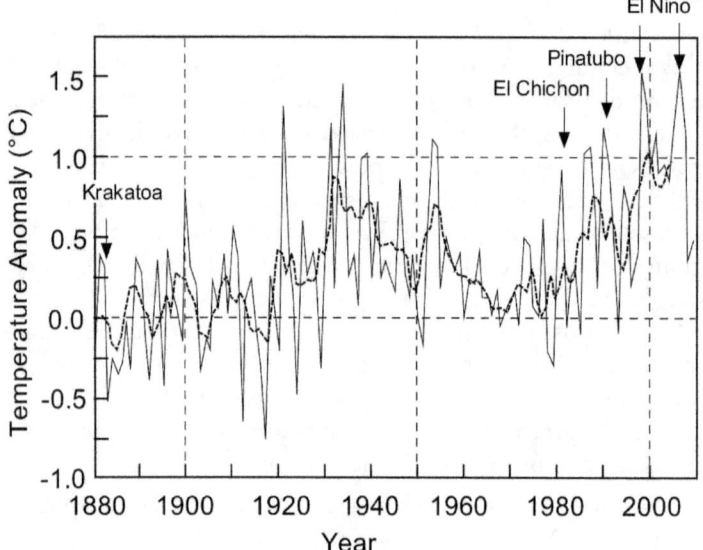

Figure 4.5. U.S. mean temperature anomalies (deviations from mean temperature). The dashed curve is a multi-year running average. Adapted from Hansen *et al.* (2010).

Kunkel *et al.* (2008) provide an analysis of weather and climate extremes for North America dating back to the mid–19th century in some cases. While the tone of this report leans toward the alarmist persuasion, the data provide mixed results. A heat wave index was defined as the occurrence of warm spells of at least 4 days in duration with mean temperature exceeding the threshold for a 1-in-10-year event. The historical occurrence of heat waves is plotted in Figure 4.6. While this index has been increasing in the past few decades, it remains far below the peaks in the 1930s. In addition, Kunkel *et al.* (2008) showed that the area over which hot daily highs and lows were reached has been increasing over the past few decades, but the percentages of total area remain low at under 0.5%.

However, there are data that suggest the warming of the past few decades may have a different character than that of the 1930s. For example, Kunkel *et al.* (2008) showed that the recent U.S. average frost-free season period was at least 10 days longer on average than the long-term average, and was significantly greater than that of the 1930s.

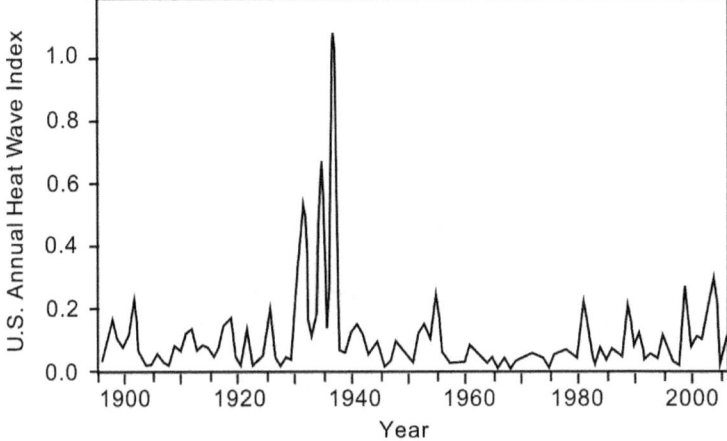

Figure 4.6. Historical variation of U.S. heat wave index (Kunkel *et al.*, 2008).

Hansen *et al.* (2012) said:

"The distribution of seasonal mean temperature anomalies has shifted toward higher temperatures and the range of anomalies has increased. An important change is the emergence of a category of summertime extremely hot outliers, more than three standard deviations (3σ) warmer than the climatology of the 1951–1980 base period. This hot extreme, which covered much less than 1% of Earth's surface during the base period, now typically covers about 10% of the land area. It follows that we can state, with a high degree of confidence, that extreme anomalies such as those in Texas and Oklahoma in 2011 and Moscow in 2010 were a consequence of global warming because their likelihood in the absence of global warming was exceedingly small."

John Christy[22] rebutted this claim showing that as usual, Hansen craftily chose to compare data and time periods that supported his claim, whereas a broader view of more data over extended time refutes his claim. One point is that Hansen chose to use data on $T_{mean} = (T_{max} + T_{min})/2$. As Christy pointed out

"T_{max} represents the temperature of a well-mixed lower tropospheric layer, especially in summer. T_{min}, on the other hand, is mostly a

<hr>

[22] *http://www.drroyspencer.com/2012/08/fun-with-summer-statistics-part-i-usa/*

measurement in a shallow layer that is easily subjected to deceptive warming as humans develop the surface around the stations."

"Since T_{max} represents a deeper layer of the troposphere, it serves as a better proxy (not perfect, but better) for measuring the accumulation of tropospheric heat, and thus the greenhouse effect. This is demonstrated theoretically and observationally in McNider *et al.* (2012). I think T_{max} is a much better way to depict the long-term temperature character of the climate."

The record highs recorded by Hansen were mostly due to highs in T_{min}, rather than T_{max}, and this can be attributed to a variety of causes other than greenhouse gas warming.

A second problem with Hansen's analysis pertains to his database and his choice of reference period. Hansen used a reference period from 1951 to 1980, and thus missed the extremely warm period of the 1930s. This would lead him to conclude that the warmth of later years was more unusual than it was.

The issue at hand is to plot year by year, the relative number of station highs that set records relative to a reference period. Christy recalculated Hansen's model for the U.S. and for the globe. For the U.S., Christy plotted the number of daily record highs in T_{max} from 1895 to 2010 as shown in Figure 4.6a. It can be seen that there were far more record highs in the 1930s than in recent years. In fact, record highs from 1910 to 1955 were generally higher than recent decades. Figure 4.6b shows the number of daily record highs for both T_{max} and T_{min}. It can be seen that T_{min} has risen more than T_{max} since 1975.

Hansen estimated that for the Northern Hemisphere, typically 10% or more of the area experienced high temperatures above the 3σ threshold for the reference period (1951-1980). Of course, choosing this short reference period ignores the relatively hot period earlier in the century. Christy estimated the results shown in Figure 4.6c. Replacing Hansen's data base with the BEST data base but retaining the short reference period, reduced the fraction of area experiencing high temperatures above the 3σ threshold by roughly a factor of two. It is not clear why this is so. But of even greater importance, when the reference period was extended, the fraction of area experiencing high temperatures above the 3σ threshold dropped even more.

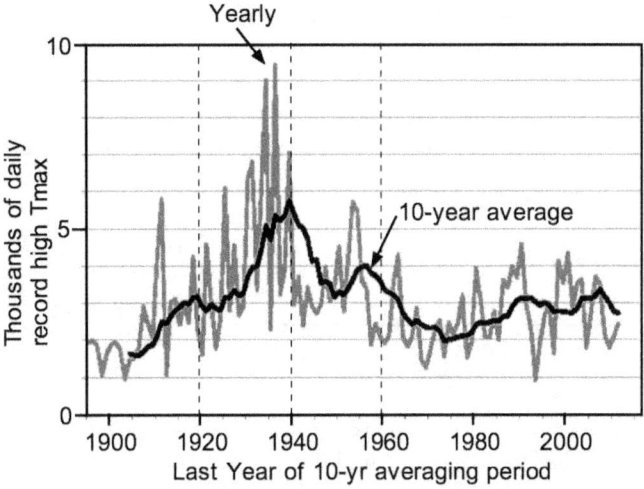

Figure 4.6a. Number of daily high U. S. T_{max} records (1895-2011) based on 970 USHCN stations with 80 years data (Christy, 2012).

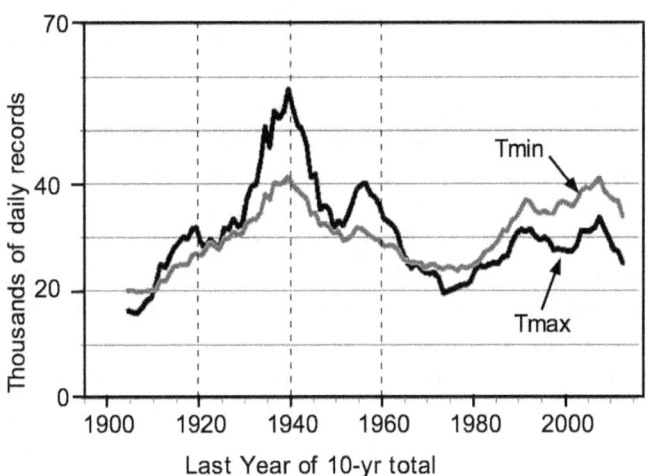

Figure 4.6b. Number of daily U. S. T_{max} and T_{min} records (1895-2011). Ten year running totals based on 970 USHCN stations with 80 years data (Christy, 2012).

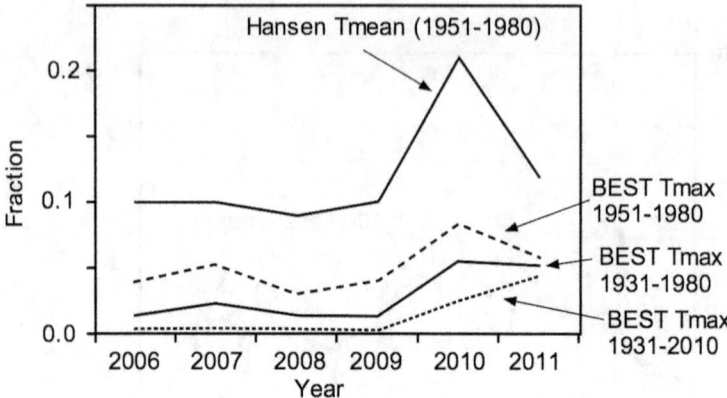

Figure 4.6c. Fraction of monitored area that exceeds 3σ threshold for reference periods and databases as indicated (Christy, 2012).

4.4.2 Global Temperatures

GISS also estimated global temperatures since 1880 as shown in Figure 4.7 (Hansen *et al.*, 2010). Steve McIntyre (*climateaudit.org*) commented briefly on this paper. He pointed out that in this update, GISS made remarkably large changes from the August 2007 version, constituting "a rewriting of history that has increased the [temperature during] 2000-6 relative to the 1930s by about 0.3 deg C". Such large changes seem suspect, especially when they significantly increase the global warming. McIntyre also emphasized that the paper by Hansen *et al.* (2010) "purported to show that urban heat islands don't matter". This paper came out shortly before Zhang *et al.* (2010) reported that UHI effects can reach 7-9°C in northeastern U.S. cities. The effects of the Krakatoa, El Chichon and Pinatubo volcanic eruptions are evident. Most recently, the effects of two major El Niños are also evident.

A breakdown of global temperature changes by latitude range is provided in Figure 4.8. This shows that by far, the greatest change in temperature was in the far north. The latitude range 28-60°N had a moderate temperature rise, and the tropics and Southern Hemisphere had minimal increases in temperature.

It is evident that as the Earth warms or cools due to various causes, the Arctic will experience greater than average warming or cooling, whereas tropical areas will undergo lesser changes. Warming in the Southern Hemisphere is limited by the large heat uptake of the Southern Ocean, leaving the Arctic as the global location with the greatest warming. Increased warming in Arctic regions is usually referred to as "Arctic amplification". Figure 4.8 shows this very clearly.

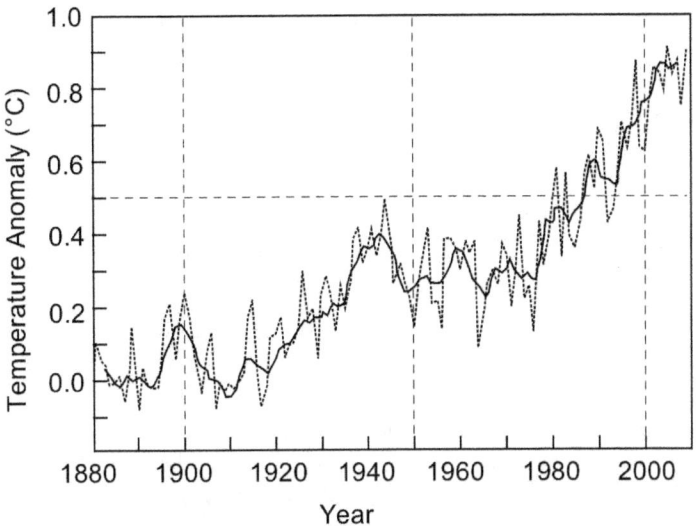

Figure 4.7. Global temperatures based on a combination of land and sea data shown with five-year moving average (solid line). Adapted from Hansen *et al.* (2010).

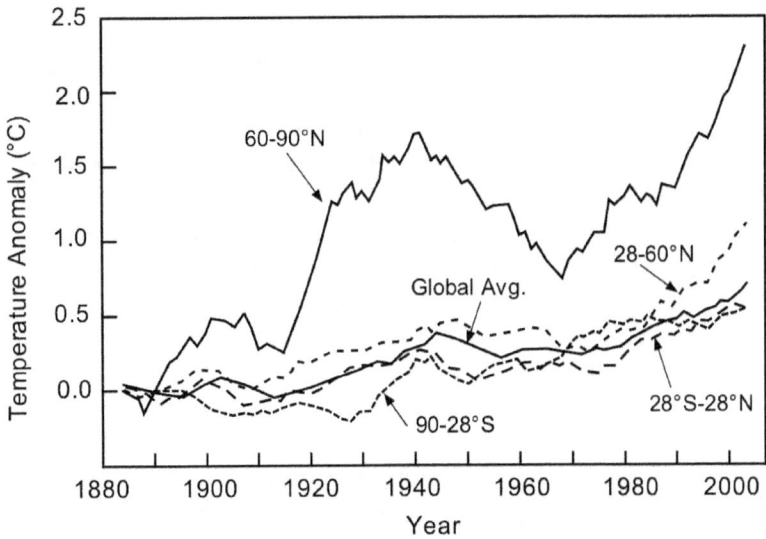

Figure 4.8. Area-weighted mean observed surface temperatures over the indicated latitude bands. The values are running 9-year means relative to the 1880-1890 mean (Shindell and Faluvegi, 2009).

Usually, Arctic amplification is regarded as the ratio of the Arctic-averaged surface air temperature change to the global average temperature change during a period when the Earth's climate changes. In the past decade, a number of studies have been conducted on Arctic amplification during the warming that occurred in the 20th century, and projections have been made of continued warming in the 21st century. Almost all of these begin by assuming that warming in the 20th and 21st centuries is predominantly due to increased concentrations of greenhouse gases, particularly CO_2, and then proceed to analyze the role that Arctic amplification will play in response to rising greenhouse gas concentrations in the 21st century. Several of these papers seem to have the intent of showing that anthropogenic emissions of CO_2 were the cause of the observed warming.

Miller *et al.* (2010, 2010a) emphasized the sensitivity of Arctic temperatures to change via various feedbacks that can act to amplify or diminish incipient climate changes. These include:

• Ice-albedo feedback – "Changes in the seasonal and areal distribution of snow and ice exert strong influences on the planetary energy balance through their impact on Earth's albedo", primarily in summer. During the winter, Arctic areas receive almost no sunlight so albedo does not matter.

• Ice-insulation feedback – "Sea ice also causes a positive insulation feedback, primarily in winter. By insulating the cold polar atmosphere in winter from the relatively warm ocean, little of the ocean's energy can be transferred to the atmosphere".

• Vegetation feedbacks – "A warming climate can cause tundra to give way to lower albedo shrub vegetation or even dark-green boreal forest. The lower albedo of shrubs and boreal forest, especially in spring when high-albedo snow cover may still bury tundra, results in earlier warming and hence exerts a positive feedback on warming".

• Permafrost feedbacks – "... poorly understood feedbacks in the Arctic involve changes in the extent of permafrost, and how changes in cloud cover interact both with permafrost and with the release of carbon dioxide and methane from the land surface. Melting allows ancient plant debris to decompose, releasing greenhouse gases (CO_2 and/or CH_4) that mix globally, amplifying the initial warming by enhancing the planetary greenhouse effect".

• Feedbacks during glacial-interglacial cycles – "slow positive feedbacks that operate on time scales of 10^4 years were important contributors to glacial-interglacial climate cycles". These included changes in albedo due to growth of ice sheets, changes in ocean volume changing land/sea areal balance, changes in vegetation, changes in water vapor, changes in dust levels, changes in greenhouse gas concentrations, etc.

Miller *et al.* (2010a) also discussed other feedbacks involving fresh water inputs to the Arctic seas and changes in thermohaline circulation.

Holland and Bitz (2003) discussed attempts to characterize Arctic amplification by means of climate models. Climate model simulations have included ice albedo feedbacks associated with variations in snow and sea-ice coverage, variations in the thickness of sea-ice, clouds, and changes in heat transported by the atmosphere and/or ocean. There is agreement among models that the Arctic warms more than subpolar regions whereas high southern latitudes exhibit a minimum warming due to changes in ocean heat uptake. Figure 4.9 shows the range of estimates of Arctic amplification by various climate models.

As we discussed previously, any change in global temperature will be amplified in the Arctic due to various feedbacks. In the last century or two, an additional factor tended to amplify warming in the Arctic, not by amplification of a global trend but by direct increases of solar energy absorbed at higher latitudes. This is due to deposition of soot on high latitude snow and ice from power plants and industry in the NH. This is discussed further in Section 6.5.

In late 2011, a team led by Richard Muller released a preliminary version of the "Berkeley Earth Surface Temperature Analysis" (BEST) that estimated global land temperatures back to year 1800. (Note that land covers only about 30% of the globe. Global temperatures will not change as rapidly as land temperatures because of thermal inertia of the oceans.) The worldwide distribution of stations was not provided, but Figure 4.4 shows the distribution for the U.S. It seems likely based on Figures 4.1 to 4.4 that station coverage for the period 1800 to 1880 was probably limited to mainly European and some U.S. locations. There are many unresolved problems with this analysis as discussed for example by Rapp (2010). It is noteworthy that one-third of the measurement stations reported that temperatures had actually dropped over the time period of measurement.

In a more recent release by the BEST Team (July 2012), the data were extended even further back in time. Rohde *et al.* (2012) reported an estimate of the Earth's average land surface temperature for the period 1753 to 2011. To address issues of potential station selection bias, they used larger sampling of stations than prior studies. For the period post-1880, their estimate was similar to those previously reported by other groups, although they claimed smaller error uncertainties. They claimed that the land temperature rise from the 1950s decade to the 2000s decade was 0.87 ± 0.05 °C (95% confidence). Both maximum and minimum daily temperatures increased during the last century. According to them, diurnal variations decreased from 1900 to 1987, and then increased; this increase is significant but not understood. Figure 4.10 shows a result of the BEST study.

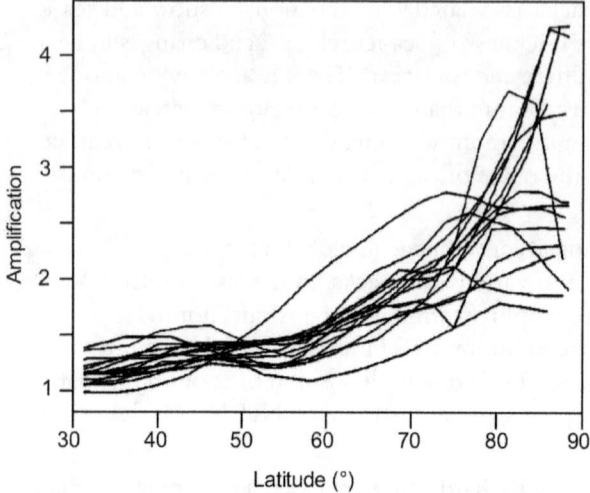

Figure 4.9. The relative amplification of air temperature 2 m above the surface as a function of latitude, as estimated by various climate models (Holland and Bitz, 2003).

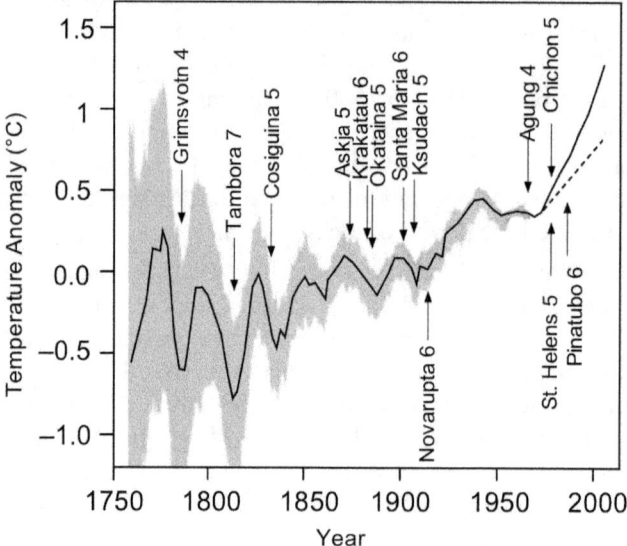

Figure 4.10. BEST 10-year average data showing dates of major volcanic eruptions with volcanic explosive indices. The dashed line shows the correction proposed by the Watts team, but this has yet to be validated.

These results are shown in Figures 4.10a and 4.10b. Figure 4.10a shows a ten-year moving average of global land temperatures. Figure 4.10b shows yearly data along with the timing of major volcanic eruptions with explosive indices shown on the graph.

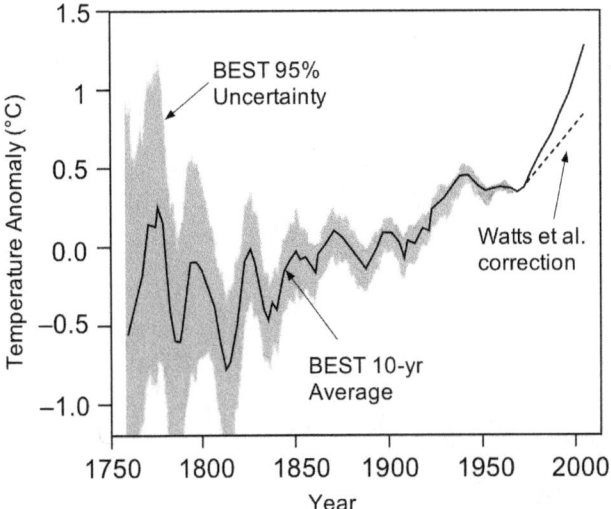

Figure 4.10a. BEST ten-year moving average of global land temperatures. The proposed correction by Watts *et al.* is under review and might not hold up under further scrutiny.

The period from 1753 to 1850 were claimed to be marked by sudden drops in land surface temperature that were coincident with volcanism; they claimed that the response function was approximately 1.5 ± 0.5 °C per 100 Tg of atmospheric sulfate. According to their interpretation, this volcanism, combined with a simple proxy for anthropogenic effects (logarithm of the CO_2 concentration), can account for much of the variation in the land surface temperature record and they claimed that the fit was not improved by the addition of a solar forcing term. Thus, for their very simple model, they concluded that solar forcing does not appear to contribute to the observed global warming of the past 250 years and the they believe that the entire change can be accounted for by a sum of volcanism and anthropogenic proxies.

The large oscillations prior to 1900 were attributed to volcanoes. A comparison of yearly data with eruptions by the largest volcanoes is given in Figure 4.10b. It is not clear why the apparent temperature effects of early volcanoes were so much greater than the effects of more recent volcanoes. Furthermore, the downward trends in the two earliest volcanoes seem to have begun prior to the eruptions. The large oscillations prior to 1850 might be artifacts if the limited number of stations were positioned in regions particularly susceptible to effects of volcanic eruptions. In particular the eruption of 1783 (*Grimsvotn*) with an explosive index of only 4 seems to have had a far greater climatic effect than more powerful eruptions that occurred later. It has been estimated that *Grimsvotn* released only 12.3 km³ whereas Tambora (1815) released 160 km³; yet they had similar climatic effects according to BEST. The volcanic explosive index is logarithmic so the volcanoes in the 19th and 20th centuries were 10 to 100 times greater than *Grimsvotn*; yet acccording to BEST, *Grimsvotn* had a much greater climatic effect. This seems unlikely to be correct.

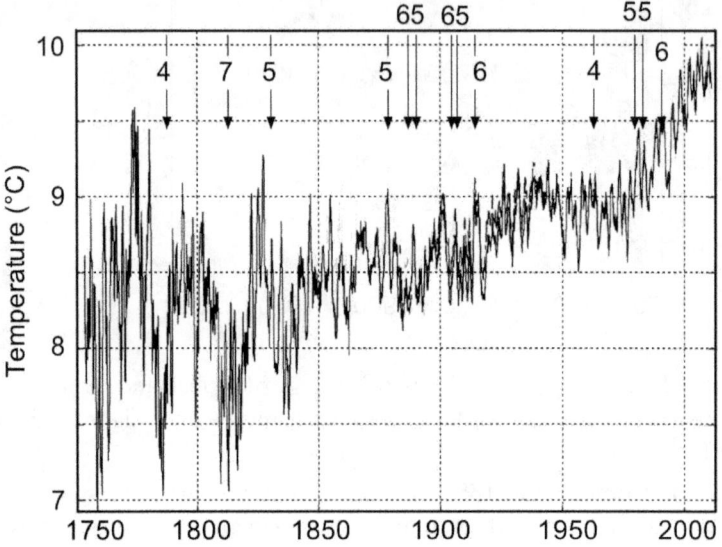

Figure 4.10b. BEST yearly land temperature data compared to volcanic eruptions with indicated explosive indices.

The folklore on the Tambora eruption is extensive. In New England, 1816 has been described as the "year without a summer". There are numerous references on the Internet to freezing and snow in June 1816, as well as crop failures. Yet the sources of these myths do not seem to be very well substantiated. In a web posting,[23] W. Eschenbach cast doubt on these myths by showing that archival data in several European cities did not seem to display unusually cold temperatures in 1816. He also showed that food commodity prices were at a low in 1816 which is contrary to what one would expect from bad weather. There were 300 responses posted to Eschenbach's posting, the overwhelming majority of which had no value. One useful response pointed to a HADCET database that showed that the summers of 1816 and 1817 were indeed relatively cold compared to the extended period 1659 to 2012 with July 1816 being the coldest July in the whole set of 353 years. A website[24] provides considerable detail on unusually cold weather in

---

[23] *http://wattsupwiththat.com/2012/04/15/missing-the-missing-summer/*
[24] *http://www.islandnet.com/~see/weather/history/1816.htm*

the summer of 1816 in the eastern United States. It is not clear where this data derives from and how valid it might be.

Another paper[25] used "recently recovered meteorological observations from 1816 onwards for three stations located in Portugal (Lisbon, Coimbra and Oporto) and also for a longer period for the Spanish stations of Madrid, Barcelona and San Fernando-Cadiz, [to produce] a better characterization of the anomalous climate for this peculiar period over southwestern Europe". They reported "... all available stations reveal a cold summer of 1816, mainly in July and August. In comparison to the 1871–1900 reference period, those two months were 2–3°C cooler, close to what has been reported for central Europe".

Another aspect of the BEST analysis of their data is the implicit assumption that the climate would have been steady, had it not been for the influences of rising CO_2, solar variations, and volcanic eruptions. Thus, they neglect internal influences (the possibility that the climate will change without external forcing). Yet we know (for example) that an internal change occurred after 1976 in the El Niño – La Niña balance that produced a heating trend from 1976 to 1998. Furthermore, as we discuss in Section 6.3, there are no reliable estimates of solar forcing dating back 250 years. It does seem likely that solar variations cannot come close to accounting for the observed data; yet we still lack reliable solar data. As we discussed previously, the team led by Watts claims that errors in the measurement stations reduce the magnitude of the overall temperature rise; however this work by the Watts team remains subject to revision.

The HadCRU group published a paper on quantifying uncertainties in global and regional temperature change (Morice *et al.*, 2012). They discuss uncertainties, and in particular, they mention "... sufficient metadata are not available and studies over small regions are too few for uncertainties in land station homogenization, urbanization and exposure biases to be adequately described on an individual grid-box level. In a similar fashion, the characterization of spatial and temporal correlations in SSTs is limited by missing ship call-sign information prior to 1981."

4.5 Tropospheric Temperatures

Marsh (2002) pointed out:

"The layer of the Earth's atmosphere from the ground up to an altitude of a few miles is called the troposphere, and the boundary between it and the rest of the atmosphere above is called the tropopause. The tropopause is about 11 miles high at the equator and only about five miles high at the poles. The troposphere is the part of the atmosphere

---

[25] Trigo, R. M. *et al.*, (2009) "Iberia in 1816, the year without a summer", *Int. J. Climatol.* **29**, 99–115.

that is responsible for the greenhouse effect, since it contains essentially all of the greenhouse gases. Because the Earth's troposphere, surface and boundary layer are closely coupled by air movement, they are considered to be a single thermodynamic system. For this reason, changes in radiative flux at the tropopause are used to express changes to the climate system."

According to the Wikipedia: "The troposphere ... contains approximately 80% of the atmosphere's mass and 99% of its water vapor and aerosols."

Most recent global climate models hindcasts and forecasts are consistent in depicting a tropical lower troposphere that warms at a rate about 1.3 times that of the surface. Thus, the models would predict that the trend for warming of the troposphere would be at a higher rate than the surface. Yet, the discussion in Section 4.10.4, particularly Figure 4.24, suggest that warming in the troposphere is even greater.

NOAA has flown about ten MSU satellites since 1979. Each satellite has a life cycle of a few years but their timing is arranged so there is overlap between successive satellites. The MSU instruments measure thermal emission from atmospheric oxygen constituting the major component of the measured brightness temperature. Initial processing of the data for the past two decades indicated significant discrepancies between temperature trends for the troposphere vs. trends observed at ground level. During the decade of the 1990s, a number of sources of error in interpreting data from the MSU satellites were eliminated. Several studies in the period 2003-2007 created greater confidence in the tropospheric measurements, and in particular, with the publication of Christy *et al.* (2007) tropospheric temperature measurements became widely accepted. There still remained some differences from ground measurements but these are believed to be real. Douglass and Christy (2009) presented evidence that past problems with calibration have been solved and they presented the latest tropospheric temperature measurements. The current data are shown in Figure 4.11.

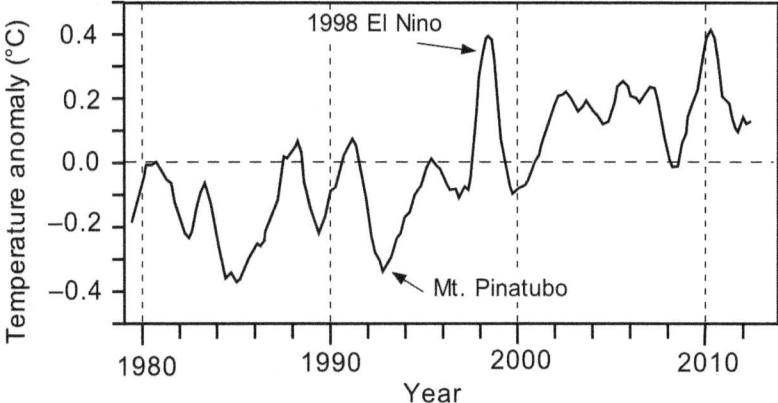

Figure 4.11. 13-month running average of tropospheric temperature over three decades. (Adapted from Christy *et al.* (2007) as updated by Spencer in 2012. (*http://www.drroyspencer.com*).

There are a number of aspects of the data on tropospheric temperatures (TT) that deserve mention. It appears that TT over the past three decades have been controlled more by volcanic eruptions and El Niño conditions than they have by increasing levels of greenhouse gases. This is illustrated in Figures 4.12 and 4.13. Obviously, these results for TT measurements are not supportive of climate models and also are contrary to the alarmist viewpoint. Hence, in keeping with recent practice over the past few years in which alarmists promptly publish rebuttals to any papers that slip through their control of which manuscripts get accepted by climate journals, it was necessary for the alarmists to publish such a rebuttal. Ben Santer took on this responsibility and the result was Santer *et al.* (2011). It is interesting, perhaps, that Santer included 16 co-authors in addition to himself; yet the nature of the work is such that it is difficult to imagine how 16 individuals could each contribute significant portions to the work. Pielke Sr. commented: "This is an unusual number of co-authors for a technical paper, but I assume Ben Santer wants to show a broad agreement with his findings".[26] In other words, many names were added to give the paper political endorsement?

Santer *et al.* (2011) were concerned with a very basic problem in climatology: how to distinguish between long-term climate change and short-term variable weather in regard to TT measurements? They treated the problem in terms of signal and noise: the signal is assumed to be a long-term linear trend of rising temperatures due increasing greenhouse gas concentrations, that is obfuscated by short-term noise. However, the climate-weather problem is innately different from a classical signal/noise problem

---

[26] *http://pielkeclimatesci.wordpress.com/category/climate-models/*

such as a radio signal affected by atmospheric activity. In that case, if the radio signal has a sufficiently narrow frequency band, and the noise has a wider frequency spectrum, the signal-to-noise ratio (S/N) can be improved with a narrow-band receiver tuned to the frequency of the radio signal. The radio signal and the noise are separate and distinct. By contrast, in the climate-weather problem, the instantaneous weather is the noise, and the signal is the long-term trend of the noise. The noise and signal are coupled in a unique way. Furthermore, there is no evidence that it is even meaningful to talk about a "trend" since there is no evidence that the variation of TT with time is linear. Remarkably, Santer *et al.* never referred to Christy *et al.* (2010) but based their analysis on older papers (e.g. Christy *et al.*, 2007).

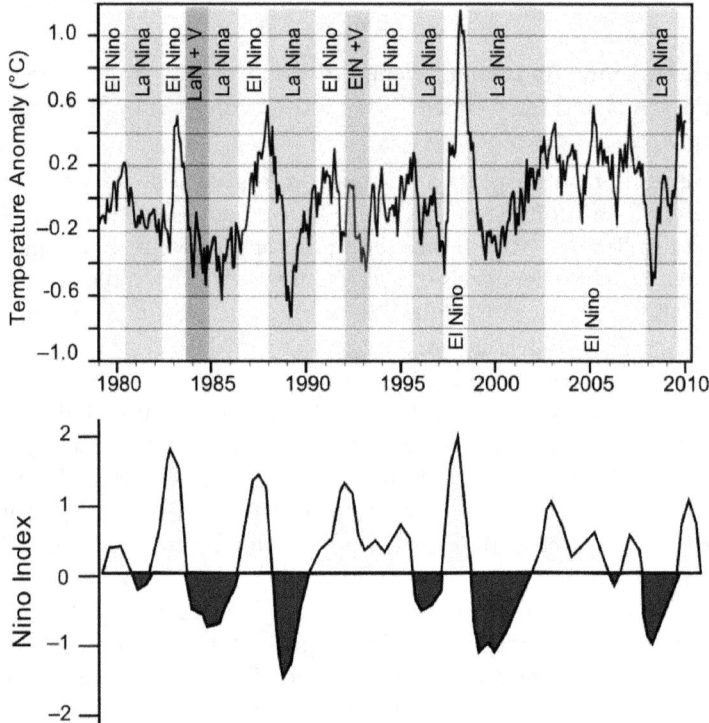

Figure 4.12. Global temperature anomaly compared to the Nino 3.4 index showing strong correlation of upticks in temperature with El Niños.

Santer *et al.* (2011) were primarily concerned with estimating how many years of data are necessary to provide a good estimate of the putative underlying linear trend. They were also intent on showing that short periods with no apparent trend do not violate the possibility that over a longer term, the trend is always there. They derived signal-to-noise (S/N) ratios for both the temperature data and the model average by means that are not clear to this writer.

As Santer *et al.* (2011) showed, one can pick any starting date and any duration length and fit a straight line to that portion of the curve of TT vs. time. They did this for various 10-year and 20-year durations. In each case, depending on the start date, they derived a best straight-line fit to the TT data for that time period. They found that the range of trends for 10-year periods was greater (-0.05 to +0.44°C/decade) than the range for 20-year periods (+0.15 to +0.25°C/decade). The trends for various start dates for ten-year trends are shown in Figure 4.14. Clearly, the trend line was steepest for a start date around 1988 (ending in the giant El Niño year of 1998). Prior to 1988 and after 1998, the trends were minimal.

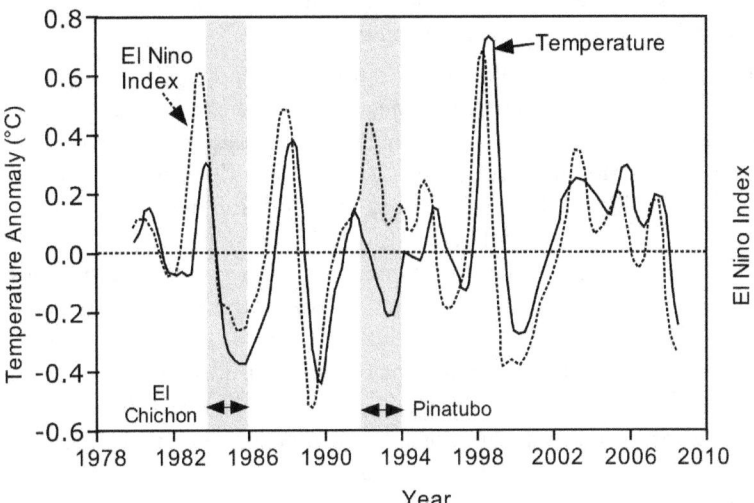

Figure 4.13. Comparison of satellite-measured temperatures with the Nino 3.4 index and volcanic eruptions.

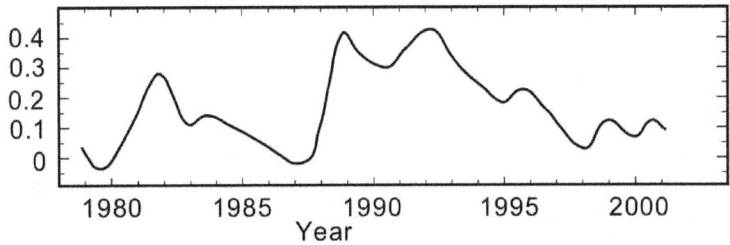

Figure 4.14. Trend (°C/decade) of TT vs. start year for ten-year durations. (Santer *et al.*, 2011).

Santer *et al.* described use of longer durations as "noise reduction", which it is, provided that one assumes the overall signal is linear in time. It still was problematic that the trend was nil after 1998 that they rationalized by saying:

> "The relatively small values of overlapping 10-year TT trends during the period 1998 to 2010 are partly due to the fact that this period is bracketed (by chance) by a large El Niño (warm) event in 1997/98, and by several smaller La Niña (cool) events at the end of the … record".

However, as Pielke pointed out, the period after 1998 was 13 years, not 10, and furthermore, as Figure 4.12 and 4.13 show, the period after 1998 had roughly equal periods of El Niño and La Niña and was not dominated by La Niñas as Santer *et al.* claimed. What Santer *et al.* (2011) implied was that an unusual conflux of a large El Niño early on and multiple La Niñas later on caused the trend to minimize for that unique period as a statistical quirk. However, that is like a baseball pitcher saying that if the opponents hadn't hit that home run, he would have won the game.

In simplistic terms, the signal-to-noise ratio can be estimated as follows. For either 10-year or 20-year durations, the signal was the mean trend derived by a straight-line fit to the TT data over that duration. The noise was the range of trends for different starting dates. For ten-year durations, the trend was $0.19 \pm 0.25°C/$decade. For twenty-year durations, the trend was $0.20 \pm 0.05°C/$decade. The signal in each case is taken as the mean trend. The distribution of trends within these ranges was similar to a normal distribution. Thus we can roughly estimate the noise as ~ 0.7 times the full width of the range. Hence, the S/N ratio for ten-year durations can be crudely estimated to be S/N $\sim 0.19/(0.7 \times 0.5) = 0.5$ and for twenty-year durations is S/N $\sim 0.2/(0.7 \times 0.1) = 2.9$. Santer *et al.* obtained S/N = 1 for ten-year durations and S/N = 2.9 for twenty-year durations. If it can be assumed that the signal varies linearly with time, one can then estimate what level of precision for the estimated trend can be obtained for any chosen duration. Santer *et al.* obviously believe that the signal is linear with time for all time. By some logic that escapes me, Santer *et al.* concluded that

> "Our results show that temperature records of at least 17 years in length are required for identifying human effects on global-mean tropospheric temperature".

This conclusion seems to be grossly exaggerated. A more proper statement might be as follows:

> *Assuming that the variability of TT is characterized by a long-term upward linear trend caused by human impact on the climate, and that variability about this trend is due to yearly variability of weather, El Niños and La Niñas, and other climatological fluctuations, the recent data suggest that the trend can be estimated for any 17-year period with a S/N ratio of roughly 2.5.*

Finally, we get to the nub of the paper by Santer *et al.* that asserted: "Claims that minimal warming over a single decade undermine findings of a slowly-evolving externally-forced warming signal are simply incorrect". Here is where Santer *et al.* attempted to dispel the notion that minimal warming for a period contradicts the belief that underneath it all, the long-term signal continues to rise at a constant rate. Pielke Sr. argued that this was an overstatement and he concluded:

> "If one accepts this statement by Santer *et al.* as correct, than what should have been written is that the observed lack of warming over a 10-year time period is still too short to definitely conclude that the models are failing to skillfully predict this aspect of the climate system" (*http://pielkeclimatesci.wordpress.com/category/climate-models/*).

However, I would go further than Pielke Sr. First of all, the period of minimal temperature rise was longer than 10 years. Second, there is no cliff at 17 years whereby trends derived from shorter periods are statistically invalid and trends derived from longer periods are valid. According to Santer *et al.* a trend derived from a 13-year period is associated with a $S/N \sim 1.5$ which though not ideal, is good enough to cast some doubt on the validity of models.

If one carries out a least squares comparison of linear and step functions to the TT , the best fits are:

Linear: $TT = -0.2 + 0.00156 (Y - 1979)$ [RMS error = 0.026]

Step: $TT = -0.14 (Y < 1998)$; $TT = 0.18 (Y \geq 1998)$ [RMS error = 0.023]

Hence, the step function fits the data slightly better than the linear function. A function based on a Niño index would undoubtedly fit much better than either of these fits.

In October 2012, the Climate Research Unit (CRU) of the University of East Anglia, one of the important organizations that monitor Earth temperatures from networks, reported on an update of their estimates of global average temperatures (including land and sea). Their results showed that during the period 1997 through 2012, the average temperature meandered up and down, but on average was unchanged over this period of 16 years. Naysayers and skeptics pounced on this announcement with glee. Presumably, Ben Santer with 20 or 30 alarmists as co-authors, will soon write a paper explaining this to be a temporary fluctuation. The CRU data, along with the Nino3.4 index, are plotted in Figure 4.14a.

The continued almost religious belief by alarmists that the temperature always rises linearly and continuously is evidently refuted. If the alarmists would only reduce their hyperbole and argue that rising greenhouse gas concentrations produce a warming force that is one of several factors controlling the Earth's climate, and there are periods during which the other factors overwhelm the greenhouse forces, perhaps we would have a rational description. Instead, the alarmists continue to find linear trends over various time periods, in some cases when they are not there.

4.6 Arctic and Antarctic Temperatures

As we pointed out in the discussion of Figures 4.8 and 4.9, the north polar region changes temperature more rapidly and with greater amplitude than other regions due to various feedback effects that amplify small changes.

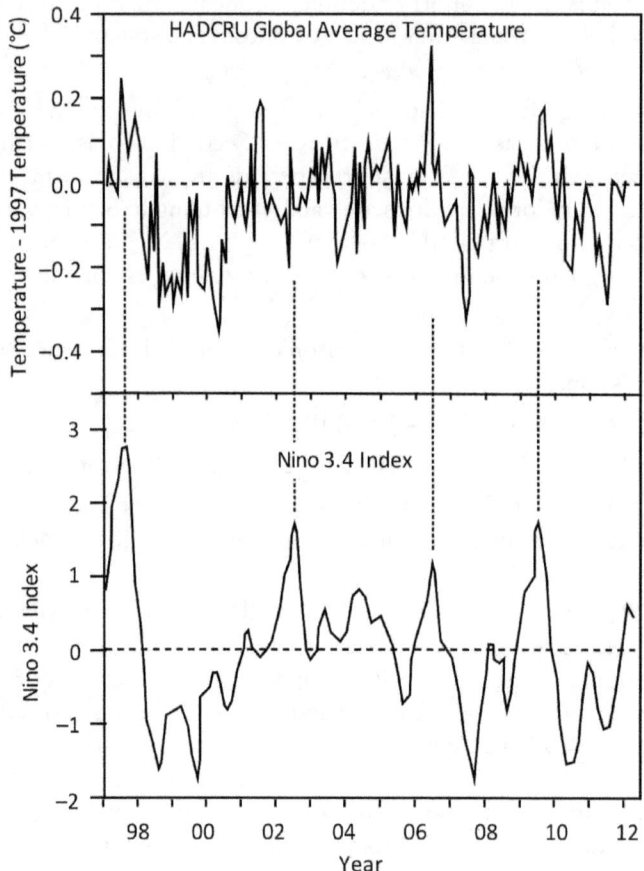

Figure 4.14a. Measured global temperatures from 1997 to 2012 compared to Nino3.4 index showing unchanged temperature average over 16 years. (*http://www.metoffice.gov.uk/hadobs/hadcrut4/data/download.html*)

4.6.1 Antarctic Temperatures

Antarctica is a very cold place. The coldest temperature recorded was –89°C. Several factors combine to make Antarctica so cold:

- Unlike the Arctic region, Antarctica is a continent whose interior areas do not benefit from the moderating influence of water.
- With 98% of its area covered with snow and ice, the Antarctic continent reflects most of the Sun's input rather than absorbing it.
- The extreme dryness of the air causes any heat that is radiated back into the atmosphere to be lost instead of being absorbed by the water vapor in the atmosphere.
- During the winter, the size of Antarctica doubles as the surrounding sea water freezes, effectively blocking heat transfer from the warmer surrounding ocean.
- Antarctica has a higher average elevation than any other continent on Earth, which results in even colder temperatures.

In discussing the Antarctic continent, it is important to distinguish between the Antarctic Peninsula, which projects to lower latitudes, and has a low elevation, vs. the Antarctic continent that lies at higher latitudes and higher elevations. Most of the climatological measurement sites are located around the continent on the coast. Only a few are inland.

Turner *et al.* (2005) analyzed 19 long-term stations over the Antarctic continent. The peninsula experienced a major warming over the last 50 years. The 14 stations around the continent were equally divided between warming and cooling, leading to an uncertain result. The data also indicated that warming trends appeared to be greater for the 1961–1990 period compared with the 1971–2000 period, suggesting that warming tailed off after 1971.

Comiso (2000) reported on surface air temperatures observed from stations in Antarctica. The results showed predominantly warming trends as high as 0.5°C/ decade along the Antarctic peninsula. Surface temperatures inferred from infrared satellite data from 1979 to 1998 were analyzed in combination with data from 21 stations that have relatively long record lengths. The surface temperatures derived from infrared data were claimed to agree well with Antarctic station data. The 45-year record of station data showed a net warming of 0.012°C/year (±67%) while the 20-year record of station data showed a net cooling of 0.008°C (±300%). The 20 year record of satellite measurements indicated a cooling of 0.042°C/year (±150%). The 20-year record length was claimed to be about the minimum length required for a meaningful trend analysis study.

Steig *et al.* (2009) estimated 50–year temperatures in the interior of Antarctica using a combination of surface data (mostly around the periphery of Antarctica), satellite data (IR radiated by the snowpack), and models to fill in missing data. It is difficult to appraise the credibility of the models used because the paper is mostly jargon and unreadable except to a very few specialists. They concluded that essentially all of Antarctica warmed in the last 50 years of the 20th century. However, several of the authors are noted alarmists, and one is the leading defender of the *hockey stick*. As always, in the process of combining data from multiple sources over extended time periods, the results are noisy and unraveling the implicit signal from noise to discover trends involves complex procedures. The procedures used by Steig *et al.* (2009) were discussed by O'Donnell *et al.* (2011). They concluded: "Though the general reconstruction concept has merit, it is susceptible to spurious results for both temperature trends and patterns". O'Donnell *et al.* investigated two alternate methods to improve the procedure. They concluded that warming over the period of 1957–2006 was concentrated in the peninsula (~0.35°C per decade). They also found that average trends for the continent, East Antarctica, and West Antarctica were half or less than those found by Steig *et al.* (2009). Subsequently, additional commentary appeared on Internet Blogs. McIntyre[27] describes efforts by the Steig team to block publication of the O'Donnell team criticism. McIntyre said: "... whatever is new in Steig *et al.* (2009) is not only incorrect, but an artifact of flawed math and whatever is valid was already known". According to McIntyre: "Steig's methodology smeared warming from the Antarctic Peninsula into other parts of Antarctica". Yet the forthcoming new edition of the IPCC Report refers generously to the Steig *et al.* paper and totally ignores the results of O'Donnell *et al.*

Frolov *et al.* (2009) presented data that suggest that temperatures in Antarctica have been meandering for many years, although the peak reached around 1980 was somewhat higher than previous peaks.

Chapman and Walsh (2007) prepared a comprehensive analysis of surface air temperature anomalies for Antarctica and the Southern Ocean at a grid resolution sufficient to capture regional variations of trends. Land surface station, automatic weather station, and gridded summarized sea surface temperature data were utilized. As Chapman and Walsh said: "The salient finding in the context of climate change is that the 45-year trends are small ..." In fact, not only were the trends small, but the trends over land were negative for the 1970s and 1980s. Most of whatever warming took place, occurred over the Antarctic Peninsula. Despite this, Chapman and Walsh interpreted the data according to the alarmist agenda, projecting strong warming over Antarctica in the 21st century based on climate models.

Tedesco and Monaghan (2010) reviewed melting of snow and ice over all of Antarctica since 1979, when routine measurement of the phenomenon via space-borne passive microwave radiometers first began" (NIPCC, 2011). Their results for the snow and ice melting trends averaged over 12 continent-wide sites over the past thirty years did not display any notable trends. In other words, they found no evidence of significant melting and the climate in

---

[27] *http://www.climateaudit.org*

Antarctica has been stable. Nevertheless, they had to pay homage to the orthodoxy by concluding "Negative melting anomalies observed in recent year do not contradict recently published results on surface temperature trends over Antarctica". They also concluded that in the future, "enhanced summer melting is likely to occur if the positive Southern Hemisphere Annular Mode (SAM) trends subside". This can be paraphrased: "enhanced summer melting is likely to occur if it occurs" – as in the Marx Brothers: "The first party is the party of the first part".

4.6.2 Arctic Temperatures

Hanna and Cappelen (2003) said:

"Different temperature and precipitation records, covering the period 1958–99, from six stations in Southern Greenland have been studied ... Despite global warming over the past few decades, the SW marginal areas of southern Greenland seem to have actually cooled, especially daytime temperatures in winter ...It will be intriguing to see if this trend continues, as it could substantially influence the mass balance (through changes in snow accumulation, surface melt runoff and iceberg calving) of the southern parts at least of the ice sheet. The overall cooling may also have caused a sharp increase in snowfall days at some of the coastal stations, where snow as a fraction of precipitation is critically dependent on temperature. These results demonstrate the regional vagaries of the global weather machine; climatic change is not a simple uniform process. Yet as one of these regions, Greenland greatly influences the surface heat budget, atmospheric circulation and (through the waxing and waning of its ice sheet) global sea level."

Przybylak (2002) provided a detailed analysis of intra-seasonal and inter-annual temperature variability for the whole Arctic for the period 1951–1990. The final conclusion was:

"The absence of significant changes in intra-seasonal and inter-annual variability of TMEAN, TMAX, TMIN and DTR is additional evidence (besides the average temperature) that in the Arctic in the period 1951–1990 no tangible manifestations of the greenhouse effect can be identified."

According to Chylek, Box, and Lesins (2004):

"A considerable and rapid warming over all of coastal Greenland occurred in the 1920s when the average annual surface air temperature rose between 2 and 4°C in less than ten years (at some stations the increase in winter temperature was as high as 6°C). This rapid warming, at a time when the change in anthropogenic production of greenhouse gases was well below the current level, suggests a high natural variability in the regional climate ...Since 1940, however, the Greenland coastal stations data have [undergone] predominantly a cooling trend. At the summit of the Greenland ice sheet the summer average temperature has decreased at the rate of 2.2°C per decade since the beginning of the measurements in 1987. This suggests that the Greenland ice sheet and coastal regions are not following the current global warming trend."

Chylek, Dubey, and Lesins (2006) found that "current Greenland warming is not unprecedented in recent Greenland history." Warming experienced from 1995 to 2005 also occurred from 1920 to 1930, except that the rate of warming from 1920 to 1930 was about 50% higher than that in 1995–2005.

Bengsston *et al.* (2004) said: "The huge warming of the Arctic that started in the early 1920s and lasted for almost two decades is one of the most spectacular climate events of the 20th century." They suggested that this was caused by enhanced wind driven oceanic inflow into the Barents Sea with an associated sea ice retreat. The magnitude of the inflow is linked to the strength of westerlies into the Barents Sea."

Two recent papers attempted to derive historical Arctic temperatures over the past 2,000 years from proxies. Kaufman *et al.* (2009) used a "compilation of proxy records from Arctic lakes, combined with complementary ice core and tree ring records, to form a new 2000-year-long, decadally resolved paleoclimate reconstruction for the Arctic" northward of 60°N. Their comparison of the proxies with temperature records during the calibration period (1860-2000) is suggestive of a relationship but lacks agreement in details. They found a slow steady decrease in Arctic temperatures from 2,000 YBP to about 1900, a sudden rise after 1900 to about 1940, and sharp decrease from 1940 to about 1970, and a subsequent rise after that. Spielhagen *et al.* (2011) derived a multidecadal-scale record of ocean temperature variations during the past 2000 years from marine sediments off Western Svalbard (79°N). Their results show a peak high temperature around year 1000 corresponding to the MWP, a rounded trough from 1400 to 1900 with a minimum around 1700 corresponding to the LIA, and a moderate rise in the 20th century. Their peak temperature in the MWP was about the same as current temperatures and their bottom in the LIA was about 0.7°C lower.

Measured land temperatures for 60°N to 90°N in the 20th century are shown in Figure 4.14b.

Kaufman *et al.* (2009) argued that the cooling prior to 1900 was due to a gradual decrease in Arctic summer insolation, and they implied (but did not overtly say) that warming after 1900 was due to greenhouse gases. They did emphasize the alarmist hyperbole that recent decades were the warmest of the past two millennia (or for greater emphasis, the past 20 decades). Spielhagen *et al.* (2011) overtly claimed that 20th century temperatures reflected global warming due to increased greenhouse gases. They said: "We find that early–21st-century temperatures of Atlantic Water entering the Arctic Ocean are unprecedented over the past 2000 years and are presumably linked to the Arctic amplification of global warming". Arrak (2012) concluded that the cause of Arctic warming in the 20th century was a relatively sudden rearrangement of the North Atlantic current system at the turn of the century that directed warm currents into the Arctic Ocean. He claimed that all observations of Arctic warming can be accounted for as consequences of these flows of warm water to the Arctic and greenhouse gases are not involved at all. It seems likely that all of these investigators had their minds made up prior to the investigations, and proceeded to interpret the data to fit their preconceived notions.

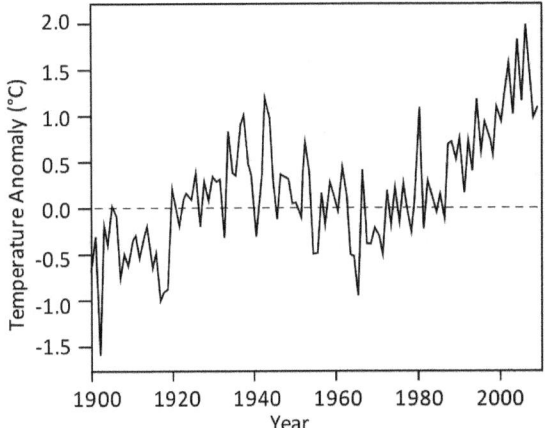

Figure 4.14b. Arctic-wide annual average surface air temperature anomalies relative to the 1961-90 mean, based on land stations north of 60°N. Data are from the CRUTEM 3v dataset (*http://www.cru.uea.ac.uk/cru/data/temperature/*).

Zwally *et al.* (2005) extended the analysis of radar altimeter data from two European remote-sensing satellites to 90.0% of the Greenland ice sheet, 77.1% of the Antarctic ice sheet, and 81.8% of the Antarctic ice shelves. The estimated changes in ice mass from elevation changes derived from 10.5 years (Greenland) and 9 years (Antarctica) of satellite radar altimetry data indicated that the Greenland ice sheet is thinning at the margins (–42 Gt/yr below the equilibrium-line altitude (ELA)) and growing inland (+53 Gt/yr above the ELA) with a small overall mass gain (+11 Gt yr). The ice sheet in West Antarctica / (WA) is losing mass (–47 Gt/yr) and the ice sheet in East Antarctica (EA) shows a small mass gain (+16 Gt/yr) for a combined net change of –31 Gt/yr. The contribution of the three ice sheets to sea level is +0.05 ± 0.03 mm/yr. The Antarctic ice shelves show corresponding mass changes of –95 Gt/yr in WA and +142 Gt/yr in EA. Expected responses of the ice sheets to climate warming are growth in thickness of the inland ice areas, due to increasing precipitation, accompanied by thinning near the margins, due to increasing surface melting.

Polyakov *et al.* (2003b) examined long-term Arctic surface air temperature and pressure data for the period 1875–2000 poleward of 62°N. The Arctic air temperature and pressure displayed substantial variability on timescales of 50–80 years. They suggested that the origin of this variability might lie in the complex interactions between the Arctic and North Atlantic. The two periods of highest temperatures in the Arctic were: in the 1930s–1940s, and in recent decades. They speculated that global warming alone cannot explain the retreat of Arctic ice observed in the 1980s–1990s. They concluded:

> "The complicated nature of Arctic temperature and pressure variations makes understanding of possible causes of the variability, and evaluation of the anthropogenic warming effect most difficult."

Howat, Joughin, and Scambos (2007) found large year-to-year variations in ice loss from ice sheets. Shepherd and Wingham (2007) reviewed the results of more than a dozen satellite-based studies of ice loss from the Greenland and Antarctic Ice Sheets. They estimated that the East Antarctica Ice Sheet (EAIS) is gaining some 25 Gt/year, the West Antarctica Ice Sheet (WAIS) is losing about 50 Gt/year, and the Greenland Ice Sheet (GIS) is losing about 100 Gt/year. These trends provide a modest contribution to sea level rise of about 0.35 mm/year. However, these short-term results should not be extrapolated because of the cyclic behavior of ice sheet loss. Frolov *et al.* (2009) presented an extensive review of centennial ice cover observations in Eurasian Arctic Shelf Seas. They emphasized:

> "The variability and state of the Arctic sea ice cover strongly depend on atmospheric conditions and on ocean dynamic and thermodynamic processes. A number of parameters influence the direction and intensity of these processes. The most significant are: the surface air temperature, wind, oceanic boundary layers and their stratification, and ocean circulation. In order to understand the causes of long-term changes in the ice cover, it is necessary to define the temporal and spatial relationships of the sea-ice cover with all [these factors]."

Frolov *et al.* (2009) utilized anomalies of mean annual surface air temperature (SAT) in the zone from 70°N to 85°N for the period from 1900 to 2003 to analyze climatic changes in the Arctic Seas throughout the last century. These authors emphasized the periodicity of cooling and warming events interpreting this figure in terms of a proposed 60-year cycle.

The sharp temperature rise in the Arctic early in the 20th century remains as a problem for those who would blame all climate change on greenhouse gases. Wood and Overland (2010) revisited this issue. They said: "the recent widespread warming of the Earth's climate is the second of two marked climatic fluctuations to attract the attention of scientists and the public since the turn of the 20th century," and the first of these: "the major early 20th century climatic fluctuation (~1920-1940)" has been "the subject of scientific enquiry from the time it was detected in the 1920s." Furthermore, they wrote that: "the early climatic fluctuation is particularly intriguing now because it shares some of the features of the present warming that has been felt so strongly in the Arctic."

These authors went back to early records to try to gain more information on the early warming episode. They said "there is evidence that the magnitude of the impacts on glaciers and tundra landscapes around the North Atlantic was larger during this period than at any other time in the historical period". They went on to say: "the ultimate cause of the early climatic fluctuation was not discovered by early authors and remains an open question," noting that "greenhouse gas forcing is not now considered to have played a major role". Thus, they suggested that: "the early climatic fluctuation was a singular event resulting from intrinsic variability in the large-scale atmosphere/ocean/land system and that it was likely initiated by atmospheric forcing". They concluded: "... thus far, human influence does not stand out relative to other, natural causes of climate change".[28]

The problem for those who attribute all climate change to greenhouse gases is that if events occasionally occur independent of greenhouse gases that produce effects greater than or equal to predicted effects of greenhouse gases, how does one distinguish between the two?

While the case can be made that recent warming in the Arctic is not unprecedented, nevertheless there are indications that in recent years, Greenland has been subject to significantly increased melting. Tedesco *et al.* (2010, 2011, 2011a) used satellite data to infer surface melting over the Greenland ice sheet. As Tedesco *et al.* pointed out, "Satellites data cannot produce estimates of runoff and liquid water content". However, these can be estimated with models. The model used by Tedesco *et al.* led to estimates that snowfall has been relatively constant over the past fifty years, although it took a dip from 2005 to 2010. Runoff increased substantially after 2000, leading to negative surface mass balances after 2000. Considering that the CO_2 concentration has been steadily increasing over this fifty-year period, it is not clear why the runoff should suddenly increase after 2000. Over the next several years, it may become apparent whether the estimate of increased runoff is real and persistent, or whether it was a temporary fluctuation.

4.7 The NH temperature dip, 1940–1978: effect of aerosols

Lefohn, Husar, and Husar (1999) was the most authoritative source on aerosol production, but Stern (2005a, b) written about six years later, appears to encompass the earlier paper as well as adding more recent data. Smith *et al.*

---

[28] The above two paragraphs were paraphrased from *http://www.sepp.org*

(2011) produced an extensive detailed update and review of global sulfur emissions by sector, and region as of 2005. They found:

"Global emissions peaked in the early 1970s and decreased until 2000, with an increase in recent years due to increased emissions in China, international shipping, and developing countries in general. An uncertainty analysis was conducted including both random and systemic uncertainties. The overall global uncertainty in sulfur dioxide emissions is relatively small, but regional uncertainties ranged up to 30%. The largest contributors to uncertainty at present are emissions from China and international shipping."

Their results for global SO_2 emissions over the past 150 years are shown in Figure 4.15. The downtrend from about 1975 to 2000 seems to be reversing due primarily to the uptrend since 1975 by China and other Asian and African regions. China is now driving the world SO_2 emission trend.

Ghan and Schwartz (2007) and Chin *et al.* (2009) provided comprehensive reviews of the status of modeling aerosols in various generations of global climate models, past, present and future. Early approaches for incorporating the effects of aerosols simply adjusted the albedo as a crude approximation, while subsequent models used much more sophisticated representations of aerosols, but even these did not adequately account for interactions of aerosols with other climate elements.

"It is now recognized that accurate representation of aerosol influences must take into account phenomena such as correlations of aerosol loading with meteorological variables and the influence of aerosol on clouds and precipitation, and hence that aerosol loading and those properties must be represented actively and interactively in climate models. It is this recognition that is driving much of the current effort to actively represent aerosol processes, properties, and effects in climate models" (Ghan and Schwartz, 2007).

The treatment of aerosols in future generations of climate models will rest on an improved understanding of the processes that control aerosol properties and the ways that they affect climate. Chin *et al.* (2009) provided a summary of unresolved issues in regard to aerosols and climate. Ghan and Schwartz (2007) provided a plan for systematically improving our understanding of aerosols in the future.

4.8 Warming in the Early 20th Century

There are several aspects of Figure 4.8 that have created difficulties for climate models oriented toward greenhouse gases as the putative cause of climate change. One factor is the rise in temperatures from 1910 to 1940 prior to large-scale CO_2 emissions. The increase in Arctic temperatures is particularly impressive, exceeding temperature changes in other regions by up to a factor of 5. As Nagashima *et al.* (2006) said,

"The warming of the Earth's near-surface temperature in the latter part of the 20th century has been mainly attributed to increases in greenhouse gases, while the warming that occurred in the early 20th century has not yet been clearly attributed to any particular climate forcing agents."

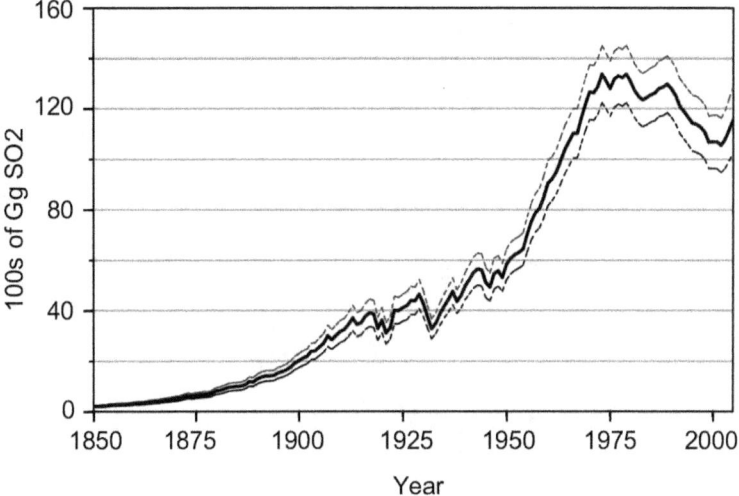

Figure 4.15. Global sulfur dioxide emissions. Dotted lines show range of uncertainty. (Smith, *et al.*, 2011).

The warming early in the century poses a major problem for the CO_2 advocates and they have no answers.

The second factor is the leveling out of temperatures from 1940 to the 1970s, including a significant drop in Arctic temperatures during this period. The third factor is the return of increasing Arctic temperatures after about 1970 and the return of rising temperatures globally during the 1970s. A fourth factor is the fact that temperature changes in NH mid-latitudes (adjacent to NH high latitudes) increased considerably more than tropical or SH temperatures. Shindell and Faluvegi (2009) pointed out that climate changes in a polar area have a significant effect on nearby mid-latitudes. Hence, Figure 4.8 suggests that the source of climate change in the 20th century might originate at least partly in the high latitudes of the NH, and spread southward with accompanying diminution.

As Bengtsson *et al.* (2004) said:

"The huge warming of the Arctic that started in the early 1920s and lasted for almost two decades is one of the most spectacular climate events of the twentieth century. During the peak period 1930–40, the annually averaged temperature anomaly for the area 60°–90°N amounted to some 1.78°C.... It was a long-lasting event commencing in the early 1920s and reaching its maximum some 20 years later. The decades after were much colder, although not as cold as in the early years of the last century. It is interesting to note that the ongoing present warming has just reached the peak value of the 1940s, and this has underpinned some views that even the present Arctic warming is dominated by factors other than increasing greenhouse gases."

Tisdale (2009) provided the reconstructed sea surface temperatures shown in Figure 4.16.

These data also show a sharp rise from about 1910 to 1940, a shallow dip from 1940 to about 1980, and a subsequent rise at roughly the same rate as that which prevailed from 1910 to 1940.

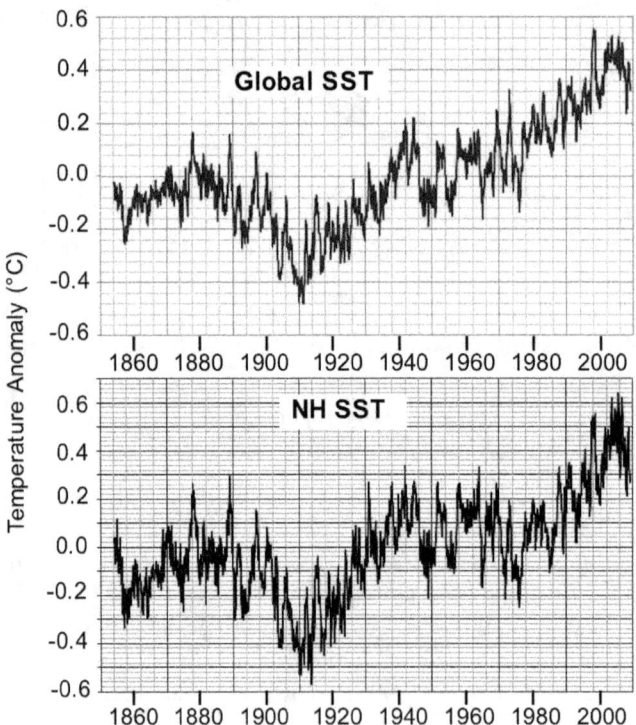

Figure 4.16. Reconstructed sea surface temperatures (SSTs) as provided by Tisdale (2009).

The sharp temperature rise in the North Polar region that occurred just prior to 1920 was remarkable. Bernaerts (2009) provides a number of historical references to this event. For example, in 1922, The Washington Post published a story: "Arctic Ocean Getting Warm; Seals Vanish and Icebergs Melt" stating that "ice conditions in the Northern North Atlantic were exceptional; in fact, so little ice has never before been noted." This article said: "The Arctic Ocean is warming up, icebergs are growing scarcer and in some places the seals are finding the water too hot." B. J. Birkeland in 1930 said the temperature rise was "probably the greatest yet known on Earth," and a few years later A. W. Ahlmann in 1946 called the event a "climatic revolution." In 1935, Jules Schokalsky received an award and in his address he said: "The branch of the North Atlantic Current which enters it by way of the edge of the continental shelf round Spitsbergen has evidently been increasing in volume, and has introduced a body of warm water so great, that the surface layer of cold water which was 200 meters thick in Nansen's time, has now been reduced to less than 100 meters in thickness."

Bernaerts mentions an oral presentation at a symposium in 1956 in which Hesselberg and Johannessen said:

"Of special interest are the data from Spitsbergen where the series of observations go back to 1912. During the first years the observations shows no conspicuous climatic change, but then comes a rapid rise of the temperature in the year 1917 to 1922. The increase of the mean temperatures in this period was about 7 degrees Celsius in the winter, 3 degrees in the spring, 3 degrees in the summer, 3 degrees in the autumn and 4 degrees for the whole year. After the year 1922 the temperature continued to rise ...but the rise was much slower ...The rise of the temperature in Spitsbergen is large compared with the rise in other parts of the world (about five times as great as in Norway)."

Bernaerts (2009) pinpointed the sharp temperature rise in the Arctic region that occurred around 1920 by studying the recorded temperatures at Spitsbergen. Spitsbergen is a remote archipelago between the North Cape in Norway and the North Pole. It is an ideal site for climate research because it is located:

- Between three huge water bodies in volume and size—the Norwegian/Greenland Sea; the Arctic Ocean; and the Barents Sea, with a modest volume (mean depth ca. 280 m) but considerable size;

- At the edge of sea ice, where regardless of the time of season at least a tiny space of the sea remains ice-free, which ensures a maritime-induced climatology, while a space covered with sea ice induces continental climatology;

- Where the Sun does not rise above the horizon for the whole winter period, from the 26th of October to the 16th of February.

As Bernaerts emphasized, the temperature rise of 1919:

- was greater in magnitude and more rapid than the recent one (1970s–2000s);

- displayed exceptionally rapid winter warming;

- had no summer signal at all.

Thus, the temperature rise from 1910 to 1940 was dominated by a single-step rise around 1919. Bernaerts (2009) provided a good deal of evidence and argument that this sudden warming was due to an influx of warm ocean waters. However, he then gave rather murky (and incredible) arguments for why he believed that naval warfare in the First World War was responsible.

Chylek *et al.* (2006) analyzed Greenland temperature records to compare the recent (1995–2005) warming period with the previous episode (1920–1930) of Greenland warming. They concluded that temperature increases in the two warming periods were of similar magnitude. However, the rate of warming in 1920–1930 was about 50% greater than that in 1995–2005.

Box *et al.* (2009) utilized a combination of meteorological station records and regional climate model output to develop a continuous 168-year (1840–2007) spatial reconstruction of monthly, seasonal, and annual mean Greenland ice sheet near-surface air temperatures. These results show a remarkable winter temperature rise just after 1920 while summer temperatures hardly changed. The post-1980 temperature increase is approaching the levels attained around 1940.

Thompson *et al.* (2008) suggested that the sharp dip of about 0.3°C in 1945 was due to uncorrected instrumental biases, implying that all temperatures after 1945 should be raised by about 0.3°C. However, Tisdale (2009) showed that this dip appears to be real because coincident changes in cloud cover and marine air temperature occurred at the same time.

The fact that there has been an increase in global temperatures in the 20th century (as demonstrated in Figure 4.8) suggests that anthropogenic activity may have contributed significantly to climate change during this time period. Anthropogenic activity includes generation of effluents from power plants, cement manufacturing, other industrial operations (carbon dioxide, black soot, sulfate aerosols), land clearing and deforestation, and large-scale irrigation). The various contributing factors act as follows:

- increased CO_2 (and other greenhouse gases) increases the absorption of outgoing IR producing a heating effect at all latitudes;
- deposition of black soot on high-latitude snow and ice increases solar absorptivity, thus producing a heating effect at higher latitudes;
- sulfate aerosols in the atmosphere produce a cooling effect by reflecting incoming sunlight;
- land clearing and deforestation increases the albedo of the Earth, producing a cooling effect.

While particles suspended in the atmosphere produce a "dimming" effect, Flanner *et al.* (2008) found that the heating effect of settled black carbon (BC) is about six times greater than the dimming effect of carbon suspended in the air, and three times greater than the equivalent forcing from CO_2. The greatest warming effect occurs in spring, when there is still ample snow and ice on the surface but solar intensity is significant.

A variety of positive feedback mechanisms amplify the trends originated by these factors. Of particular importance are (1) the increase in humidity expected from a warming trend which adds water vapor greenhouse warming, and (2) the positive feedback of albedo changes as high-latitude snow and ice expands or contracts.

Estimates exist for the rate of emission of CO_2, BC, and sulfate aerosols in the 20th century. CO_2 emissions have increased steadily throughout the 20th century and are still increasing, even though emission per unit energy generated has decreased with time. BC emissions increased sharply toward the end of the 19th century and peaked in the 1920s. However, emissions continued through the remainder of the century, but the primary emission regions changed from the U.S., Europe and the former U.S.S.R in the first half of the century to developing nations (particularly China) in the second half of the century. Aerosol emissions rose throughout the 20th century but peaked in the early 1980s. Nevertheless, aerosol emissions remained significant after the peak.

In addition to these anthropogenic influences, there seem to be natural cycles in the climate that produce significant climate changes even in the absence of anthropogenic influences. Of particular interest is the behavior of the oceans that retain a great deal of heat energy. The Pacific Ocean underwent a major change in the mid-1970s that produced a predominantly El Niño condition for almost three decades. These warm surface waters undoubtedly contributed to the climate in the NH after the 1970s.

The task of resolving anthropogenic influences from natural cycles is rendered difficult by the variety of anthropogenic influences and the uncertainties in historical climate changes. Many climatologists have focused on the role of increased CO_2 as the putative driver of global warming via the greenhouse effect with amplification by increased water vapor. All climate models lead to significant amplification of temperature increases in polar areas, as discussed by Serreze *et al.* (2008) and Holland and Bitz (2003). While the majority of climatologists have focused on greenhouse gases as the putative source of 20th-century global warming with amplification in polar areas, it is also possible that an important source of global warming was located in the Arctic with consequent spillover into neighboring latitudes.

Greenhouse gases cannot possibly explain the sharp rise in Arctic temperatures prior to 1940 because carbon emissions were relatively low during that period. A more likely explanation centers on the role of BC from the U.S., Europe, and the former U.S.S.R in the first half of the century falling on Arctic snow and ice. By about 1940, BC emissions from mid–north latitudes diminished but sulfate aerosol emissions increased steeply after 1940. It seems possible that aerosols were dominant in producing a cooling effect from about 1940 to the mid-1970s. BC played a secondary role during this period, being overwhelmed by the sharp rise in sulfate aerosol emissions. It is not so clear what was responsible for the renewed global warming that began in the 1970s. While BC emissions began to rise again during this period, the region of predominant emission moved from more northerly locations to Asian regions at lower latitudes. It is not clear how efficiently such emissions can be transported to higher latitudes. CO_2 emissions rose sharply after the 1970s. But another factor is the state change of the Pacific Ocean in the 1970s. There is a close correspondence between surface air temperatures and El Niño indices over the past 30 years.

Bengtsson *et al.* (2004) suggested that "four possible mechanisms, individually or in combination, could have contributed to the early twentieth century warming: anthropogenic effects, increased solar irradiation, reduced volcanic activity, and internal variability of the climate system." They concluded: "It seems unlikely that anthropogenic forcing on its own could have caused the warming, since the change in greenhouse gas forcing in the early decades of the twentieth century was only some 20% of the present." However, in considering anthropogenic effects, they dealt only with greenhouse gases, and did not consider the deposition of BC. They pointed out the uncertainties in reconstructing past solar irradiance, and they dismissed volcanic activity as the cause of this warming. Therefore, they sought an answer in terms of the atmospheric flow pattern that drives ocean circulation and results in the advection of warm water into the northeastern North Atlantic. Johannessen *et al.* (2004) concluded that the warming of the 1920s and 1930s was due to "natural fluctuations internal to the climate system." Reductions in albedo due to decreasing sea ice induced by wind changes were attributed as the cause of this early warming. However, they claimed that more recent warming in the 1980s and 1990s was due to greenhouse gases. However, Polyakov *et al.* (2003) concluded that the Arctic is subject to natural oscillatory variations the principal driver for climate change, and that "[greenhouse] warming alone cannot explain the retreat of Arctic ice observed in the 1980s–90s." Their final conclusion was:

> "The complicated nature of Arctic temperature and pressure variations makes understanding of possible causes of the variability, and evaluation of the anthropogenic warming effect most difficult."

In Section 6.5 another possible factor in Arctic climate change is explored: deposition of BC on Arctic snow and ice.

4.9. Climate in the 20th Century and El Niños

We have already seen a strong connection between climate and the El Niño index over the past thirty years in Figures 4.11 to 4.13.

Maasch (2009) provides a good description of the ENSO cycle:[29]

"It has long been recognized that when sea-surface temperatures warm during El Niño, typically in December, rainfall also increases and the success of fisheries decreases along the north Peruvian coast. But not every summer sea-surface warming is the same. Some years are characterized by especially warm water that remains until May or even strong warm intervals. Local warming of waters off Peru coincides with positive SST (sea-surface temperature) anomalies over a much larger domain, namely the eastern half of the equatorial Pacific. La Niña is the opposite phenomenon, referring to abnormally cold SST in the eastern half of the equatorial Pacific. We now know that these changes are part of a much larger climate pattern and that the atmosphere and ocean are coupled in the equatorial Pacific".

In the "normal" state, the sea surface temperature (and oceanic heat content) along the equator in the Pacific is warm in the west (near Asia) and cold in the east (near South America). The ocean surface waters are well mixed by wind stirring. Along the equator in the Pacific, this surface mixed layer is usually 150 m deep or deeper in the west, but it becomes shallower to the east until it essentially disappears near the South American coast. Sea level is also higher in the west. The trade winds, driving currents westward along the equator, feed and maintain the build-up of excess warm water on the western side (Cane, 1983).

It has been observed that quasi-periodic variations occur about this normal state every few years, in which the Pacific climate changes from the so-called El Niño conditions to the so-called La Niña conditions. NOAA has described these episodes as follows. El Niño episodes reflect periods of exceptionally warm sea surface temperatures across the eastern tropical Pacific. La Niña episodes represent periods of below-average sea surface temperatures across the eastern tropical Pacific. These episodes typically last approximately 9–12 months. During a strong El Niño, ocean temperatures can average 2°C–3.5°C above normal between the Date Line and the west coast of South America. These areas of exceptionally warm waters coincide with the regions of above-average tropical rainfall. During La Niña, temperatures average 1°C–3°C below normal between the Date Line and the west coast of South America. This large region of below-average temperatures coincides with the area of well below–average tropical rainfall. For both El Niño and La Niña the tropical rainfall, wind, and air pressure patterns over the equatorial Pacific Ocean are most strongly linked to the underlying sea surface temperatures, and vice versa, during December– April. During this

---

[29] In 2012, Tisdale produced a remarkably extensive compendium of data, description and analysis of El Niño phenomena in the form of a lengthy book: *http://bobtisdale.wordpress.com/*

period the El Niño and La Niña conditions are typically strongest, and have the strongest impacts on U.S. weather patterns. El Niño and La Niña effect episodes typically last approximately 9–12 months. They often begin to form during June–August, reach peak strength during December–April, and then decay during May–July of the next year. However, some prolonged episodes have lasted 2 years and even as long as 3–4 years. While their periodicity can be quite irregular, El Niño and La Niña typically occur every 3–5 years on average.

El Niños were originally recognized by fishermen off the coast of South America as the appearance of unusually warm water in the Pacific Ocean, occurring near Christmas and hence the name "El Niño" referring to the holy child. El Niño and La Niña are the warm and cold phases of an oscillation referred to as El Niño /Southern Oscillation, or ENSO, which has typically had a period of about 3–7 years. El Niño is thus one phase of a natural mode of oscillation that results from unstable interactions between the tropical Pacific Ocean and the atmosphere.

Although ENSO originates in the tropical Pacific Ocean–atmosphere system, it affects weather patterns over the entire world. According to a NOAA website, there is evidence that ENSO has been occurring for at least 125,000 years. El Niños vary in intensity and duration. The 1982 and 1997–1998 El Niños were particularly strong. In such a strong El Niño, the accumulation of excess heat in the eastern Pacific is about 10^{16} kWh, a very large amount of energy (total consumption of energy in the U.S. is about 10^{13} kWh per year). This affects the climate of a good part of the world, not merely the Pacific. For example, realclimate.org reports: "El Niños typically perturb the winter Northern Hemisphere jet stream in a way that favors anomalous warmth over much of the northern half of the U.S., the typical amplitude of the warming is about 1°C."

Changnon (2000) described the remarkable El Niño that lasted from May 1997 to May 1998 as "the climate event of the century". From August 1997 to December 1997, sea surface temperatures (SSTs) rose by 1°C for the dateline to the South American coast, by 4°C east of 140°W, and by more than 5°C near the Galapagos Islands. These temperatures persisted until May 1998. As a result, during the winter of 1997–1998, temperatures across the northern states of the U.S. were 3°F to 10°F warmer than normal. Precipitation across the southeastern states was 40% to 50% above normal. Temperature and precipitation records were broken in many areas. In the spring of 2008, precipitation in Los Angeles and San Francisco was triple the normal level. Similarly for Tampa, Florida. Spring 1998 temperatures in a corridor of 22 U.S. states ranging from the Midwest to the northeast were the warmest in 100 years. This created a field day for alarmists who trumpeted the proclamation that 1998 was the hottest year in a hundred years, which though true, had nothing to do with carbon dioxide. However, Al Gore insisted it was due to greenhouse gases.

The atmospheric signature, the Southern Oscillation, reflects the monthly or seasonal fluctuations in the air pressure difference between Tahiti and Darwin. The Southern Oscillation Index (SOI) is a measure of the large-scale fluctuations in air pressure occurring between the western and eastern tropical Pacific during El Niño and La Niña episodes. Traditionally, this index has been calculated based on the differences in air pressure anomaly between Tahiti and Darwin, Australia. In general, smoothed time series of the SOI correspond very well with changes in ocean temperatures across the eastern tropical Pacific. The negative phase of the SOI represents below-normal air pressure at Tahiti and above-normal air pressure at Darwin. Prolonged periods of negative SOI values coincide with abnormally warm ocean waters across the eastern tropical Pacific typical of El Niño episodes. Prolonged periods of positive SOI values coincide with abnormally cold ocean waters across the eastern tropical Pacific typical of La Niña episodes.

Estimates of the SOI have been made that date back to the mid-19th century (*http://www.cgd.ucar.edu/cas/catalog/climind/soi.html*). These results show roughly equal positive and negative fluctuations prior to about 1976 (see Figure 4.17). However, in the last 25 years of the 20th century, the trend has been predominantly negative (McLean, 2007a, b). This is demonstrated in Figure 4.18.

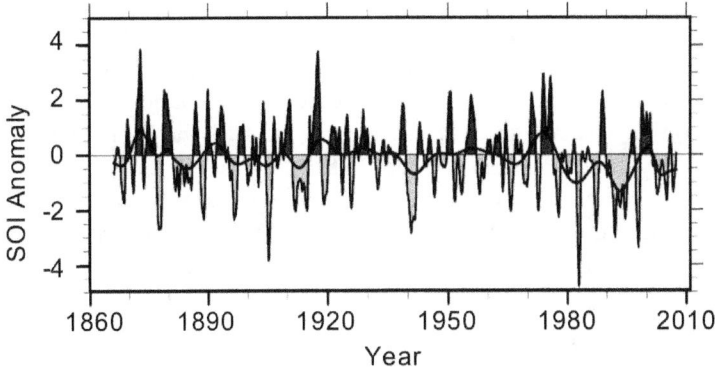

Figure 4.17. Long-term SOI index based on monthly values. Also shown are smoothed curves filtered to remove fluctuations of less than 8 months duration. Values prior to 1935 should be used with caution. Adapted from *http://www.cgd.ucar.edu/cas/catalog/climind/soi.html*

This trend in the SOI has led to some controversy in connection with global warming. On the one hand, some climatologists have used computer simulations to infer that this trend in SOI is due to anthropogenic-induced global warming (e.g. Vecchi, 2006). Power and Smith (2007) also suggest human activity as a cause of negative SOI but they admit: "there is currently no consensus amongst climate models concerning change in the behavior of ENSO in response to global warming."

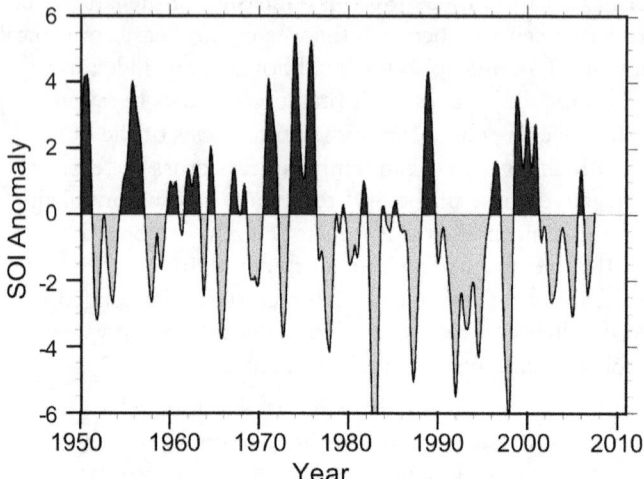

Figure 4.18. SOI Index since 1950 showing predominantly negative values after 1976. Adapted from

http://www.cgd.ucar.edu/cas/catalog/climind/soi.html

On the other hand, two notable naysayers (John McLean and Bob Foster) have suggested that a rather sudden and decisive change in the circulation patterns and upwelling characteristics in the Pacific began around 1976 and has continued to this day. As can be seen from Figure 4.18, the Pacific has been predominantly under El Niño conditions for the past 30 years. According to McLean and Foster, this predominance of warm surface waters in the Pacific has heated the Earth, particularly in the NH, and generated a rather abrupt upturn in global warming after 1976 (see Figure 4.7). According to McLean:

> "The abruptness of this change in upwelling appears likely to be related to some cataclysmic event in the region. Scientists would surely have noticed any shift in winds that was strong enough to cause a semi-permanent 25% reduction in the upwelling of eastern Pacific cold water so the answer is probably hidden in the ocean itself. The only cataclysmic event in the general region at that time was the Guatemala earthquake of February 1976 in which 250,000 people were killed, but any link is purely speculative at the moment." (*http://mclean.ch/climate/global_warming.htm*)

This is illustrated more dramatically in Figure 4.19. As is the case in most issues in climate change, there are diverse opinions. Has something changed in the Pacific Ocean that is heating the Earth, independent of greenhouse gases? Or has the effect of greenhouse gases produced a change in the Pacific Ocean?

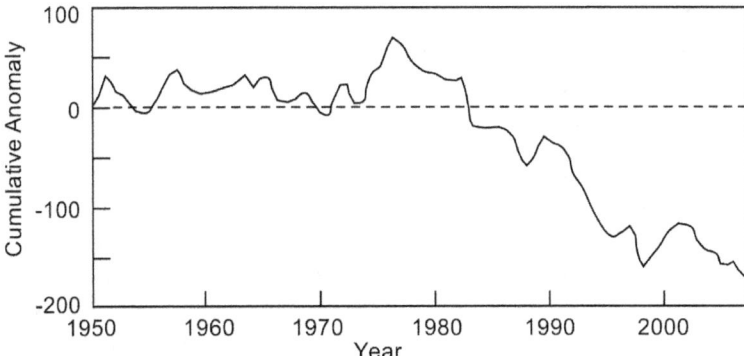

Figure 4.19. Integral of SOI anomaly since 1950. Adapted from McLean (2007a, b)

Further corroboration of the abrupt change in the El Niño–Southern Oscillation behavior around 1976/1977 was provided by other studies. For example, Meehl and Washington (1996) investigated possible causes for the effect:

> "Sea-surface temperatures in the tropical Pacific Ocean increased by several tenths of a degree during the 1980s and early 1990s, contributing to the observed global warming during this period."

Trenberth and Hoar (1997) said: "... the tendency for more El Niño and fewer La Niña events since the late 1970s is highly unusual and very likely to be accounted for solely by natural variability."

Frauenfeld *et al.* (2005) developed "a uniquely interdecadal Pacific signal [IPS] embedded in the Pacific Ocean–Northern Hemisphere atmosphere system that, by its statistical construction, is representative of the interaction between the large scale atmospheric circulation of the Northern Hemisphere and SSTs of the Pacific Ocean." They said "the time series of the IPS from 1949–2000 [was] dominated by the Pacific Climate Shift with negative anomalies prior to 1976/77 and almost exclusively positive anomalies since..." Their plots of the time series of the IPS show a sharp step function around 1977.

Guilderson and Schrag (1998) said:

> "Several studies have noted that the pattern of El Niño–Southern Oscillation (ENSO) variability changed in 1976, with warm (El Niño) events becoming more frequent and more intense. This '1976 Pacific climate shift' has been characterized as a warming in SSTs through much of the eastern tropical Pacific."

Desler, Alexander, and Timlin (1996) said:

"A prominent decade-long perturbation in climate occurred during the time period [1970–1991] in which surface waters cooled by 1°C in the central and western North Pacific and warmed by about the same amount along the west coast of North America from late 1976 to 1988."

DiLorenzo *et al.* (2007) said:

"Particularly dramatic physical and biological excursions occurred during the 1976–77 change in the Pacific Decadal Oscillation."

Hare and Mantua (2000) said:

"It is now widely accepted that a climatic regime shift transpired in the North Pacific Ocean in the winter of 1976–77. This regime shift has had far reaching consequences for the large marine ecosystems of the North Pacific. Despite the strength and scope of the changes initiated by the shift, it was 10 to 15 years before it was fully recognized. Subsequent research has suggested that this event was not unique in the historical record but merely the latest in a succession of climatic regime shifts."

Wu, Lee, and Liu (2005) said:

"The 1970s North Pacific climate regime shift is marked by a notable transition from the persistent warming (cooling) condition over the central (eastern) North Pacific since the late 1960s toward the opposite condition around the mid 1970s...This large-scale decadal climatic regime shift has produced far-reaching impacts on both the physical and biological environment over the North Pacific and downstream over North America."

Kim and Miller (2007) studied "the 1976/1977 climate regime shift." They concluded that the thermocline warmed but did not deepen. Power and Smith (2007) emphasized that "the lowest 30-year average value of the June–December SOI just occurred in 1977–2006" along with "the highest tropical sea-surface temperatures on record [in] what appears to be a concurrent period of unprecedented El Niño dominance."

According to NOAA:

"El Niños are not caused by global warming. Clear evidence exists from a variety of sources (including archaeological studies) that El Niños have been present for hundreds, and some indicators suggest maybe millions, of years. However, it has been hypothesized that warmer global sea surface temperatures can enhance the El Niño phenomenon, and it is also true that El Niños have been more frequent and intense in recent decades. Recent climate model results that simulate the 21st century with increased greenhouse gases suggest that El Niño-like sea surface temperature patterns in the tropical Pacific are likely to be more persistent."

"A rather abrupt change in the El Niño–Southern Oscillation behavior occurred around 1976/77 and the new regime has persisted... However, it is unclear as to whether this apparent change in the ENSO cycle is caused by global warming."

McLean *et al.* (2009) concluded that the El Niño index

"... is a dominant and consistent influence on mean global temperature. Shifts in temperature are consistent with shifts in the [El Niño index] that occur about 7 months earlier. The relationship weakens or breaks down at times of volcanic eruption in the tropics... Since the mid-1990s, little volcanic activity has been observed in the tropics and global average temperatures have risen and fallen in close accord with the [El Niño index] of 7 months earlier. Finally, this study has shown that natural climate forcing associated with ENSO is a major contributor to variability and perhaps recent trends in global temperature, a relationship that is not included in current global climate models."

According to their estimates, changes in the El Niño index could account for about 70% of the variance in the global tropospheric temperature over the past 50 years. The results of McLean *et al.* (2009) would seem to be a major stumbling block for alarmists who attribute most of the warming of the 20th century to greenhouse gases. It is therefore not surprising that the alarmists struck back with members of the cabal publishing Foster *et al.* (2010), that claimed that the results of McLean *et al.* (2009) "are seriously in error" and concluded "In fact, the general rise in temperatures over the 2nd half of the 20th century is very likely predominantly due to anthropogenic emissions of greenhouse gases". Foster *et al.* (2010) fell back on climate models that attribute only 15-30% of temperature variation in the 20th century to variability of the El Niño index. McLean *et al.*[30] attempted to rebut the criticism by Foster *et al.* (2010), but the *Journal of Geophysical Research* (JGR) refused to publish it. There are several important aspects of this episode that require further elaboration. These include (1) technical aspects, (2) attitudes and collusion amongst cabal alarmists, and (3) collusion of the JGR with cabal alarmists.

In regard to technical aspects, the issue revolves about methods used for filtering in statistical processing of data. Foster *et al.* (2010) appear to have made some valid criticisms of specific details, but these do not negate the strong correlation of the El Niño index with climate change. Perhaps the contribution of the El Niño index is less than the 70% claimed by McLean *et*

---

[30] Their rebuttal, which will be referred to as "McLean2010" is available at either of these websites:

icecap.us/images/uploads/McLeanetalSPPIpaper2Z-March24.pdf
http://scienceandpublicpolicy.org/originals/censorship_at_agu.html

al. (2009), but clearly the El Niño index is an important factor in climate change as evidenced by Figures 4.11 to 4.13. It seems doubtful that climatology has sufficient data and analytical insight to pin down its quantitative share in influencing climate change. A number of authors, even members of the *alarmist cabal* have admitted that climate models do not adequately account for El Niño effects.

McLean (2010) presented excerpts from the *climategate* emails that clearly show that the *alarmist cabal* regarded McLean *et al.* (2009) as a threat to their orthodoxy, and they colluded together to disparage McLean *et al.* (2009). The cabal regards itself as a police force to eradicate any contrary evidence or analysis that would refute their overemphasis on greenhouse gases.

The original paper by McLean *et al.* (2009) was approved by three independent reviewers for the JGR. One reviewer said:

> "I found the paper to be well-organized, well-written, and clear on the importance of the research. The abstract is informative, reference section is excellent, and the graphics are of high quality. The findings are likely to be of interest to a wide variety of readers."

The JGR published Foster *et al.* (2010), a rather vicious criticism of McLean *et al.* (2009), but refused to publish McLean's response. Evidently, the JGR is acting in collusion with the alarmist cabal, and probably regrets that McLean *et al.* (2009) "slipped through". McLean (2010) provides all the details.

Stammerjohn *et al.* (2008) showed that Antarctic annual sea ice retreat and advance is related to variability of the El Niño–Southern Oscillation and Southern Annular Mode.

Douglass (2010) pointed out that commonly used El Niño indices contain an unwanted effect from the annual cycle that can be reduced by digital filtering. He then developed an improved El Niño index dating from 1856. He analyzed the occurrences of major El Niño and La Niña events in some detail, and pointed out the existence of a basic asymmetry favoring El Niños in recent times. While Douglass (2010) restricted his attention to peak values of the index, it is also instructive to integrate the El Niño index over time. The result is shown in Figure 4.20. This shows that from ~1860 to ~1920, the El Niño and La Niña events were quite balanced. There was a predominance of El Niño events from about 1920 to ~1940, followed by a predominance of La Niña events from ~1940 to the late 1970s. Starting around 1980, El Niño events have been strongly dominant. The integral of the Niño index is more or less parallel to global temperature variability, suggesting that El Niño and La Niña events have been associated with most of the global temperature change in the past century and a half (see Figure 4.21). The index itself seems to correlate well with the tropospheric temperature.

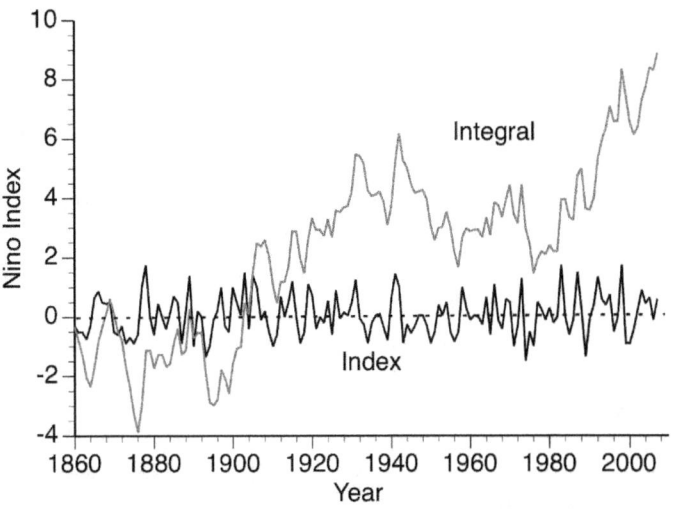

Figure 4.20. Integral of Douglass' modified El Niño index.

Additional discussion and references on ENSO effects are given in NIPCC (2011) and particularly in Tisdale's book at *http://bobtisdale.wordpress.com/*.

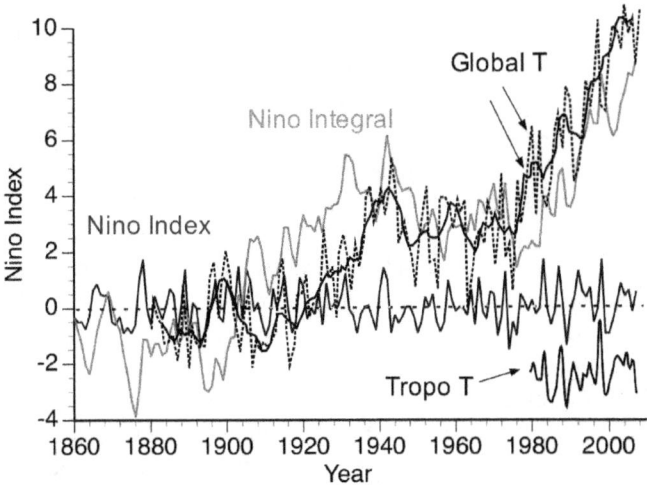

Figure 4.21. Comparison of Douglass' modified El Niño index and its integral with measured global surface temperature anomalies and tropospheric temperature anomalies.

4.10 Warming of the Oceans
4.10.1 Introduction

While many climate studies focus on temperature changes in the atmosphere at the surface or in the troposphere (e.g. BEST) the climate of the Earth is ultimately determined by the temperatures of the oceans. The oceans have a heat capacity about 1,000 times greater than the atmosphere and land surface. Although air temperatures may change much more rapidly than ocean temperatures, it is the ocean temperature distribution that will ultimately determine the climate of the Earth.

The "holy grail" of climatology that many climatologists are searching for, is the expected rise in global average temperature if the CO_2 concentration is doubled from the pre-industrial value of roughly 280 ppm (*aka.* climate sensitivity). Since 99.9% of the Earth's heat capacity lies in the oceans, the temperatures of the oceans ultimately controls the Earth's climate. If we could predict the increase in global average ocean temperature resulting from a doubling of the CO_2 concentration from the pre-industrial value, we could infer the climate sensitivity of the Earth. Unfortunately, like almost every issue in climatology, it is difficult to reduce uncertainties in estimates of future trends.

There is considerable evidence that sea surface temperatures (SST) have warmed significantly over the course of the 20th century (*e.g.* Tisdale, 2012). There is also evidence that on average, the bulk oceans have warmed over this time period but the temperature gains were smaller (*e.g.* Levitus *et al.*, 2012). Nevertheless, the oceans are so vast that this represents a very large amount of heat. While the data on ocean warming leave much to be desired, there seems to be little doubt that the oceans have acquired a significant amount of heat over the last five decades. Aside from the still unsettled issue of specifying the ocean warming to higher accuracy, the question arises as to what factors caused the ocean warming. Had there been long-term systematic increases in solar intensity and/or decreases in cloud cover, that would certainly have contributed significantly to ocean warming. As we discuss in Section 6.3, there are several models that attempt to estimate how the solar intensity has varied during the 20th century, but they all make assumptions that cannot be validated and therefore they remain highly speculative. Estimates of global variation of cloud cover have been made by a number of investigators. A slightly upbeat review was provided by Dai *et al.* (2006) who found that cloud cover decreased over the past several decades. But even these authors admit "large inadequacies in monitoring long-term changes in global cloudiness with surface and satellite observations" and the sub-title of their paper is "A Tale of Monitoring Inadequacies". Eastman *et al.* (2011) concluded that long-term secular changes in cloud cover over the oceans may have occurred during the period 1954-2008. Norris (2005) reported decreases in cloud cover over the oceans from 1952 to 1997. However, Evan *et al.* (2007) pointed out problems with cloud data and suggested that the International Satellite Cloud Climatology Project (ISCCP) data set of cloud amounts "may not be appropriate". Norris and Slingo (2009) discussed the inadequacies of cloud observation systems. It should be noted that cloud cover over the oceans exceeds that over land by a considerable amount (~68% vs. ~54 %).

Spencer and Braswell (2010, 2011) pointed out that "internal factors" operating near the Earth's surface may affect global heat flows. These might include random variations in cloud cover in the Earth's climate system, "brought about through circulation, induced changes in tropospheric wind shear, frontal system behavior, precipitation efficiency, trade wind inversion strength, or any other of the myriad processes that can potentially affect cloud formation other than feedback upon temperature" and non-radiative forcing of temperature change such as tropical intra-seasonal oscillations in the rate of heat transfer from the ocean to the atmosphere".

While we are remain uncertain about past variability of solar intensity and cloud cover, and variability due to internal factors remains speculative, one thing we do know is that over the past ~ 120 years, the CO_2 concentration in the atmosphere has risen fairly steadily. Climate models show that there is a resulting decrease in upward flux of long wave radiation at the top of the atmosphere (TOA) that produces an imbalance in the Earth's energy balance. The Earth then warms to regain equilibrium. Concurrently, the net downward back radiation from increased levels of greenhouse gases in the troposphere increases, due to the increase in CO_2 concentration as well as the increase in tropospheric temperature. It is widely theorized that much of this heat flux ends up in the oceans as ocean warming. Some estimates of the Earth's imbalance indicate that it exceeds the gains in heat content of the oceans, thus giving rise to discussions in the blogosphere about "the missing heat". At the other end of the scale, Tisdale (2012) claims that since the IR radiation is absorbed in the top few microns of the ocean, it cannot heat the ocean and merely leads to greater evaporation. While we cannot claim to provide a full explanation of the warming of the oceans due to all factors, we think it is possible to at least partly clarify the contribution of the greenhouse effect to ocean warming.

In this section, we are concerned with three aspects of the warming of the oceans:

(i) The excess flux of down-welling back radiation IR that impinges on the ocean surface for several levels of CO_2 concentration compared to the pre-industrial value (i.e. the *forcing* at the surface).

(ii) The heating effect in the oceans produced by increased down-welling back radiation IR flux from the sky.

(iii) A comparison of the estimated rate of warming of the oceans from (i) and (ii) with the measured change in ocean heat content over the past few decades.

The effects of back radiation from the sky on oceans were studied as early as the 1920s and continued in the 1960s to the 1980s by a number of investigators. Since then, the ocean heating process has been incorporated into massive climate models and the physics of the process has been buried in these models. Several climatologists, notably James Hansen, have emphasized their estimates of a energy flux imbalance at the top of the atmosphere (TOA). It has then been argued that this net energy flux into the Earth must end up somewhere, and the only place that could be is the oceans. There has been some consternation because measured increases in ocean heat content appear to be lower than predictions based on the estimated energy flux imbalance at the TOA. Meanwhile, the process by which energy accumulates in the ocean may require further elucidation.

A number of websites currently claim that back radiation from greenhouse gases cannot heat the oceans. For example, one website[31] asserts: "since the LWIR re-radiation from increasing 'greenhouse gases' is only capable of penetrating a minuscule few microns (millionths of a meter) past the surface and no further, it could therefore only cause evaporation (and thus cooling) of the surface 'skin' of the oceans". Another web page[32] says "It is impossible for a 1.7 W/m^2 increase [predicted by the IPCC due to man-made greenhouse gases] in downward 'clear sky' atmospheric LWIR flux to heat the oceans." Another website[33] says: "Infrared radiation from 'greenhouse gases' causes evaporative cooling of the oceans rather than heating". Yet another website[34] ridicules Livermore scientists for claiming that rising greenhouse gases warm the oceans. Finally, another website says: "Since the ocean is on average warmer than the atmosphere, the energy flux across the ocean/atmosphere interface is on average carrying heat from the ocean to the air.... So given the general direction of the motion of the energy, how can infrared energy be pushed into the ocean, when it can't penetrate the surface further than its own wavelength?"[35]

If these claims were correct, then any warming of the oceans would have to be attributed to increases in solar intensity or decreases in cloud cover. In

[31] http://hockeyschtick.blogspot.com/2010/08/why-greenhouse-gases-wont-heat-oceans.html
[32] http://hockeyschtick.blogspot.com/2010/08/energy-environment-full-special-issue.html
[33] http://theantislave.wordpress.com/2012/09/04/man-made-co2-is-not-the-driver-of-global-warming/
[34] http://stevengoddard.wordpress.com/2012/06/11/livermore-moves-reality-out-of-the-physical-world/
[35] http://tallbloke.wordpress.com/2011/03/03/tallbloke-back-radiation-oceans-and-energy-exchange/

the sections that follow, we attempt to clarify (to the extent possible) the role of forcing due to increased CO_2 in the atmosphere on ocean warming.

4.10.2 Brief Overview of Model

In our model, we start with an ocean in equilibrium with the air above it. The profile of temperatures below the surface at a tropical location is shown schematically in Figure 4.22. Initially, there is a temperature profile (A) for the equilibrium state before applying a forcing to the surface. The short-term curve (B) shows the initial response to the forcing. The initial temperature rise at the surface is ΔT_i. As a result, the mixed layer of the ocean begins warming. As it warms, there is a progression of temperature profiles until a new equilibrium is established at a new ΔT. Figure 4.23 shows the same information in greater detail, with curve B representing an intermediate stage in the passage from no forcing to ultimately a new equilibrium under forcing with curve D. The ultimate temperature rise of the mixed layer of the ocean is ΔT in this figure.

The ultimate new equilibrium established after passage of sufficient time is shown as (C). This graph is not to scale. The difference in temperatures between the mixed layer and the ocean surface was exaggerated to make the graphic clearer.

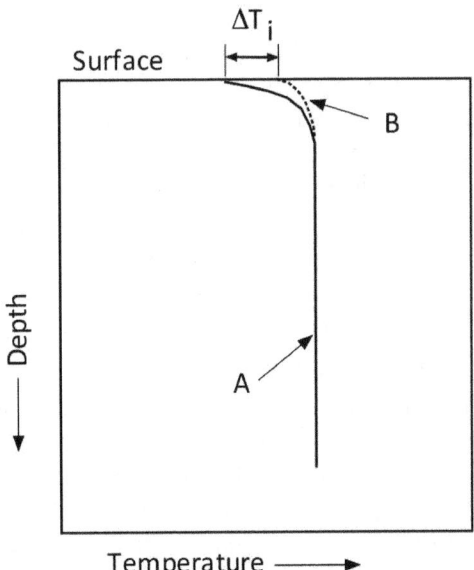

Figure 4.22. Schematic temperature profile in a tropical ocean. The initial curve (A) refers to equilibrium before applying a forcing to the surface. The short-term curve (B) shows the initial response to the forcing. This graph is

not to scale. The difference in temperatures between the mixed layer and the ocean surface was exaggerated to make the graphic clearer.

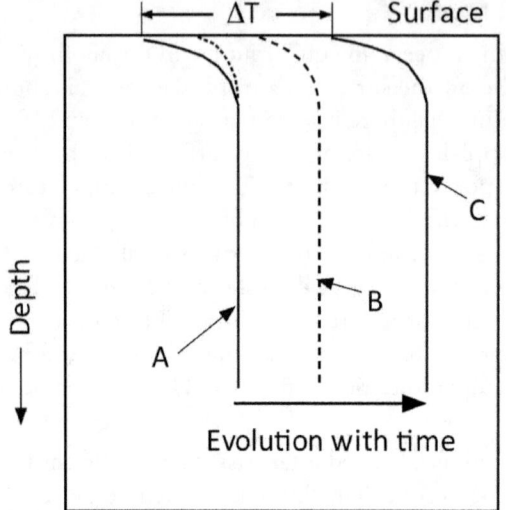

Figure 4.23. Schematic temperature profile in a tropical ocean. The initial curve (*A*) refers to equilibrium before applying a forcing to the surface. The response (*B*) shows an intermediate stage in the process as time progresses.

As these changes take place, changes are likely to occur in the air above the ocean surface. In some zero'th order models, it has been assumed that the air remains unchanged in temperature and humidity, even as the ocean surface warms. At the other extreme, one can assume that the air temperature tracks the ocean surface temperature and the relative humidity in the air remains constant (the absolute humidity increases). Reality probably lies between these extremes.

We begin with a model for the initial response to an increase in long wave back radiation to the surface. This radiant flux is absorbed in the surface resulting in a temperature increase at the surface.

4.10.3 The Initial Response to Forcing at the Ocean Surface

Professor Anthony Mills (private communication) carried out an energy balance about the ocean surface, in which he modeled the ocean in two layers. The upper layer is a thin surface layer at T_S below which is a mixed layer at T_L. As Mills pointed out, on average T_L always exceeds T_S by a small amount due to absorption of short wave radiation in the mixed layer below the ocean surface while the surface is losing energy to the air. When back radiation is increased, say by 1 W/m², the initial effect will be an increase in T_S while T_L will hardly change. As a consequence of the rise in T_S, the convective, latent and radiative losses to the atmosphere will increase. But in addition, the ongoing heat transfer from the mixed ocean layer to the ocean surface will be reduced because $T_L - T_S$ will be reduced compared to what it was before the forcing was applied to the surface. As time progresses, T_L will slowly rise and eventually establish a new equilibrium value, slightly greater than the increased value of T_S. This is illustrated in Figures 4.22 and 4.23. As a result, the mixed ocean layer will gradually warm due to the influence of an increase in long wave back radiation impinging on the ocean.

As we shall see, the effect of an increase in back radiation due to increased CO_2 results in an increase in the surface temperature T_S that reduces the convective heat loss from the well mixed water layer to the surface. In this sense, the effect of increased back radiation due to increased CO_2 is to heat the ocean because the ocean will lose heat to the atmosphere less effectively, resulting in an increase in the bulk temperature of the ocean.[36]

An energy flux balance about the ocean-atmosphere interface, prior to application of a forcing is:

$$(1 - \alpha)\, Q_S + Q_L - Q_{LH} - Q_{SH} - Q_B = 0$$

in which:

Q_L = Rate of heat gain by the ocean mixed layer per unit area

Q_{LH} = Rate of latent heat loss from surface to air per unit area

Q_{SH} = Rate of sensible heat loss from surface to air per unit area

Q_B = Rate of net back radiation from surface to air per unit area

Q_S = solar intensity falling on ocean surface

α = reflectivity of ocean surface

F = increase in back radiation from the sky to the ocean surface due to increasing CO_2 per unit area

At night, we can discard the solar intensity term. The flux balance at night is

$$Q_L - Q_{LH} - Q_{SH} - Q_B = 0$$

When we apply a forcing F to the ocean surface at night, this equation becomes:

$$F + q_L - q_{LH} - q_{SH} - q_B = 0$$

in which the lower case q's represent values after the forcing is turned on. As we pointed out above, T_S will increase initially while T_L will lag. The increase in T_S will cause q_{LH}, q_{SH}, and q_B to increase, while q_L will decrease. Subtracting the two flux balance equations, we obtain:

$$F = -\Delta Q_L + \Delta Q_{LH} + \Delta Q_{SH} + \Delta Q_B$$

---

[36] This concept was presented on the *realclimate.org* website at
http://www.realclimate.org/index.php/archives/2006/09/why-greenhouse-gases-heat-the-ocean/.
However no quantitative estimates were given.

where each $\Delta Q_i = q_i - Q_i$. The T_S before and after forcing are designated as T_{S1} and T_{S2} (same for T_A).

Approximations for Q_{LH}, Q_{SH}, and Q_B were given by Newell and Dopplick (1979). From these, we find:

Latent Heat Loss = $Q_{LH} = 6.08\ (h_S - h_A)\ w$

Sensible Heat Loss = $Q_{SH} = 2.51\ (T_S - T_A)\ w$

Net Back Radiation = $Q_B = 0.94\ [\sigma\ T_S{}^4\ (0.56 - 0.08\ e_A{}^{1/2})]$

It should be noted that there are quite a number of empirical formulas for net back radiation as discussed in 4.10.7. The formula used by Newell and Dopplick is one of several such formulas that are similar.

e_A = water vapor pressure in the air (product of relative humidity and saturation water vapor pressure) (mb)

R = relative humidity of air (%)

h_S = saturation specific humidity (g/kg) at T_S

$h_A = R\ h_S$ at T_A

R = relative humidity at T_A

T_A = air temperature (K)

T_S = ocean surface temperature (K)

w = wind speed (m/s)

The expressions for the changes in energy flux after applying the forcing are:

$$\Delta Q_{LH} = 6.08\ (3)\ (0.25)\ [h_S(T_{S2}) - h_S(T_{S1})]\ \text{W/m}^2$$

$$\Delta Q_{SH} = 22.59\ [(T_{S2} - T_{A2}) - (T_{S1} - T_{A1})]\ \text{W/m}^2 = 22.59\ (T_{S2} - T_{S1})\ \text{W/m}^2$$

$$\Delta Q_B = 5.33 \times 10^{-8}\ \{T_{S2}{}^4\ (0.56 - 0.08 \times [e_A(T_{S2})]^{0.5}) - T_{S1}{}^4\ (0.56 - 0.08 \times [e_A(T_{S1})]^{0.5})\}\ \text{W/m}^2$$

where $e_A(T_{S2})$ means e_A at T_{S2}, the relative humidity was fixed at 75%, and the wind speed was taken as 3 m/s.

In the zero'th approximation, one assumes that the air temperature does not change and the humidity of the air does not change.

At night, $Q_L = h_L(T_L - T_S)$ where h_L (W/m^2-K) is a convective heat transfer coefficient controlling heat transfer from the well mixed layer at temperature T_L, to the interface at temperature T_S. The heat transfer coefficients were estimated as shown below:

Ocean Condition	h_L W/m^2-K
1. Still – natural convection	216
2. Wind speed 3 m/s	420
3. Wind speed 5 m/s	1050
4. Wind speed 10 m/s	3600

If one assumes that initially, T_L is unchanged by the forcing, it follows that

$$\Delta Q_L = h_L(T_{S1} - T_{S2})$$

This term is negative. The coefficient h_L is very large. As a result, the dominant term is $-\Delta Q_L$ in the expression:

$$F = -\Delta Q_L + \Delta Q_{LH} + \Delta Q_{SH} + \Delta Q_B$$

For example, consider a case where $T_{S1} = 294$ K and $F = 1$ W/m^2. We can use trial and error, varying a guessed value of T_{S2} until F is equal to $-\Delta Q_L + \Delta Q_{LH} + \Delta Q_{SH} + \Delta Q_B$.

When this procedure is carried out, we find that a temperature rise ΔT_S equal to about 0.0045°C produces the following changes in flux (assuming T_A does not change):

$\Delta Q_{LH} = 0.08$ W/m^2

$\Delta Q_{SH} = 0.04$ W/m^2

$\Delta Q_B = 0.02$ W/m^2

$\Delta Q_L = 0.986$ W/m^2

With a higher wind speed, ΔQ_L would be amplified. Clearly, the overwhelming effect of a rise in T_S is to reduce the heat flux from the mixed layer to the ocean surface. Therefore, most of the effect of an increase in back radiation is to heat the mixed layer. This conclusion is independent of any assumptions regarding changes in the atmosphere that result from warming of the ocean surface. If we repeated the calculation assuming that the air temperature tracks the ocean surface temperature and the relative humidity of the air remains constant, that would decrease energy loss from the ocean surface to the air, and amplify the importance of energy loss from the mixed layer to the ocean surface.

4.10.4 Establishment of a New Equilibrium

In the previous section, we found that the initial response of the ocean to a forcing applied to the ocean surface is a rise in the surface temperature and a decrease in energy flux from the mixed layer to the ocean surface. This is illustrated as curves B in Figures 4.22 and 4.23.

As time progresses, the mixed layer continues to lose energy at a lower rate than prior to application of the forcing to the surface. As a result, the mixed layer gradually warms (as shown in curve C in Figure 4.23). However, as the mixed layer warms, the rate of energy transfer from the mixed layer to the surface increases, and thus the rate of warming of the mixed layer decreases as time progresses. Eventually, the rate of energy transfer from the mixed layer increases enough to establish a new equilibrium at a higher mixed layer temperature (see curve C in Figure 4.22 or curve D in Figure 4.23). When this new equilibrium is established we can treat the upper ocean as a mixed layer and ignore the small difference in temperature between the mixed layer and the surface. We then carry out an energy balance about the ocean surface.

4.10.4.1 Zero'th Order Model

Newell and Dopplick (1979) used an idealized model to estimate the effect of increased back radiation on the ocean surface. They started with an atmosphere and ocean in equilibrium. In their model, they kept all parameters of the atmosphere constant (*e.g.* temperature, absolute humidity) but allowed the temperature of the upper mixed layer of the ocean to rise as a consequence of the additional back radiation. The back radiation is absorbed into the top few microns of the ocean surface. This additional energy flux to the ocean surface can end up in four possible responses:

(1) Warming of the mixed layer of ocean as expressed in an increase in $T_S \approx T_L$.

(2) Sensible heating of the atmosphere. If the ocean warms and the atmosphere is unchanged, the ocean will transfer some heat to the atmosphere. Sensible heat transfer depends on wind speed.

(3) Latent heat loss. If evaporation from the ocean surface increases, some energy will be transferred from the ocean to the atmosphere in the form of water vapor. Latent heat loss depends very sharply on wind speed.

(4) Radiant heat loss. If the ocean surface warms, it will radiate more energy back into the atmosphere. Since we assume that the air does not change as the ocean warms, the back radiation from the air does not increase, and therefore the net back radiation from the ocean to the air increases.

The temperature of the mixed layer, $T_S = T_L$, will rise until the sum of sensible, latent heat and radiative losses increase by an amount equal to the increase in back radiation forcing due to increasing CO_2. As we pointed out previously, Newell and Dopplick (1979) used the following equations:

Latent Heat Loss = $Q_{LH} = 6.08 \ (h_S - h_A) \ w$

Sensible Heat Loss = $Q_{SH} = 2.51 \ (T_S - T_A) \ w$

Net Back Radiation $= Q_B = 0.94 \, [\sigma \, T_S^4 \, (0.56 - 0.08 \, e_A^{1/2})]$

Prior to forcing, an energy balance on the mixed layer is:

$$(1 - \alpha) \, Q_S - Q_{LH} - Q_{SH} - Q_B = 0$$

in which heat transfer to the deep ocean is ignored. Since IR is absorbed in the top few microns of the ocean surface, we assume that any forcing (X) W/m^2 injects (X) W/m^2 into the ocean surface. Since convective heat transfer between the surface and the top mixed layer of ocean is very efficient, this warms the upper ocean layer. As the ocean warms, energy losses to the atmosphere via convection, radiation and latent heat take place. Ultimately, a new equilibrium is reached at a higher value of T_S. In the zero'th order model used by Newell and Dopplick, the atmosphere does not change as the ocean warms. To proceed numerically, we add an arbitrarily chosen increment of temperature to T_S. For this value of ΔT_S, we calculate the terms in the energy balance and determine the increase in total heat loss to the atmosphere. We can then repeat this by trial and error until the increase in heat loss is equal to the forcing, and this increase in T_S is due to the forcing.

As an illustration, let $T_S = T_L = T_A$. Further assume that $R = 75\%$, and choose a wind speed $= 3$ m/s. For example, if we arbitrarily choose $T_S = T_L = T_A = 294$ K, and use the saturation vapor pressure of water @ 21°C = 24.79 mb, we obtain prior to forcing:

Latent Heat Loss $= Q_{LH} = 6.08 \, (b_S - b_A) \, w = 6.08 \, (15.70 - 11.78) \, 3 = 6.08 \, (3.92) \, (3) = 71.59$ W/m^2

Sensible Heat Loss $= Q_{SH} = 2.51 \, (T_S - T_A) \, 3 = 0$

Net Back Radiation $= Q_B = 0.94 \, [\sigma \, T_S^4 \, (0.56 - 0.08 \, e_A^{1/2})] = 0.94$ x $(5.67$ x $10^{-8})$ x $(294)^4$ x $(0.56 - 0.08$ x $[0.75$ x $24.79]^{.5}) = 85.64$ W/m^2

Further discussion of the origin of the formula for Q_B is given in Section 4.10.7. According to this simplistic model for the unforced case, $T_S = T_L = T_A$ will adjust to provide a total energy flux from the ocean surface to match an incoming energy flux of 157.23 W/m^2. Had we used a lower relative humidity for the air, the latent heat loss would have been higher, and *vice versa*.

When forcing is introduced in this simple model, $T_S = T_L$ will increase while we assume for simplicity that T_A remains unchanged. With forcing,

Latent Heat Loss $= Q_{LH} = 6.08 \, (b_S - 11.78) \, 3 = 18.24 \, (b_S - 11.78)$ W/m^2

Sensible Heat Loss $= Q_{SH} = 2.51 \, (T_S - 294) \, 3$ W/m$^2 = 7.53 \, (T_S - 294)$ W/m^2

Net Back Radiation $= Q_B = 0.94 \, [\sigma \, T_S^4 \, (0.56 - 0.08 \, e_A^{1/2})] = 0.94$ x $(5.67$ x $10^{-8})$ x $(T_S)^4$ x $(0.56 - 0.08$ x $[0.75$ x $24.79]^{0.5})$ W/m^2

Total Heat Loss = Latent heat loss + Sensible heat loss + Net back radiation

We can now calculate these values as a function of $T_S = T_L$. For each assumed value of $T_S = T_L$, we subtract 157.23 W/m² from the calculated total heat loss and attribute this difference to the forcing that produced the increase in $T_S = T_L$. As it turns out, the total heat loss is a very sensitive function of $T_S = T_L$. Increasing $T_S = T_L$ by 0.02°C increases the total heat loss by about 0.39 W/m². Hence, according to this simple model, adding say, 0.36 W/m² of forcing to the back radiation will only increase $T_S = T_L$ by about 0.02°C. Adding 1.0 W/m² of forcing to the back radiation will only increase $T_S = T_L$ by about 0.05°C. Some calculations are shown in Table 4.1.

As we shall show in Section 4.10.4, the forcing at the surface resulting from a doubling of CO2 from the pre-industrial value, is about 1 W/m². From Table 4.1, it follows that a surface forcing of about 1 W/m² will only warm the oceans by about 0.05°C in this simple model in which it is assumed that the atmosphere does not change as the ocean warms. Had we used a relative humidity of say 60% instead of the value 75% used for Table 4.1, only small changes in the numbers would result.

Table 4.1. Change in total heat loss as a function of T_S when the atmosphere is unchanged (W/m²).

T_S (K)	Latent Heat Loss	Sensible Heat Loss	Net Back Radiation	Total Loss	Change in Total Loss
294	71.59	0.00	85.64	157.23	0.00
294.01	71.77	0.08	85.65	157.43	0.20
294.02	71.96	0.15	85.66	157.62	0.39
294.03	72.14	0.23	85.68	157.82	0.59
294.04	72.32	0.30	85.69	158.01	0.78
294.05	72.50	0.38	85.70	158.20	0.97
294.06	72.69	0.45	85.71	158.40	1.17

4.10.4.2 First Order Model

Watts (1980) commented on the paper by Newell and Dopplick (1979) – to be denoted ND1. Watts pointed out:

> "... under the assumption that the temperature and humidity ratio of the atmospheric air over the sea remain constant as the radiative flux downward at the ocean surface increases. If the ocean and atmosphere temperatures change together, the change in sensible heat flux will be zero according to the author's equation. The temperature of the atmosphere will increase, of course, because the latent heat that leaves the ocean must end up in the atmosphere after condensation occurs. The specific humidity of the atmosphere will also increase."

He went on to say "Estimating how the latent heat, sensible heat and back radiation change in response to long-term changes in the ocean temperature is a difficult problem indeed." Watts (1980) suggested that there is a "constant value of relative humidity when the temperature changes" However, as we discuss in Section 5.5.2, although this is a widely used approximation, the basis for it is actually quite weak.

Newell and Dopplick (1981) replied to Watts. One point they made was "a good fraction of the additional energy received by the surface air from the warmer sea could be radiated away by the additional water vapor". However, they admitted: "Clearly, we should go further than our 'first iteration' and additional computations [to] include consideration of the variation of sea surface temperature with air temperature and moisture". They also said: "We agree that the additional energy used in evaporation will eventually appear in the free atmosphere above the surface layer. It is by no means obvious that it will cause an increase in temperature. First of all part of the increase may be offset by additional cooling to space from the increased water vapor. Second, there are a myriad of water-cloud-circulation feedback processes that may come into play".

Watts (1982) provided a rebuttal to this note. He claimed that the method used in ND1 is merely a first iteration. In this simple model, one fixes atmospheric conditions to the initial value and calculates the forcing corresponding to any change in $T_S \approx T_L$. According to Watts, one must take into account changes in the atmosphere above the ocean that result from an increase in $T_S \approx T_L$, and use these (with the increased $T_S \approx T_L$) as the starting point for a second iteration. The problem is how to assess the changes in atmospheric parameters resulting from an increase in $T_S \approx T_L$? Watts (1982) made the assumption that T_A remains equal to $T_S \approx T_L$ as $T_S \approx T_L$ rises, and furthermore the relative humidity in the atmosphere remains constant as T_A rises. These assumptions are qualitatively in the right direction. Surely, one expects T_A to rise as $T_S \approx T_L$ increases, but why should it remain equal to $T_S \approx T_L$? One also expects the absolute humidity to rise, but why should the relative humidity remain constant? In fact, we don't really know much about the relative humidity. It was assumed to be 75% for purposes of calculating Table 4.1, but that was merely a wild guess. Nevertheless, if we make the assumptions suggested by Watts (1982) and set $T_A \approx T_S \approx T_L$ and assume constant relative humidity = 75% as T_A rises, for purposes of a second iteration, we must use

Latent Heat Loss = Q_{LH} = 6.08 [h_S – 0.75 h_S] 3 = 18.2 [0.25 h_S] W/m²

Sensible Heat Loss = Q_{SH} = 2.51 (T_S – T_A) 3 W/m² = 0

Net Back Radiation = Q_B = 0.94 [σ T_S^4 (0.56 – 0.08 $e_A^{1/2}$)] = 0.94 x (5.67 x 10⁻⁸) x (T_S)⁴ x (0.56 – 0.08 x [e_A]⁰·⁵) W/m² = 5.33 x 10⁻⁸ T_S^4 (0.56 – 0.08 x [e_A]⁰·⁵) W/m²

Total Heat Loss = Latent heat loss + Sensible heat loss + Net back radiation

We obtain the results shown in Table 4.2.

In this approximation, since $T_A \approx T_S \approx T_L$, there is no sensible heat loss. The latent heat loss increases as $T_S \approx T_L$ increases. However, the net back radiation decreases significantly as $T_S \approx T_L$ increases. The ultimate result is that if one assumes $T_A \approx T_S \approx T_L$ and the relative humidity remains constant, the modeled temperature rise increases significantly compared to the model of Newell and Dopplick. For example, $T_S \approx T_L$ rises by about 0.6°C due to a forcing of 1 W/m². For a forcing of 0.4 W/m², characteristic of the average for the past 55 years, the temperature gain is estimated to be about 0.25°C.

Table 4.2. Change in total heat loss as a function of T_S when we set $T_A \approx T_S \approx T_L$ (W/m²).

T_S (K)	Saturation Humidity (mb)	Sensible Heat Loss	Net Back Radiation	Latent Heat Loss	Total Loss	Change in Total Loss
294.0	24.79	0.00	85.64	71.59	157.23	0.00
294.1	24.92	0.00	85.38	72.05	157.43	0.20
294.2	25.07	0.00	85.08	72.50	157.59	0.36
294.3	25.22	0.00	84.79	72.96	157.75	0.52
294.4	25.37	0.00	84.49	73.42	157.91	0.68
294.5	25.51	0.00	84.23	73.87	158.10	0.87
294.6	25.66	0.00	83.94	74.33	158.26	1.03
294.7	25.80	0.00	83.64	74.78	158.42	1.19
294.8	25.95	0.00	83.35	75.24	158.59	1.36
294.9	26.10	0.00	83.05	75.70	158.75	1.52
295.0	26.25	0.00	82.76	76.15	158.91	1.68
295.1	26.41	0.00	82.42	76.61	159.03	1.80
295.2	26.56	0.00	82.13	77.06	159.19	1.96
295.3	26.72	0.00	81.81	77.52	159.33	2.10
295.4	26.88	0.00	81.50	77.98	159.47	2.24
295.5	27.03	0.00	81.19	78.43	159.62	2.39
295.6	27.19	0.00	80.87	78.89	159.76	2.53

However, Watts (1982) pointed out:

"... the surface energy-balance model gives results that are extremely sensitive to slight model-parameter changes".

Small changes in the assumed value of the humidity in the atmosphere produce large changes in the temperature rise. For example, if Table 4.2 is repeated with a wind speeds of 5 m/s, the temperature rise for 1 w/m² forcing is reduced to 0.25°C. If the wind speed is reduced to 2 m/s one finds the calculation to be unstable. As Watts emphasized, "The surface heat flux calculation is very, very sensitive to small changes in the components of the heat balance". He pointed out that a heat balance at the tropopause does not suffer from this problem. However, a heat balance at the tropopause does not provide us with direct insight as to the effect of back radiation on the oceans.

It is evident from these calculations that a crucial unknown is the change in the air above the ocean as the ocean surface warms. In the zeroth order model, it was assumed that the air did not change as the ocean surface warmed. As a result, when the ocean surface warms in this model, although the net back radiation increases extremely slowly, while the latent heat loss and sensible heat loss increase at significant rates. This allows the ocean surface to rid itself of excess energy that would have accumulated due to a reduction in flux from the mixed layer to the ocean surface. Hence a new equilibrium is achieved with a rather small increase in ocean temperature. In the first order model, the air temperature is set equal to the ocean temperature and the relative humidity is assumed to be constant. Thus the sensible heat loss is zero. The latent heat loss increases with temperature as before. However, the net back radiation now decreases sharply as the ocean and temperatures increase, due to the increased humidity in the air. Thus, the ocean surface is less able to lose energy to the air, and the temperature rise is greater. The problem is that the total heat loss from the ocean depends on three terms, one of which is assumed to be zero, a second decreases with temperature, and the third increases with temperature. The term that decreases with increasing temperature depends critically on the humidity in the air above the ocean. The term that increases with temperature depends critically on wind speed. Depending on assumptions made about these variables, the final equilibrium temperature of the mixed layer of the ocean can be almost any value.

4.10.4.3 Analysis by Ramanathan (1981)

Ramanathan (1981) argued that as a consequence of doubling CO_2 in the atmosphere, not only is there an increase in back radiation at the surface of about 1.2 W/m^2, but there are two additional feedback fluxes that affect the surface.

One additional factor is direct radiative heating warming of the troposphere that increases the IR emission by the radiatively active constituents of the tropospere (clouds, H_2O, CO_2, O_3 and other trace gases). This amplifies the back radiation flux. However the transmission efficiency for the tropospheric emission to penetrate to the surface is unclear from Ramanathan's paper.

The second process "concerns the interactions between ocean surface temperature, the hydrological cycle and tropospheric convective adjustment processes". According to Ramanathan (1981):

> "The surface warming due to [increased back radiation and radiation from the warm troposphere] enhances H_2O evaporation into the troposphere, which indirectly amplifies the surface warming in two ways: (i) The latent heat released within the troposphere (resulting from the enhanced evaporation) warms the troposphere, thus enhancing tropospheric IR emission; and (ii) the enhancement in the evaporation also increases the absolute humidity of the troposphere which, of course, increases tropospheric IR emission. A fraction of the increase in tropospheric IR is emitted upward to space and the remainder is emitted downward to the surface. The downward fraction of the enhanced emission amplifies the surface warming by [increased back radiation and radiation from the warm troposphere]. This feedback between temperature, H_2O evaporation and IR emission is primarily controlled by ocean-atmosphere interactions since the world oceans are the primary source for atmospheric H_2O. The magnitude of the amplification is strongly determined by tropospheric convective adjustment processes and its subsequent effect on tropospheric lapse rates. This dependence arises because the partitioning of the enhanced IR emission between upward and downward components is controlled by lapse rate changes".

However, the latent heat release seems to have been taken into account in the first process, and one wonders whether this effect was counted twice? Furthermore, it is not clear how much effect a supposed increase in tropospheric humidity would have on back radiation at the surface, considering that air above the ocean is already quite humid and clouds act as a barrier to IR transmission. Ramanathan seems to have assumed a low opacity for the atmosphere.

According to Ramanathan a back radiation flux of 1.2 W/m^2 leads to increases of 2.3 W/m^2 due to increased radiation from the warming troposphere, and 12.0 W/m^2 from increases in radiation from increases in atmospheric humidity. Thus, he argued that the total back flux is about 15.5 W/m^2, rather than 1.2 W/m^2. He claimed "Newell and Dopplick's approach underestimates surface warming by about a factor of 30-40". As stated above, this writer suspects that the feedback is inhibited by clouds and low lying humidity of air over the ocean, so Ramanathan's estimates seem grossly exaggerated. It should be noted that Trenberth et al. (2009) provide a wide range of estimates of variable sign and this value has considerable uncertainty.

Using a multi-layer global climate model in one dimension, Ramanathan estimated the effect of the three factors mentioned above on ultimate equilibrium ocean temperature due to a doubling of CO_2:

	Increased Back Radiation	Warming of Atmosphere	Increased Water Vapor	Total
Surface Flux (W/m^2)	1.2	2.3	12.2	15.5
ΔT_S (°C)	0.17	0.33	1.7	2.2

There are several aspects of Ramanathan's treatment that are difficult to comprehend. One is that when he compared his calculation (omitting feedbacks) with that of Newell and Dopplick, he agreed that a 1.2 W/m^2 forcing produces a temperature rise without feedbacks is about 0.04°C. Yet in the above table he indicates a rise of 0.17°C without feedback.

Ramanathan's estimates for the effects of warming of the atmosphere and increased humidity depend on assumptions regarding changes to the atmosphere (temperature and humidity) that result from increased ocean temperature and apparently, he assumed a relatively transparent atmosphere. One wonders if he has properly taken into account clouds over the oceans. Despite the length of his article, it is written in such a confusing manner that it is difficult to determine exactly what he assumed for these changes.

Nevertheless, one thing stands out from Ramanathan's calculation. Even with his seemingly exaggerated large feedback terms, he found that the oceans only warm by about 2.2°C. Since ocean temperatures will ultimately control the Earth's climate, this would seem to limit the future rise in global temperature due to a doubling of CO_2 from the pre-industrial value.

4.10.4.4 Summary of Models

In our model, the initial response of the ocean to an increase in back radiation flux is an increase in the ocean surface temperature, T_S, while the mixed layer of ocean remains at its original temperature, T_L. Although the increase in T_S tends to increase heat loss from the surface, this increase is far less than the increase in back radiation flux. But the effect of an increase in T_S is a decrease in $T_L - T_S$, resulting in a large reduction in energy flux transported from the mixed layer to the surface. The new (increased) value of T_S is that obtained when the sum of increased heat losses upward and decreased ocean energy flux to the surface balances the increase in back radiation flux. As time progresses, the energy flux form the mixed layer to the surface remains less than it was before the forcing was applied, and therefore it warms (T_L increases with time). As T_L increases, the ocean energy flux to the surface gradually increases. After passage of sufficient time, a new equilibrium is eventually established in which T_S and T_L are both higher than before the forcing was applied. To model this new equilibrium it is sufficient to use a lumped ocean model for the mixed layer and surface by assuming $T_S \approx T_L$ because the difference between T_S and T_L is expected to be smaller than the temperature change resulting from the forcing. One then takes an energy flux balance about the ocean surface, and varies $T_S \approx T_L$ until the calculated heat loss from the surface equals the forcing. This is the new equilibrium value of $T_S \approx T_L$.

In order to utilize the model quantitatively, we need to estimate the increase in back radiation flux at the ocean surface due to increased CO_2, and we also need to estimate the changes that occur in the air above the ocean (in order to estimate heat losses from the ocean to the air).

The basic forcing at the ocean surface due to increased CO_2 is well understood, as described in Section 4.10.4. The forcing at the surface due to doubling of CO_2 from the pre-industrial value is about 1 to 1.2 W/m^2. What is not clear is to what extent this basic forcing is augmented by radiation from a warmer troposphere. Ramanathan (1981) concluded that this effect is very large but his results seem exaggerated.

The effect of increased ocean surface temperature on the air above can only be conjectured. If radiation from a warmer troposphere is neglected, and one makes the zero'th order assumption that the atmosphere remains unchanged as the ocean warms, one finds a very small increase in ocean temperature ($\sim 0.05°C$) due to an increase in back radiation of 1 W/m^2. On the other hand, if radiation from a warmer troposphere is neglected, and one assumes that the air temperature remains equal to the ocean surface temperature and the relative humidity of the air remains constant, the calculated increase in ocean temperature due to an increase in back radiation of 1 W/m^2 is about $0.6°C$. Unfortunately, this calculation is very sensitive to assumptions made about the condition of the air, and even more sensitive to assumptions about the wind velocity. Hence it is not possible to obtain precise quantitative estimates of the ultimate equilibrium temperature of the mixed ocean layer when subjected to a forcing at the surface.

One thing we can assert however, is that warming of the oceans by an increase in back radiation to the surface is an efficient process, and the initial rate of heat loss by the mixed layer is only slightly less than the magnitude of the imposed forcing. As we shall see in Section 4.10.4, the average surface forcing due to increased CO_2 over the past 55 years was roughly 0.4 W/m^2. It is therefore not unreasonable to expect that the upper mixed level of the ocean would have warmed by an input of roughly this amount over that time period. The equilibrium increase in temperature due to this forcing depends on the level of additional radiation from a warmer troposphere at the surface, as well as the characteristics of the air above the oceans (temperature, humidity, wind speed, cloud cover).

4.10.5 Excess flux of down-welling back radiation IR impinging on the ocean surface due to rising CO_2 concentrations

Quite a number of studies have estimated the forcing (additional downward back IR radiation) due to an increase in the CO_2 concentration from pre-industrial levels. Most of these estimated the forcing at the top of the atmosphere (TOA) for a doubling of CO_2. However, we are interested in the downward back IR radiation at the surface, in order to estimate the heat flux impinging on the oceans. The estimates given in this section do not include feedbacks.

Newell and Dopplick (1979) estimated that the IR flux change at the surface due to increasing the CO_2 concentration from 330 ppm to 600 ppm would be latitude dependent, and would vary from roughly 0.8 W/m² at low latitudes to roughly 1.5 W/m² at high latitude. When the calculation was repeated for clear skies, this range increased to 1.1 to 2.6 W/m².

Ramanathan (1981) estimated the forcing as a function of altitude. He pointed out:

"The troposphere as a whole is subjected to a net radiative heating of about 3.5 W/m² (and ~3 W/m² on a global average basis) for a doubling of CO_2 , which is roughly a factor of 3 larger than the surface heating." (Note: He added an increase in the downward emission from the stratosphere by about 1.2 W/m² to obtain the total forcing at the tropopause). His profile for forcing from doubling CO_2 is shown in Figure 4.24.

As Ramanathan (1981) put it:

"It is commonly stated that CO_2 absorbs upwelling radiation and then re-emits it to the surface as back radiation. The CO_2 bands overlap with water vapor bands whose opacity is so large that most of the back radiation from CO_2 is absorbed by the intervening layer of H_2O. As a result, the CO_2 back radiation at the surface increases by only 1.2 W/m² as opposed to the 4.3 W/m² tropopause radiative forcing."

Lindzen (2007) analyzed several general circulation models and showed that "warming is strongly peaked in the tropical troposphere". He went on to conclude: "Roughly speaking, the warming [near the tropopause] in the tropics is ... twice to about three times larger than near the surface regardless of the sensitivity of the particular model. This is, in fact, the signature (or fingerprint) of greenhouse warming. Stated somewhat differently, if we observe warming in the tropical upper troposphere, then the greenhouse contribution to warming at the surface should be between less than half and one third the warming seen in the upper troposphere.... The modeling studies establish that the ratio of upper tropospheric tropical warming to surface warming is approximately 2.5:1 regardless of the model sensitivity."

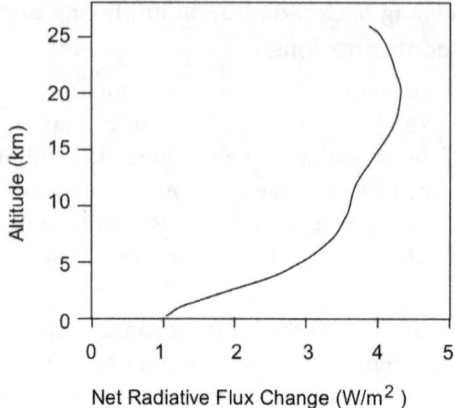

Figure 4.24. Forcing due to doubling CO_2 from pre-industrial value as a function of altitude (Ramanathan, 1991).

Collins *et al.* (2006) asserted "The interaction of short wave and long wave radiation with an (idealized) atmosphere free of clouds and aerosols can be calculated to a very high degree of accuracy". They said "the introduction of clouds would greatly complicate the ... exercise and therefore clouds are omitted from [the calculations]". They said: "Flux is defined as flux for clear-sky and aerosol-free conditions and forcing" and was without stratospheric adjustment. They also said "the effects of adjustment on forcing are approximately -13% for CO_2." They estimated clear sky forcing at (1) the top of the model, (2) at a pressure of 200 mb (surrogate for tropopause) and (3) at the surface. Their results are given in Table 4.3.

Table 4.3. Clear sky forcing at various altitudes for various changes in greenhouse gas concentrations (Collins *et al.*, 2006).

Change	Forcing (W/m^2)		
	Top of model	200 mb	surface
CO_2 goes from 287 ppm to 369 ppm	0.9	1.8	0.4
CO_2 goes from 287 ppm to 574 ppm	2.5	5.2	1.3
All greenhouse gases go from year 1860 to year 2000	2.2	3.0	1.3

In a later paper, Iacono *et al.* (2008) improved the previous clear sky estimates with better models that included more layers in the model atmosphere. Their results are shown in Table 4.4. In both cases (Collins *et al.* and Iacono *et al.*) the forcing at the surface is considerably less than at the troposphere or top of the model.

Table 4.4. Clear sky forcing at various altitudes for various changes in greenhouse gas concentrations (Iacono *et al.*, 2008).

Change	Forcing (W/m^2)		
	Top of model	200 mb	surface
CO_2 goes from 287 ppm to 369 ppm	1.1	2.0	0.6
CO_2 goes from 287 ppm to 574 ppm	3.0	5.7	1.7
All greenhouse gases go from year 1860 to year 2000	2.1	3.0	1.1

The effect of clouds can be surmised from the work of Schmitt and Randall (1991) who pointed out: "Clouds influence the surface energy budget through cloud shadows, by downward emission of infrared radiation from cloud base and by blocking downward infrared radiation emitted above the level of the cloud. Through these various effects, the clouds can modulate the CO_2 forcing". They included clouds in their climate model. They "... evaluated the CO_2 forcing, [by running] the radiation code twice; once with a CO_2 mixing ratio of 330 ppm, and a second time with 660 ppm. The CO_2 forcing is then obtained as the difference in the long wave radiation fields between these two cases". Schmitt and Randall (1991) divided the atmosphere into 11 layers and assumed a distribution of clouds. They used a global climate model to estimate the forcing due to a doubling of CO_2 from the 330 ppm level. Their test runs indicate that the forcing due to doubling CO_2 maximizes near the tropopause and decreases at higher and lower altitudes. Surface forcing is considerably lower than that at the TOA. Averaged over all latitudes, they found that forcing at the surface was of the order of ~ 1.5 W/m^2 for clear skies and 1.0 W/m^2 when clouds were included. At the tropopause, forcing was of the order of ~ 5.1 W/m^2 for clear skies and 4.4 W/m^2 when clouds were included. These results for clear skies are comparable to those of Iacono *et al.* (2008).

From this work, we may conclude that the forcing at the surface due to a doubling of CO_2 from the pre-industrial value of ~ 280 ppm would be about 1.0 to 1.2 W/m^2 when clouds are included. The forcing at the surface in going from the pre-industrial level of CO_2 to the present value (~ 395 ppm) is about 0.6 W/m^2. The average forcing over the past 55 years was roughly 0.4 W/m^2.

4.10.6 Some Basic Quantities

Before proceeding to discuss ocean warming, it is useful to present some basic quantities:

- area of oceans ~ 361 million km^2
- volume of oceans ~ 1.3 billion km^3
- average depth of oceans $\sim 1.3 \times 10^9 / 3.61 \times 10^8 = 3.6$ km $= 3,600$ m

- Heat capacity of ocean water: 3,993 J/kg/°C
- Density of ocean water ~ 1.025×10^{12} kg/km³
- Heat capacity of oceans: $1.025 \times 10^{12} \times 3,993 \times 1.3 \times 10^9 = 5.3 \times 10^{24}$ J/°C

If you added 1.0 W/m² to the oceans, the total heat input would be $1.0 \times 361 \times 10^6 \times 10^6 = 3.6 \times 10^{14}$ Watts (equivalent to J/s). Note: We used the area of the oceans, not the area of the Earth.

In the course of a year, there are $3,600 \times 24 \times 365 = 3.2 \times 10^7$ s

In a year, the heat input to the oceans would be:

3.6×10^{14} J/s $\times 3.2 \times 10^7$ s $= 1.2 \times 10^{22}$ J

The temperature rise of the oceans in one year would then be:

$1.2 \times 10^{22}/5.3 \times 10^{24} = 2.3 \times 10^{-3} = 0.0023°C$

Of course, the ocean is heated in layers and more heat would appear in the upper layers than the lower layers. The distribution of ocean volume vs. depth is shown in Figure 4.25.

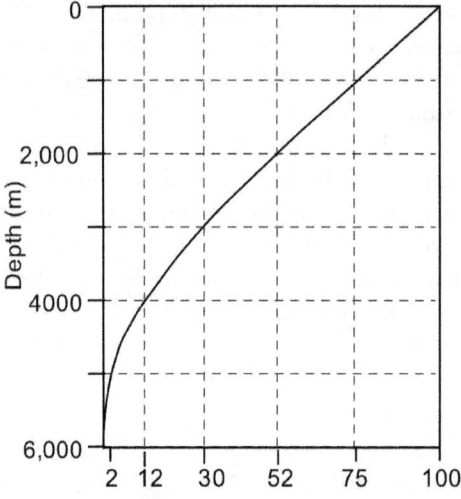

Figure 4.25. Percent of world ocean volume below any depth.

4.10.7 Measured Heat Content of the Oceans

The Earth's atmosphere is very tenuous compared to the oceans. As Pielke (2003) discussed, the energy balance of the Earth can be estimated by either calculating the various forcings that act on the Earth at the top of the atmosphere (TOA) and thereby estimating the net flux (W/m²) acting to heat the Earth, or by estimating the heat deposited into the Earth system (e.g. Joules/yr). In the former case, the argument is that the imbalance in energy flux at the TOA must be stored in the Earth system, and the only place it can go is the oceans; therefore it must heat the oceans. Since the heat capacity of the Earth lies mainly in the oceans, one can estimate the heat gain of the Earth by measuring the heat gain of the oceans. There are difficulties in both approaches. Of particular importance in regard to forcings are the unknowns relative to changes in humidity, aerosols and clouds as the Earth warms and industrialization proceeds. The problem in estimating ocean heat content is that the oceans are so diverse, and the collective mass of the oceans is so huge that even with a large heat flux imbalance, the temperature rise is small and difficult to measure.

Palmer and Haines (2009) estimated the global heat absorbed by the ocean to a depth of 220 m from 1970 to 2000 to be 0.25 W/m^2 (based on ocean area, not global area).

According to Levitus *et al.* (2012), over the 55-year period from 1955 to 2010, the heat content of the oceans down to a depth of 2000 m increased by 2.4 x 10^{23} J. They claim that this corresponds to an average temperature rise for the 0-2000 m layer of 0.09°C. Since the 0-2000 m layer occupies 48% of the volume of the oceans (see Figure 4.23), its volume is about 0.48 x 1.3 x 10^9 km³ = 6.3 x 10^8 km³.

The average temperature rise over 55 years for this layer is thus estimated to be

2.4 x 10^{23} J /(6.3 x 10^8 km³ x 1.025 x 10^{12} kg/km³ x 3993 J/kg/°C) = 0.08°C

which is close to the value given by Levitus *et al.* (2012). The same calculation can be repeated for the 0-700 km layer. According to Levitus *et al.* (2012), over the 55-year period from 1955 to 2010, the heat content of the ocean down to depth 700 m increased by 1.67 x 10^{23} J. They claim that this corresponds to an average temperature rise for the 0-700 m layer of 0.18°C. Since the 0-700 m layer occupies 16% of the volume of the oceans, its volume is about 0.16 x 1.3 x 10^9 km³ = 2.1 x 10^8 km³. The average temperature rise over 55 years for this layer is thus estimated to be

1.67 x 10^{23} J /(2.1x 10^8 km³ x 1.025 x 10^{12} kg/km³ x 3993 J/kg/°C) = 0.19°C

By subtraction, we can infer that for the 700-2000 m layer, the increase in heat content was 0.73 x 10^{23} J. The volume of this layer is roughly 4.2 x 10^8 km³. The average temperature rise over 55 years for this layer is thus estimated to be

0.73 x 10^{23} J /(4.2 x 10^8 km³ x 1.025 x 10^{12} kg/km³ x 3993 J/kg/°C) = 0.04°C.

The average rates of temperature increase per year over the 55-year span were:

0-700 m layer: 0.19°C/55 = 0.0035°C/yr

700-2000 m layer: 0.04/55 = 0.00073°C/yr

On average, the heat input to the oceans per unit area can be estimated as follows:

Over the 55-year period, the heat content of the ocean down to depth 2000 m increased by 2.4 x 10^{23} J. The volume of ocean below 2,000 m is 52%. The average temperature rise over the 55-year span below 2,000 m is not known but was probably around 0.01°C or less. Adopting as a guess the 0.01°C figure, the total heat gain of the entire ocean over the 55-year period is estimated to be 3.0 x 10^{23} J.

Since the area of the oceans is 3.61 x 10^8 km², the total heat gain per unit area of ocean was

3.0 x 10^{23} J/3.61 x 10^8 km² = 8.3 x 10^{14} J/km². Over the 55-year period there were

55 x 3.2 x 10^7 = 1.76 x 10^9 seconds

Therefore on average, the rate of heat gain by the oceans per unit area of ocean was

8.3 x 10^{14}/1.76 x 10^9 = 4.7 x 10^5 W/km² = 0.47 W/m²

It is noteworthy that Hamon *et al.* (2012) made corrections to the world ocean database and concluded that the results "show a fairly prominent trend in 0–700 m ocean heat content of 0.39 × 10^{22} J/yr between 1970 and 2008". Levitus *et al.* found a trend of 1.67 x 10^{23} J/55 years = 0.30 × 10^{22} J/yr between 1955 and 2010 for the 0-700 m layer.

Loeb *et al.* (2012) reviewed more recent data on ocean warming. Their results are shown in the Table 4.5.

Table 4.5. Rate of heat absorption for the 0-700 m ocean layer (W/m²) per unit area of earth for two time periods. Note that the area used to calculate the heat flux is the area of the Earth, rather than the area of the oceans. The area of the earth is 1.41 times the area of the oceans. Therefore, to compare these data with the previous given calculations, one should multiply the figures in the table by 1.41. (Loeb *et al.*, 2012)

Time Period	Measured W/m² in top 700 m of ocean		
	PMEL/JPL/JIMAR	NODC	Hadley
1993-2003	0.66±0.17	0.48±0.23	0.40±0.21
2004-2008	0.18±0.60	0.10±0.60	0.31±0.57

The error bars for the 2004-2008 data are excessively large.

Church *et al.* (2011) estimated the heat gain by the oceans for the periods 1972 – 2008 and 1993-2008. By subtraction, we can also estimate the heat gain for 1972-1992. Their reported heat gains are shown in Table 4.6. From these data and the volumes of the layers, we can (as before) calculate the estimated temperature rise for each layer for each time period. This is given in Table 4.7. The rate of heat gain by the various layers for various time periods per unit area of ocean is given in Table 4.8.

Table 4.6. Heat gains in units of 10^{21} J by time period and ocean layer according to Church *et al.* (2011).

	1972-2008	1993-2008	1972-1992
0 – 700 m	112.6	45.9	66.7
700 – 3,000 m	49.7	20.7	29.0
3,000 m to bottom	30.7	12.8	17.9
Total	193.0	79.4	113.6
Years	27	18	11
Seconds	8.6×10^8	5.8×10^8	3.5×10^8

Table 4.7. Total temperature rise (°C) by time period and ocean layer based on data of Church *et al.* (2011).

	Volume (km³)	1972-2008	1993-2008	1972-1992
0 – 700 m	2.1×10^8	0.131	0.053	0.078
700 – 3,000 m	7.0×10^8	0.017	0.007	0.010
3,000 m to bottom	3.9×10^8	0.019	0.008	0.011
Total	1.3×10^9	0.036	0.015	0.021

Table 4.8. Rate of heat absorption by time period and layer (W/m²) per unit area of ocean based on data of Church *et al.* (2011).

	1972-2008	1993-2008	1972-1992
0 – 700 m	0.36	0.22	0.53
700 – 3,000 m	0.16	0.10	0.23
3,000 m to bottom	0.10	0.06	0.14
Total	0.62	0.38	0.90

Trenberth (2010) described ocean heat measurements prior to buildup of the Argo profiling float system in 2003-2005:

"Before then, the bulk of the observations of the ocean were from expendable bathythermographs (XBTs) dropped from ships along their tracks as opportunities arose. As a result, coverage was spotty and irregular, and missing over many regions such as the Southern Ocean. The XBTs recorded temperatures, but the exact depth they were at was an estimate based on an assigned drop rate, which turned out to be sensitive to the exact design and character of the XBT probe. Recent careful comparisons with calibrated probes deployed from research vessels have shown the need for corrections. The severe under-sampling of the ocean until about five years ago, along with the variety of methods used to correct for problems and biases, has led to many estimates of how the temperatures in the ocean have changed over time. Of particular interest for climate is the vertically integrated ocean heat content. The reprocessing of XBT and Argo observations has resolved some issues ... but there remains a surprisingly large spread among different estimates of ocean heat content as discussed in the paper by Lyman *et al.* (2010). They delved into the origins of these differences, and compiled and reprocessed a common data set for the upper 700 m of the ocean."

Trenberth concluded that for the 0-700 m ocean layer, the measured heat gain was 0.90 W/m^2 over the time interval 1994 to 2008 and 0.77 W/m^2 over the time interval from 2004 to 2008. His conclusions suggest a slowdown after 2003 when the data network was better. These results are per unit area of the oceans and are considerably higher than those obtained by Levitus *et al.* and Church *et al.* It is not clear why his values are so much higher than those who analyzed the data.

There seems to be considerable uncertainty in the measured rates of heat gain by the oceans over the past few decades. The data appear to suggest a heat gain of about 0.5 W/m^2 (area of ocean).

Climate models indicate that a doubling of CO_2 produces a forcing at the surface of about 1.1 W/m^2 when clouds are included. Current CO_2 forcing is estimated to be about 60% of the forcing due to a doubling. Over the past 55 years, the average CO_2 forcing at the surface is estimated to be roughly 40% of the forcing due to a doubling. Hence, the estimated average forcing at the surface with clouds included over the past 55 years is roughly estimated to be about 0.4 W/m^2. The simple heat transfer model suggest that the effect of increase in back radiation is to reduce ocean heat loss, leading to ocean warming. As it turns out, the magnitude of the reduction in ocean heat loss is roughly 95% of the magnitude of the increase in back radiation. Measurements of heat gain by the oceans suggest a heat gain over this period averaging about 0.5 W/m^2. The heat gain by the oceans appears to have been of roughly the same magnitude as that predicted by CO_2 forcing without feedbacks, although the data and models are highly approximate. This is in contrast to some claims in the literature that since the long wave radiation is absorbed in the top few microns, it ends up mainly causing evaporation rather than ocean warming. If strong feedbacks are included as indicated by Ramanathan (1981) the predicted rate of ocean warming would be 15 time greater, and the problem of what happened to the "missing heat" would arise.

Some climatologists have argued that aside from the absolute rate of gain of heat of the oceans, the rate of heat gain has been accelerating in recent years. Some of this is tied to the rate of sea level rise, but the recent discovery that ground water depletion is contributing significantly to sea level rise adds a new wrinkle to this issue.

At a 2012 ASME Colloquium on energy and climate change held at the Disney Hotel in Anaheim, CA, Josh Willis made a presentation in which he claimed that there was a "hockey stick" in a plot of sea level vs. time due to an acceleration late in the 20th century from 2 mm/yr to 3 mm/yr. Closely related to this is the recently issued paper by Levitus *et al.* (2012). This paper was commented on by Pielke, Sr. (2012)[37] and Eschenbach (2012)[38] amongst others.

The Levitus *et al.* paper presents integrated heat content over the past 55 years, although the monitoring system for the oceans to depth 700 m was not fully in place until the mid-1990s. If one differentiates the integral curve of Levitus *et al.*, an approximate curve for the annual increase in heat content is obtained (units of 10^{22} J/yr). There is no evidence of acceleration or a "hockey stick" in these data. As Eschenbach pointed out, the temperature changes corresponding to these changes in heat content amount to roughly 0.0012°C/yr. He questioned whether such measurements can be made accurately. Since the integral curve used five-year smoothing, these annual changes are not actually year-to-year changes in the change in heat content, but rather, year-to-year changes in the five-year average. Therefore, one presumes that the annual changes in heat content were considerably greater than that shown in the figure. This would make sense considering how small the temperature changes actually were. As Eschenbach said: "Why not show the actual annual data? What are the averages hiding?"

Another relevant curve was proved by Pielke, Sr. (2008) showing no acceleration.

4.10.8 Effective Back Radiation at the Ocean Surface

Sverdrup *et al.* (1942) discussed the effective back radiation over the ocean surface. According to these authors:

> "The sea surface emits long-wave heat radiation, radiating nearly like a black body, the energy of the outgoing radiation being proportional to the fourth power of the absolute temperature of the surface. At the same time the sea surface receives long-wave radiation from the atmosphere, mainly from the water vapor. A small part of this incoming long-wave radiation is reflected from the sea surface, but the greater portion is absorbed in a small fraction of a centimeter of water, because the absorption coefficients are enormous at long wave lengths. The effective back radiation from the sea surface is represented by the difference between the 'temperature radiation' of the surface and the long-wave radiation from the atmosphere, and this effective radiation depends mainly upon the temperature of the sea surface and the water-vapor content of the atmosphere [a few meters above the surface].... According to Ångstrom (1920), the latter is proportional to the local vapor pressure,

---

[37] *pielkeclimatesci.wordpress.com/2012/.../the-overstatement-of-certaint...*
[38] *http://wattsupwiththat.com/2012/04/23/an-ocean-of-overconfidence/*

which can be computed from the relative humidity if the air temperature is known. Over the oceans, the air temperature deviates so little from the sea-surface temperature that the vapor pressure can be obtained with sufficient accuracy from the sea-surface temperature and the relative humidity of the air at a short distance above the surface."

Estimates of the back radiation as a function of temperature and cloud cover date back to fairly early in the 20th century. Several empirical formulas were proposed in the first half of the 20th century and these were widely used to analyze ocean-atmosphere interactions. As the years went by, remembrance of the basis for the formulas gradually faded while use of the formulas expanded.

Ångstrom (1920) presented the graph shown in Figure 4.26.

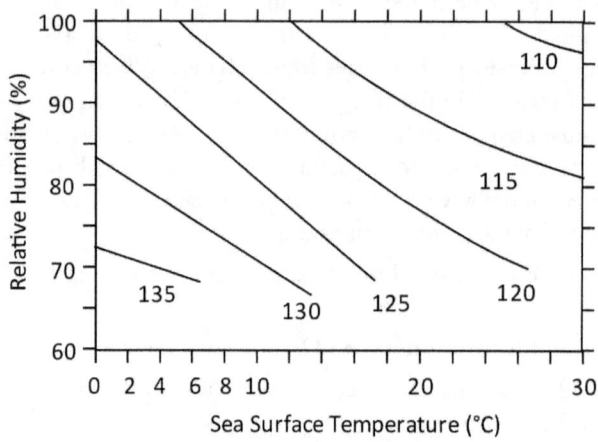

Figure 4.26. Effective back radiation in W/m^2 from the sea surface to a clear sky as a function of sea-surface temperature and relative humidity of the air at a height of a few meters (Sverdrup *et al.*, 1942).

Sverdrup *et al.* went on to say:

"In the presence of clouds the effective back radiation is cut down because the radiation from the atmosphere is increased. The empirical relation can be written

$$Q = Q_0 \, (1 - 0.083 \, C)$$

where Q_0 is the back radiation for a clear sky and where C is the cloudiness on the scale 1 to 10. A diurnal or annual variation in the cloudiness will lead to a corresponding variation in the effective back radiation. On an average, the diurnal variation of cloudiness over the oceans is very small and can be neglected, but the annual variation is in some regions considerable. The above equation is applicable to average conditions only, because the reduction of the effective back radiation due to clouds depends upon the altitude and the density of the clouds. Thus, if the sky is completely covered by cirrus, alto-stratus, or stratocumulus clouds, the effective radiation is about $0.75 \, Q_o$, $0.4 \, Q_o$, and $0.1 \, Q_o$, respectively."

Assuming cloud cover over the ocean averages roughly 0.6, the back radiation would be roughly half of the values given in Figure 1 for clear skies.

Sverdrup *et al.* cite "Brunt's empirical formula":

$$Q_B = Q \, (1 - 0.44 - 0.08 \, e_A^{0.5})$$

where Q is the radiation of a black body having the temperature of the sea surface and e_A is the vapor pressure of water vapor in the air in millibars.

These formulas were used by a number of investigators (*e.g.* Koto (1966) and Hasse (1971) used the graph provided by Sverdup *et al.* together with the correction for cloudiness given above).

Baldwin (1970) used results of Anderson (1954) and James (1966) to estimate the effective back radiation. He used Anderson's equation:

$$Q_B = (4.75 \times 10^{-9}) \, T_S^4 \, (1 - a + b \, e_A)$$

where

$a = 0.74 + 0.025 \, C \, \exp(-0.0584 \, h)$

$b = 0.0049 - 0.00054 \, C \, \exp(-0.060 \, h)$

in which:

e_A = Water vapor pressure in the air at T_S (mb)

C = Cloud amount in tenths

h = Average cloud height in thousands of feet.

Haney (1971) estimated the net upward flux of long wave radiation, using an empirical relationship due to Brunt, which was also used by others:

$$Q_B = Q^* \, \sigma \, T_S^4$$

where

$$Q^* = 0.985 \, [0.39 - 0.05 \, (e_A)^{0.5}](1 - 0.6 \, C^2),$$

Dorman (1974):

$$Q_B = [\sigma \, T_S^4 \, (a - b \, e_A^{1/2})](1 - c \, C) + d \, T_S \, (T_S - T_A)$$

with constants given in a book by Wyrtki.

Lane (1989) provided an empirical equation for the effective back radiation, "that takes into account the complex atmospheric absorption and radiation":

$$Q_B = 0.96 \, \sigma \, T_A^4 \, (11.7 - 0.23 \, e_A)(1 - c \, C) + 3.84 \, \sigma \, T_A^3 \, (T_S - T_A)$$

in which

C = fractional cloud cover

c = A coefficient that varies with latitude and is roughly 0.5 at tropical latitudes ranging to 0.6 at 30° latitude

Gill (1982) suggested use of the formula:

$$Q_B = 0.985 \, [\sigma \, T_S^4 \, (0.39 - 0.05 \, e_A^{1/2})(1 - 0.6 \, C^2)$$

Maughan (1966) measured the outgoing and incoming IR levels over a body of water and subtracted these to obtain the back radiation flux.

Huang and Park (1975) attempted to use insolation measurements from ocean buoys to infer cloud cover, based on correlations of dependence of back radiation on cloud cover in the mid-latitude North Pacific Ocean (43°N). They referenced a number of earlier publications that utilized the so-called Berliand formula:

$$Q_B = Q_o \, (1 - a \, C - b \, C^2)$$

where C is cloud cover, Q_o is the clear sky back radiation and a and b are constants. At the latitude of measurements, a ~ b ~ 0.38. They mention a formula attributed to Berliand and Berliand in 1952 as follows:

$$Q_B = 0.985 \, [\sigma \, T_S^4 \, (0.39 - 0.05 \, e_A^{1/2})(1 - k \, C^2) + + 2.91 \, \sigma \, T_A^3$$
$$(T_S - T_A)$$

Sopkin (2008) used a similar formula attributed to Berliand and Berliand in 1952.

Kraus and Booth (1961) attributed the following formula to Brunt in 1944 and Dorsey in 1940:

$$Q_B = 0.985 \, \sigma \, T_S^4 \, (0.39 - 0.0504 \, e_A^{1/2})(1 - k \, C^2)$$

Siegel and Dickey (1986) measured the net long wave radiation at the sea surface over the eastern North Pacific Ocean for 22 days during the fall of 1982. They referred to a number of previous attempts to correlate simple formulas to measurements of net back radiation, and then went on to compare their measurements with a number of these formulas (Table 4.9). The mean value of T_S was 297.8 K.

Table 4.9. Clear Sky Formulas for Back Radiation (Siegel and Dickey, 1986)

Reference	Formula for Back Radiation	(W/m^2)
Berliand and Berliand [1952]	$Q_o = \varepsilon\sigma\,T_S^4\,(0.39 - 0.05\,e_A^{1/2}) + 4\varepsilon\sigma\,T_A^3\,(T_S\text{-}T_A)$	85.2
Brunt [1932]	$Q_o = \varepsilon\sigma\,T_S^4\,(0.39 - 0.05\,e_A^{1/2})$	81.1
Efimova [1961]	$Q_o = \varepsilon\sigma\,T_S^4\,(0.254 - 0.00495\,e_A)$	73.0
Bunker [1976]	$Q_o = \varepsilon\sigma\,T_S^4\,(0.257 - 0.005\,e_A) + 4\varepsilon\sigma\,T_A^3\,(T_S\text{-}T_A)$	78.9
Anderson [1952]	$Q_o = \varepsilon\sigma\,[T_S^4 - T_A^4(0.74 + 0.0049e_A)]$	81.0
Swinbank [1963]	$Q_o = \varepsilon\sigma\,[T_S^4 - 9.36 \times 10^{-6}\,T_A^6]$	88.5
Clarke et al. [1974]	$Q_o = \varepsilon\sigma\,T_S^4\,(0.39 - 0.05\,e_A^{1/2}) + 4\varepsilon\sigma\,T_S^3\,(T_S\text{-}T_A)$	86.5

The various formulas accounted for clouds using a formula of the type

$$Q_B = Q_o\,(1 - B\,C^{1/N})$$

where is the clear sky formula and B and N are constants. The predictions of the formulas for $C \sim 0.72$ are given in Table 4.10. The measured mean effective back radiation was 52.0 W/m^2.

Table 4.10. Simple formulas for back radiation including clouds.

	N	B	Q_B
Berliand and Berliand [1952]	1	0.55	52.3
	2	0.63	55.3
Brunt [1932]	1	0.49	51.4
	2	0.57	53.6
Efirnova [1961]	1	0.44	50.2
	2	0.52	51.3
Bunker [1976]	1	0.50	51.3
	2	0.57	53.5
Anderson [1952]	1	0.52	51.6
	2	0.60	53.9
Swinbank [1963]	1	0.58	52.0
	2	0.66	54.8
Clark et al. [1974]	1	0.56	52.6
	2	0.64	55.8

Newell and Dopplick (1979) claimed that estimates of Q_B were "discussed extensively in the literature" and based on this, they used:

$$Q_B = 0.94\,[\sigma\,T_S^4\,(0.56 - 0.08\,e_A^{1/2})]$$

which presumably includes clouds and pertains to tropical latitudes. Their estimate at 297.8 K is 47.5 W/m².

Zapadka *et al.* (2007) simultaneously measured data regarding the long wave radiation of the sea surface and its contiguous air layer, the water vapor pressure in the air above the water, and the cloud cover. These data were gathered during numerous research cruises in the Baltic in 2000–03 and were supplemented by satellite data characterizing the cloud cover over the whole Baltic. From this, they derived an improved formula for the back radiation:

$$Q_B = 0.985 \ \sigma \ T_S{}^4 - \sigma \ T_A{}^4 \ (0.685 - 0.00452 \ e_A)(1 - B \ C^N)$$

They provide values of B and N for various types of clouds.

For purposes of estimating the temperature rise when the oceans are exposed to a long-term increase in back radiation, the absolute value of the back radiation is not very important. What is important is the temperature dependence of the back radiation. The temperature dependence of the back radiation depends on two terms: T_S and e_A. Using the Brunt formula (for example) we have

$$dQ_o / dT_S = 4\varepsilon\sigma \ T_S{}^3 \ (0.39 - 0.025 \ e_A{}^{-1/2} \ de_A / dT_S)$$

4.11 The Climate Debate Revisited

We have shown that the global average temperature rose in the 20th century. However the global network for temperature measurement leaves much to be desired. It seems likely that the measured rise in temperature may be exaggerated by urban heat island effects. The recent work by Watts *et al.* suggests that this effect is far greater than has previously been assumed. However, this work is being revised and the final results are not available. The precision to which we know the global average temperature decreases significantly toward the early part of the 20th century and even more so in the 19th century. It is noteworthy that according to the BEST study, 1/3 of land measurement sites had local temperatures that actually decreased in the 20th century. Furthermore the temperature did not rise uniformly in lock step with the rise in CO_2, but rose sharply early in the 20th century, flattened out in the middle of the 20th century and then rose again late in the 20th century. By far, the greatest temperature rise occurred at high northern latitudes. Some of this was undoubtedly due to deposition of black carbon on ice and snow, not greenhouse gases. Measurements of tropospheric temperature over the past three decades show that temperatures seem to vary closely with El Niño indices, rather than CO_2 concentration. The temperature seems to have been relatively flat from 1979 to 1997 and flat from 1999 to 2012, with a single step jump upward during the great El Niño of 1998. The alarmist view of various aspects of the climate debate as embodied in Table 1.1 may now be judged in the context of 20th century temperatures. See Table 4.11.

Table 4.11 Comparison of alarmist view of climate change with 20th century observations.

Aspect	Observations
Temperature rise over past century vs. natural fluctuations	Whether the 20th century warming was within the scope of past natural fluctuations depends on which models are used for reconstruction of millennial temperatures. The answer here is not clearcut. It seems likely that there was a significant *Little Ice Age* and the rise in temperature from 1880 to 1940 was due to emergence from the LIA.
Rising CO_2 was the cause of rising temperatures in 20th century	There are a number of observations that cast doubt on the consensus belief that rising CO_2 was the predominant cause of temperature rise in the 20th century. These include: (1) Most of the temperature rise in the 20th century occurred at high northern latitudes – with relative temperatures increasing faster than climate models predict for Arctic amplification; (2) 1/3 of land measurement sites reported a decrease in temperature; (3) The sharp temperature rise from 1920 to 1940 was prior to major buildup of CO_2 in the atmosphere; (4) The decrease in temperature from 1940 to 1976 was counter to the buildup of CO_2 although it could be attributed to the reverse effect of aerosols; (5) Temperatures appear to be controlled by El Niño indices rather than CO_2 concentration; and (6) Temperatures over the past 30 years were flat except for a single step up in the great El Niño of 1998 – while CO_2 was rising to new heights. On the other hand, there is evidence that the oceans have warmed over the past five decades and this can logically be attributed to the increases in greenhouse gases over that period.
Climate models provide reasonable estimates of future warming in BAU scenario	Climate models do not seem to properly account for the close tie of climate to El Niño indices. Since they do not work in recent past, we have no basis for believing they can predict the future.

5 CARBON DIOXIDE THROUGH THE AGES

One of the most pressing issues of our time is the possibility that rising CO_2 concentrations in the atmosphere might lead to significant global warming in the future that could produce deleterious impacts on humankind. Hence, the relationship between CO_2 concentration and climate has become a very central and critical scientific issue. However, in addition to being a scientific issue, rising CO_2 has also become a political issue as well. This is due to several factors:

(1) In the process of consuming fossil fuels, cement production, and other industrial activities, the world produces large amounts of CO_2. If the world continues in a business-as-usual scenario, CO_2 production will continue to rise in the 21st century, leading to higher CO_2 concentrations in the atmosphere. The potential cure for too much CO_2 requires a draconian change in the way that energy is generated and used by the world, and this change may not be technically feasible, and even if it turns out to be technically feasible, it will likely be extremely costly. Indeed, it is possible that it may not be possible to provide the people of the world with energy to run the industrialized world if CO_2 emissions must be cut as dramatically as alarmists claim.

(2) A significant number of climatologists have adopted an "alarmist" view in which they believe that continuation of business-as-usual energy policies in the 21st century will be disastrous to humankind. Many of them have voiced this viewpoint via the Internet, meetings and media. Furthermore, this bias has crept into scientific publications published in peer review journals. A smaller number of climatologists have been skeptical of the certainty expressed by alarmists. Liberal politicians have been swayed by alarmists into

enacting severe constraints on future CO_2 emissions. These constraints require that by such and such a future year, we must emit considerably less CO_2. It is not clear that these constraints can be met technically or financially. Indeed, the benchmark used by several governments is an 80% reduction in CO_2 emissions by 2050 – a goal that almost certainly cannot, and will not be met. Conservative politicians tend to lean toward the skeptical view, more from a political perspective than from any scientific understanding. It ashould be noted that governments have not been clear whether this means an 80% reduction from present emission levels or an 80% reduction from that expected on the basis of a business-as-usual scenario. If, as seems likely, it is an 80% reduction from present levels, that is equivalent to approximately an 88% reduction from a business-as-usual scenario.

(3) Quite a few prominent climatologists in their zeal to save the world from overheating (and possibly to secure more funding for climate research) have engaged in unprofessional activities in an attempt to exclude the skeptics from science publications. They have also manipulated data to exaggerate the threat of rising CO_2 and they have presented their results in a biased and one-sided manner. Some have prevented others from checking their results by holding their data in secret. In many cases, they have drawn conclusions from sparse and noisy data, yet made bold claims of certainty in their conclusions. The exposure of these shenanigans has hurt their credibility in some quarters; nevertheless, the science questions remain regarding the impact of rising CO_2.

(4) Under auspices of the United Nations, the *Intergovernmental Program on Climate Change* (IPCC) has been co-opted by alarmists regarding the effect of CO_2 on climate, and they have widely promulgated the belief that "the debate is over" regarding the impact of rising CO_2, yet considerable uncertainty remains in all the issues.

5.1 Recent and Future Times

There is considerable evidence that the CO_2 concentration in the atmosphere has varied over a very wide range during geological time. Nevertheless, despite this wide-ranging past, a seemingly repeatable pattern has emerged over the past few hundred thousand years. The Earth has alternated between ice ages and interglacial periods at roughly 100,000-year intervals. Ice cores reveal that during interglacial periods, typical CO_2 concentrations were about 280 ppm while at the height of glaciation during ice ages the CO_2 concentration dropped to roughly 190 ppm. About 20,000 years ago, the Earth was at its Last Glacial Maximum (LGM) and the CO_2 concentration was roughly 190 ppm. As the ice age waned, the CO_2 concentration rose, and reached a plateau of roughly 280 ppm where it has remained for the past ~8,000 years. As a result, we are accustomed to think of 280 ppm as the "normal" pre-industrial level of CO_2 in the atmosphere.

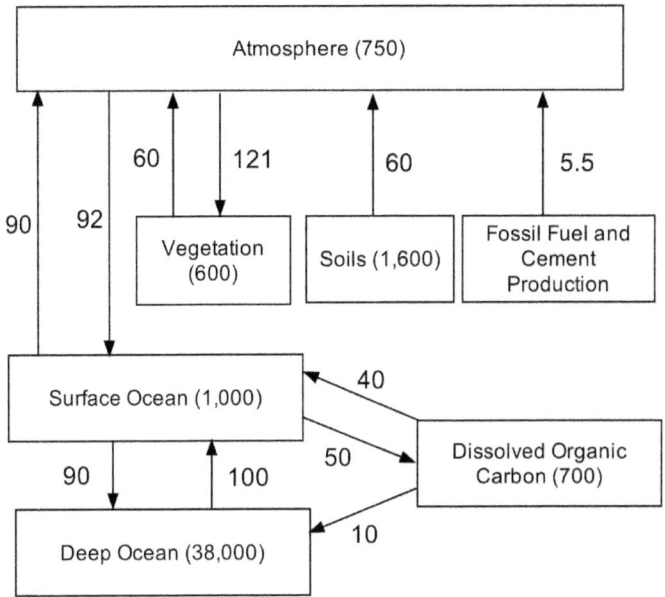

Figure 5.1. Rough estimates of carbon storage and annual carbon fluxes as of about year 2000. Storage is in gigatons (Gt) of carbon and fluxes are in Gt/year of carbon. Ocean sediments are not shown, nor are many smaller contributors. Adapted from Northrup (2004). As of year 2012, the fossil fuel and cement emissions are closer to 8.0 Gt/yr vs. 5.5 Gt/yr in the figure.

Knowledge of carbon exchange between the atmosphere, land, and the oceans is important in understanding the rate of build-up of CO_2 in the atmosphere due to human intervention in the Earth environment. However, the Earth's carbon cycle is complex. The oceans, the biosphere, and land exchange very large amounts of CO_2 each year, and the contributions from fossil fuel and cement production are relatively small. Nevertheless, the human contributions (from fossil fuel and cement production) appear to be enough to upset the delicate Earth balance, leading to a buildup of CO_2 in the atmosphere. Figure 5.1 provides a very rough simplified version of the carbon content and annual carbon fluxes in the Earth's ecosystem. Sabine *et al.* (2004) estimated the long-term (1800–1994) and recent (1980–1999) transfers of CO_2 between the biosphere, the atmosphere, and the oceans. They estimated that about 58% ended in the atmosphere and 42% ended in the oceans. It has been suggested that over the past two decades, about half of net emissions remained in the atmosphere and half was taken up by the oceans. Douglass (2005) pointed out that the common approach is to treat nature as unchanging, and attribute all increases in atmospheric CO_2 concentration to anthropogenic sources. However, he argued that natural emissions of CO_2 increased in the latter part of the 20th century due to prevalence of El Niños that raised surface ocean temperatures, and some major volcanic eruptions that emitted large amounts of CO_2. When these factors were taken into account, he found that some of the increase of the late twentieth century atmospheric CO_2 was due to natural causes. After taking this into account, he concluded that the rate of increase of the late twentieth century atmospheric CO_2 concentration was approximately constant at around 44% of nominal anthropogenic emissions. In addition, he argued that massive coal fires in northern China and biomass burnings have added large amounts of CO_2 to the anthropogenic side of the ledger. Thus, his estimate of anthropogenic CO_2 emissions is greater than other estimates, and he concluded that only about 30% of late twentieth century CO_2 emissions from anthropogenic sources (other than biomass burnings) ended up in the atmosphere. This is considerably lower than the nominal figure 50% that is often used. A rough estimate is that for each Gt of carbon[39] added to the atmosphere, the concentration of CO_2 rises by about 0.4 ppm to 0.5 ppm. Anthropogenic emissions of carbon have steadily increased since the 19th century, and currently run about 8 Gt/yr. If roughly half of were to end up in the atmosphere, the concentration would rise at the rate of about 1.8 ppm per year. The estimated CO_2 concentration over the past two millennia is shown in Figure 5.2. As of 2011, the concentration reached about 395 ppm and is increasing at around 2 ppm per year.

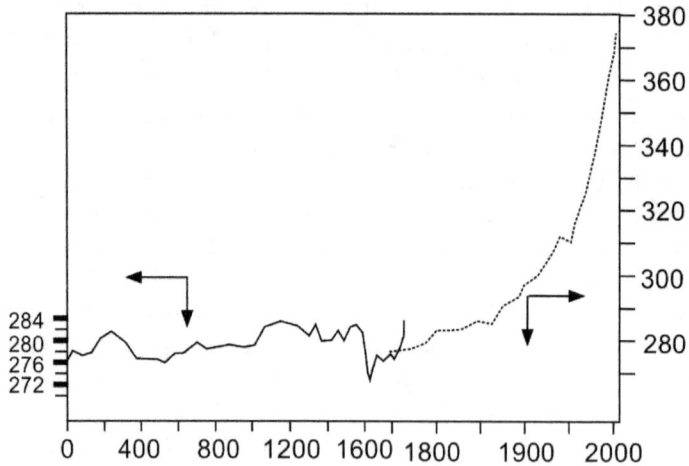

Figure 5.2. Atmospheric CO_2 concentration over the past 2,000 years (ppm). Adapted from Etheridge *et al.* (1996), and extended with recent data from NOAA.

The actual variation of CO_2 concentration goes through an annual sawtooth cycle reflecting a seasonal fluctuation in CO_2 of up to about 5 ppm. Plants draw carbon dioxide from the atmosphere to make food. In the

---

[39] One Gt of carbon is equivalent to $44/12 = 3.67$ Gt of CO_2.

Northern Hemisphere, the CO_2 concentration peaks in early spring just before plant growth begins and falls off in October when the growing season ends. See Figure 5.2a.

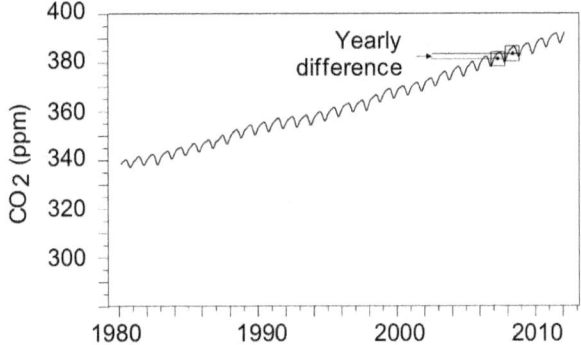

Figure 5.2a. Annual and long-term variation of CO_2 concentration (Humlum, *et al.*, 2012).

One website[40] said this is an "important paper [that] strongly suggests man-made CO_2 is not the driver of global warming" and "CO_2 released from use of fossil fuels have little influence on the observed changes in the amount of atmospheric CO_2, and changes in atmospheric CO_2 are not tracking changes in human emissions". Such conclusions are somewhat exaggerated.

Humlum *et al.* analyzed in detail the annual variations in CO_2 as illustrated in Figure 5.2a. While both CO_2 and temperature generally increased during this 31-year period, the rates of change varied significantly during the period. They showed that changes in CO_2 correlated somewhat with changes in sea surface temperature (SST) but the CO_2 change lagged the SST change by about 11-12 months. They concluded that "A main control on atmospheric CO_2 appears to be the ocean surface temperature". They mentioned a possible connection to the giant 1998 El Niño but did not elaborate on the relationship of the entire sequence of data to El Niño indices.

Consider Figure 5.2b. The uppermost curve shows the NINO3.4 index from 1980 to 2011. Peak El Niños are labeled with letters A to F. The middle curve shows the change in CO_2 concentration per year plotted on a monthly basis. The peaks in this curve are also subjectively labeled A to F. The average change in CO_2 concentration per year can be interpreted either as a ramp or a step-function. Arbitrarily adopting the step function, the average change in CO_2 concentration per year varied from year to year by about 1.5 ppm/yr prior to the 1998 El Niño, and varied from year to year by about 2.0 ppm/yr after the 1998 El Niño. These are depicted as horizontal dashed lines x and y.

---

[40] *http://wattsupwiththat.com/2012/08/30/important-paper-strongly-suggests-man-made-co2-is-not-the-driver-of-global-warming/*

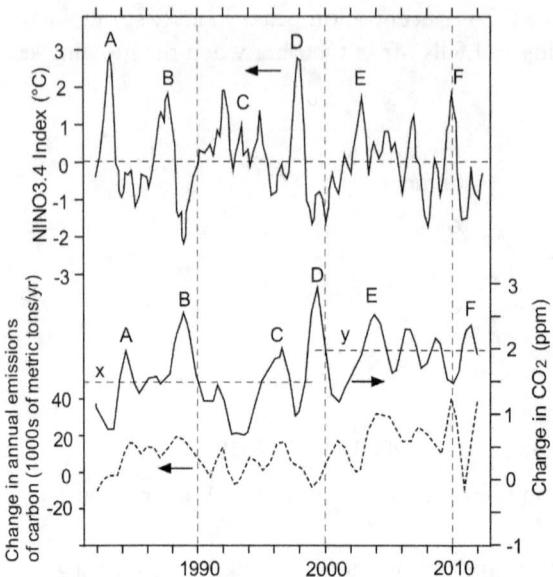

Figure 5.2b. CO_2 emissions and concentration and Niño index.

The lowermost curve shows the annual change in anthropogenic CO_2 emissions plotted on a per month basis. A rough rule of thumb is that each Gt of carbon (3.67 Gt of CO_2) produces the equivalent of about 0.5 ppm of CO_2 in the atmosphere if none of it is absorbed. Figure 5.2b below shows that annual variations in global emissions of carbon are typically about 2×10^4 metric tons per year which if unabsorbed, would produce annual changes in CO_2 that are far too small to account for the observed variations in the average change in CO_2 concentration per year.

The point made by Humlum *et al.* is that the average change in CO_2 concentration per year lags the change in ocean temperature by about 11-12 months. As Tisdale (2012) showed in his book, El Niños leave behind them a pool of warm surface waters. As a result, the average change in CO_2 concentration per year tends to lag the NINO3.4 index by a bit more than a year. This correlation is far from perfect but it seems to have some validity, particularly for the major El Niño that started toward the end of 1997. The data suggest that the ability of the oceans to absorb CO_2 emitted by human activity responds to the state of the NINO3.4 index with a delay of a bit over a year.

Human activity is presently emitting roughly 8 Gt/yr of carbon, which if unabsorbed, would be sufficient to increase the atmospheric concentration of CO_2 by about 4 ppm per year. Over a period of years, we might assume that (very) roughly half of that CO_2 is absorbed by earth systems (oceans, biosphere, ...) and the other very rough half ends in the atmosphere raising the atmospheric concentration by about 2 ppm. However, on a year-by-year basis, the proportion of emitted CO_2 that is absorbed by the Earth systems varies considerably, mainly due to the presence of warm surface waters in the Pacific produced quasi-periodically by El Niños. According to the graphical data in Figure 5.2b, the annual increase in CO_2 concentration can be as high as 3 ppm (following the 1998 El Niño) or as low as 1 ppm (between peaks B and C). During the most recent period after the 1998 El Niño, variations in annual increase in CO_2 concentration seem to have varied roughly as 2 ± 0.5 ppm or $\pm 25\%$. These results seem to suggest that while roughly half of emissions might end up in the atmosphere over an extended period, annual variations in the distribution of emitted CO_2 between the atmosphere and the earth system are significant, and strongly dependent on prevalence of El Niños.

Tisdale showed that from 1976 to about 2005, there was a pronounced prevalence of El Niños over La Niñas. He argued that this could account for all of the warming of the Earth during that period without invoking the greenhouse effect. However, it seems likely that during this period, a greater proportion of emitted CO_2 ended up in the atmosphere due to prevalence of El Niños, and this might have amplified the natural El Niño warming effect via greenhouse gas forcing. McLean et al. (2009) estimated that 70% of warming over this period was due to El Niños while Foster et al. (2010) fell back on climate models that attribute only 15-30% of temperature variation in the 20th century to variability of the El Niño index. As is usual in climate matters, one has only to glance at the authors to know in advance what spin the results are likely to show. The Foster paper included the crème de la crème of climategate characters while the Mclean paper was written by skeptics.

The proportion of global heating from 1976 to 2005 due to prevalence of El Niños over La Niñas vs. greenhouse gas forcing remains uncertain. Nevertheless, the state of the Pacific Ocean is clearly important, not only for its impact on the atmospheric temperature, but also because it regulates the annual rise in CO_2 concentration.

In any event, the paper by Humlum et al. does not "strongly suggest man-made CO_2 is not the driver of global warming" but it does suggest (as Tisdale, 2012 has emphasized) that CO_2 is not the entire story for global warming, and variation of El Niños vs. La Niñas also has a significant effect.

Many projections have been made of the future use of fossil fuels during the remainder of the 21st century and the consequent future rise in CO_2 concentration. This will depend on a number of factors: world prosperity, industrialization of developing countries, rate of introduction of renewable energy, and gains in energy efficiency. These projections vary widely in their assumptions. Some assume business-as-usual with no special efforts to reduce carbon emissions, leading to rising carbon emissions as high as 25 Gt/yr late in the 21st century. Others assume severe reductions in carbon emissions ramping down quickly as the 21st century proceeds. One middle-of-the-road scenario that has been widely used is the so-called IS92a projection that calls for increased energy efficiency and gradually increasing contributions from renewable energy, but involves heavy use of coal, and allows for significant growth in total energy usage in the 21st century. For any arbitrary future scenario of emissions, one can roughly estimate how the CO_2 concentration will vary in the 21st century.

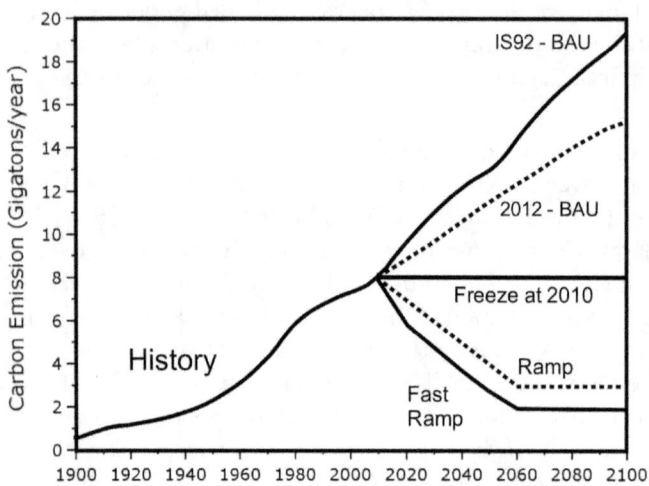

Figure 5.3. Annual emissions of carbon for five future scenarios in the 21st century.

Figure 5.3 shows five conceivable future scenarios for future emissions. One is the widely used middle-of-the-road "business as usual" scenario from the IPCC known as "IS92a." In this scenario, annual CO_2 emissions continue to increase through the 21st century. However, it seems likely that a significant part of the predicted coal usage might be replaced by natural gas, which would reduce the level of future emissions. Hence, another scenario in which natural gas replaces part of the coal in IS92a is also shown as 2012-BAU. Three other hypothetical future scenarios are shown in this figure. In one scenario, the CO_2 emission rate is held constant at the 2010 rate (estimated to be about 8 Gt/yr of carbon) for the remainder of the 21st century. The 8 Gt/yr of carbon emissions consists of about 2 Gt/yr from land clearing and about 6 Gt/yr from fossil fuel burning and cement production. The expectation in "business as usual" is that the land use figure will not change markedly but the fossil fuel combustion will increase significantly in the "business as usual" scenario. In the other scenarios, there are downward ramps to lower emission rates as the 21st century wears on. It should be noted that these latter three scenarios require draconian modifications to the way that industrialized societies produce and consume energy. These five scenarios lead to the buildups of CO_2 in the atmosphere as shown in Figure 5.4, assuming that half of the CO_2 emitted ends up in the atmosphere. It was roughly estimated that each Gt of carbon emitted leads to a ~ 0.25 ppm increase in CO_2 concentration assuming half of the carbon ends up as atmospheric CO_2.

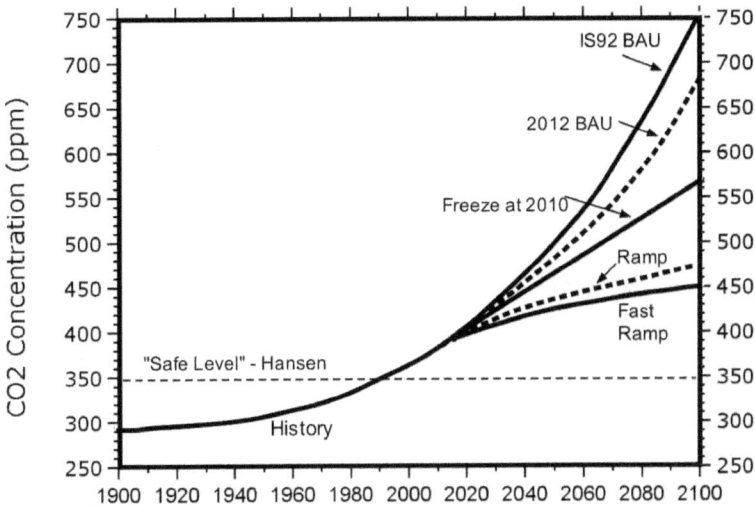

Figure 5.4. Buildup of CO_2 in the atmosphere corresponding to the five scenarios in Figure 5.3.

While this 50% assumption was representative of the past, it is not clear whether it will hold in the future. One model predicts that the 50% distribution will continue through at least 2040 (Mackenzie *et al.*, 2001). Sokolov *et al.* (2009) believe that the uptake by the Earth system will become saturated and reach a limit in the future. For any particular scenario for future CO_2 emissions, one must estimate the consequent rise in CO_2 concentration in the atmosphere. The common folklore is that about half the CO_2 emitted in the late 20th century was absorbed by the Earth system and the remaining half remained in the atmosphere. Knorr (2009) and Curtin (2009) reviewed data regarding the percentage of anthropogenically produced CO_2 that remains in the atmosphere as reported by studies such as that of Jones *et al.* (2005). The so-called "airborne fraction" of CO_2 remaining in the atmosphere "is known to have stayed remarkably constant over the past five decades" at "around 40%" (Knorr, 2009). Curtin (2009) estimated that:

"... since 1958 on average 56 percent of total global emissions of CO_2 have been absorbed by the oceans (both by dissolving and by biotic uptakes) and by the terrestrial biosphere's vegetation, so that only 44 percent have remained 'aloft', and thereby increasing the atmospheric concentration of CO_2".

Absorption by the Earth system in the future is uncertain. Several studies (e.g. Solomon *et al.*, 2009; Meinshausen *et al.*, 2009; Sokolov *et al.*, 2009; Schuster and Watson, 2007) claim that the capacities of the terrestrial and ocean sinks to uptake CO_2 are limited and will reach saturation levels of around 5 Gt C/year as CO_2 continues to be generated anthropogenically in the future. Sokolov *et al.* (2009) considered four levels of future CO_2 emissions and in all four cases the Earth system becomes saturated by around 2050. As a result, they claim that the percentage of emitted CO_2 taken up by the Earth system will decrease with time, and in the particular case of unrestricted future emissions, the percent of uptake by the Earth system will drop from the present value (estimated by them to be ~ 40%) to ~ 10% by 2100. If this proves to be correct, the projections of future CO_2 concentrations in Figure 5.4 (based on constant uptake at ~50%) could be very low.

Curtin (2009) challenged this conclusion. He argued that this assumption was based on old data that suggested that while plant growth rates increased with increasing CO_2 concentration at low to moderate CO_2 levels, plant growth rates would saturate when the CO_2 concentration becomes sufficiently high. The best data on world food production support the conclusion that there is no evidence of a slowing down in the ability of the Earth system to absorb CO_2 as CO_2 increases. He pointed out that over a 50-year period, the growth rate of atmospheric CO_2 has been slower than the growth rate of CO_2 emissions while the growth rate of CO_2 emissions has increased significantly. He argued that if the Earth system continues to absorb CO_2 in the future at its present rate, and if CO_2 emissions are subjected to draconian reductions, this could result in a significant reduction in world food production – which is dependent on the present high CO_2 concentration.

Ballantyne *et al.* (2012) analyzed 50 years of global carbon dioxide measurements and found that the processes by which the planet's oceans and ecosystems absorb the greenhouse gas are not yet at capacity. They concluded:

> "Globally, these carbon dioxide 'sinks' have roughly kept pace with emissions from human activities, continuing to draw about half of the emitted CO_2 back out of the atmosphere. However, we do not expect this to continue indefinitely."

The paper suggests that "we do not yet understand well enough the processes by which ecosystems of the world are removing CO_2 from the atmosphere, or the relative importance of possible sinks: regrowing forests on different continents, for example, or changing absorption of carbon dioxide by various ocean regions". Over the 50-year period 1960-2010, they found that cumulative emissions of carbon were 350 Gt. Of this, about 45%

accumulated in the atmosphere and 55% was absorbed by the Earth's carbon sinks.

In the "business as usual" IS92a scenario, the CO_2 concentration reaches 750 ppm by the end of the 21st century. With replacement of some coal by natural gas, this drops to under 700 ppm. If the CO_2 emission rate is frozen at the 2010 level, the CO_2 concentration reaches about 570 ppm by the end of the 21st century. Even the down-ramps leads to some elevation of the atmospheric CO_2 concentration. Hence, it seems likely that the CO_2 concentration in the atmosphere will reach at least 560 ppm before the end of the 21st century.

5.2 CO_2 and the Greenhouse Effect

A simplistic explanation for why increased CO_2 concentration can lead to increased temperature of the Earth is illustrated in Figure 5.5.

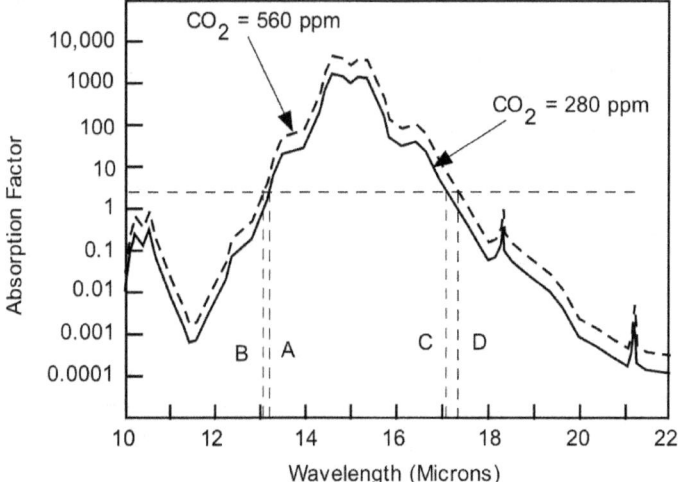

Figure 5.5. Absorption factor (absorptivity concentration integrated over vertical path through atmosphere) for CO_2 vs. wavelength. The horizontal dashed line corresponds to an absorption factor of 3 (essentially complete absorption along a vertical path through the atmosphere).

Carbon dioxide absorbs outgoing IR radiation emitted by the Earth primarily in the absorption band in the wavelength band between 13 mm and 17 mm. The absorption of any wavelength in the atmosphere is dependent on the integral of the absorptivity times the concentration on a vertical path through the atmosphere. This integral is called the absorption factor. Since the absolute amount of absorption depends on the exponential function of the absorption factor, an absorption factor of 3 corresponds to about 99% of complete absorption. As Figure 5.5 shows, with the pre-industrial level of 280 ppm of CO_2, the entire absorption band from 13 mm to 17 mm is fully saturated. Adding more CO_2 to the atmosphere does not increase the absorption significantly within the saturated region. Only at the "wings" of the absorption band is there any significant increase in absorption by the atmosphere when the CO_2 concentration is increased. Thus, with a CO_2 concentration of 280 ppm, absorption is saturated between vertical dashed lines A and C in Figure 5.5.

If in the future, the CO_2 concentration is doubled (from the pre-industrial value) to 560 ppm, the absorption curve moves up by a factor of 2 on the vertical log scale, and the saturated region expands to the region between vertical lines B and D, producing a net heating effect. However, the additional heating effect (from absorption in regions between A and B, and between C and D) is much smaller than the original heating effect in going from 0 ppm to 280 ppm (region between A and C). Thus we see that as more and more CO_2 is added to the atmosphere, the heating effect decreases per unit amount of CO_2 added.

Lindzen (2007) described the effect of additional absorption by CO_2 and other greenhouse gases in the atmosphere. He pointed out that the simplistic discription of the greenhouse effect is that greenhouse gases and clouds "inhibit cooling of the Earth by thermal radiation, and serves as a blanket which causes the Earth to be warmer than it otherwise would be". Lindzen objected to this description because as he said:

> "... the surface of the Earth does not cool primarily by thermal radiation".... There is so much greenhouse opacity immediately above the ground that the surface cannot effectively cool by the emission of thermal radiation. Instead, heat is carried away from the surface by fluid motions ranging from the cumulonimbus towers of the tropics to the weather and planetary scale waves of the extratropics. These motions carry the heat upward and poleward to [altitudes] where it is possible for thermal radiation emitted from these levels to escape to space."

The lower atmosphere is relatively opaque to IR radiation but as the altitude increases, the density decreases and transmission of IR improves. Lindzen defines an altitude sufficiently high that the optical depth for IR is about 1 so that transmission of IR is attenuated roughly as e^{-1}. This is a region of the atmosphere that can radiate energy from the Earth to space. He provided a description of the effect of adding greenhouse gases to the atmosphere as shown in Figure 5.5a.

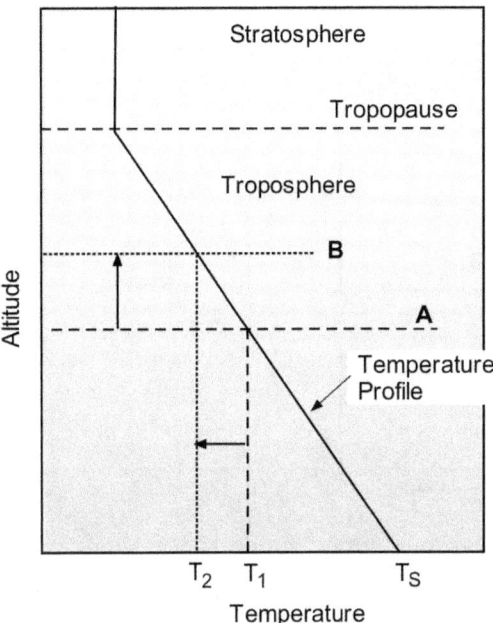

Figure 5.5a. Lindzen's picture of how the greenhouse effect works.

The temperature of the atmosphere decreases with altitude to some level known as the tropopause. The height of the tropopause varies from about 16 km in the tropics to about 12 km at 30° latitude, and to about 8 km in polar regions. Before adding greenhouse gases to the atmosphere, the altitude at which the optical depth for IR is around 1 is denoted **A** in Figure 5.5a. When greenhouse gases are added to the atmosphere, the level at which the optical depth is around 1 is raised in altitude due to the additional absorption by the greenhouse gas. As Lindzen points out, "because the temperature of the atmosphere decreases with altitude, the new characteristic altitude for emission is colder than the previous altitude". The new altitude for emission is **B**, and $T_1 > T_2$. Since IR emission is proportional to the fourth power of the temperature, emission from altitude **B** is reduced compared to what it was at altitude **A**. Thus the Earth cannot cool as effectively. The Earth will therefore warm until the temperature at altitude **B** approaches the original temperature. This is shown in Figure 5.5b. The original temperature at the surface is T_S, leading to the vertical temperature profile shown as the slanted dotted line. Before the temperature rises, the temperature at the altitude for emission is T_2 at point 2. After the surface temperature rises to $T_{S'}$, the new vertical temperature profile is the slanted solid line and the temperature at altitude **B** rises back to T_1. The Earth is now back in thermal balance.

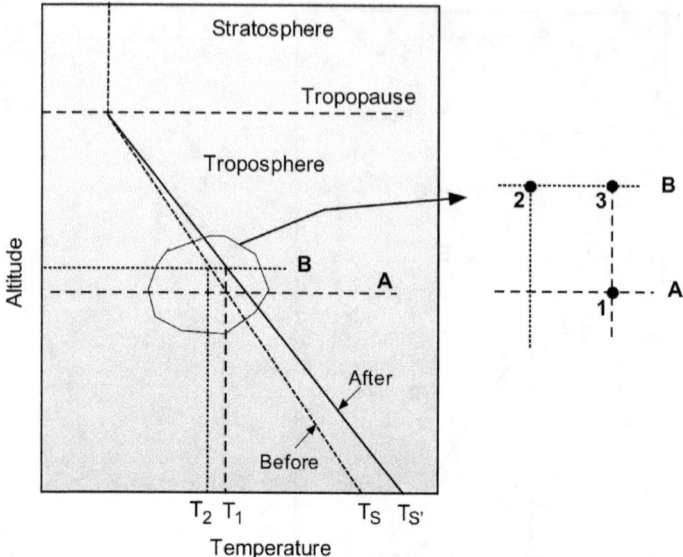

Figure 5.5b. Lindzen's picture of how the greenhouse effect works; part two.

Lindzen also pointed out that climate models indicate that when the Earth warms due to the greenhouse effect, warming is strongly peaked in the tropical troposphere near the altitude where the optical depth $\tau \sim 1$. (See for example, Figure 4.24). He concluded:

"Roughly speaking, the warming at $\tau = 1$ in the tropics is about two to three times larger than near the surface regardless of the sensitivity of the particular [climate] model. This is, in fact, the signature (or fingerprint) of greenhouse warming. Stated somewhat differently, if we observe warming in the tropical upper troposphere, then the greenhouse contribution to warming at the surface should be between less than half and one third the warming seen in the upper troposphere."

Climatologists describe the effect of a greenhouse gas such as CO_2 in absorbing outgoing IR radiation as exerting a "forcing" at the top of the atmosphere. This forcing is a hypothetical heat flow downward measured in W/m^2. Hansen *et al.* (2000) have estimated the forcing on the climate due to changes in CO_2 concentration as shown in Figure 5.6.

Estimating the quantitative shape of the curve in Figure 5.6 is of great importance for understanding the effect of future CO_2 emissions on future climate change. In Figure 5.6, vertical lines represent:

A = typical CO_2 at glacial maximum in an ice age;

B = typical CO_2 during an interglacial period between ice ages;

C = current CO_2 level due to human impact on environment; and

D = CO_2 level after it doubles compared to pre-industrial levels. The estimated forcing are shown as vertical double arrows:

F1 = forcing in the transition from a glacial maximum to an interglacial period (~ 3 W/m²)

F2 = forcing due to CO_2 rise since before the industrial period to the present (~2 W/m²)

F3 = forcing due to change from pre-industrial levels to doubled CO_2 (~ 3.7 W/m²).

According to these calculations, the rise in CO_2 from pre-industrial times to the present has already produced about half the forcing that will result from doubling CO_2 from pre-industrial times. According to basic theory, when a forcing of ~ 3.7 W/m² is applied to the top of the atmosphere (via doubling CO_2 from ~280 to ~560 ppm) the Earth will warm until it radiates outward an additional 3.7 W/m² to compensate for the downward forcing. Simple radiative equilibrium requires that the Earth warm up approximately 1.2°C in this case. Hence, if there were no other changes, and the Earth remained exactly as before except for an increase in global average temperature, doubling CO_2 would produce a temperature increase of about 1.2°C.

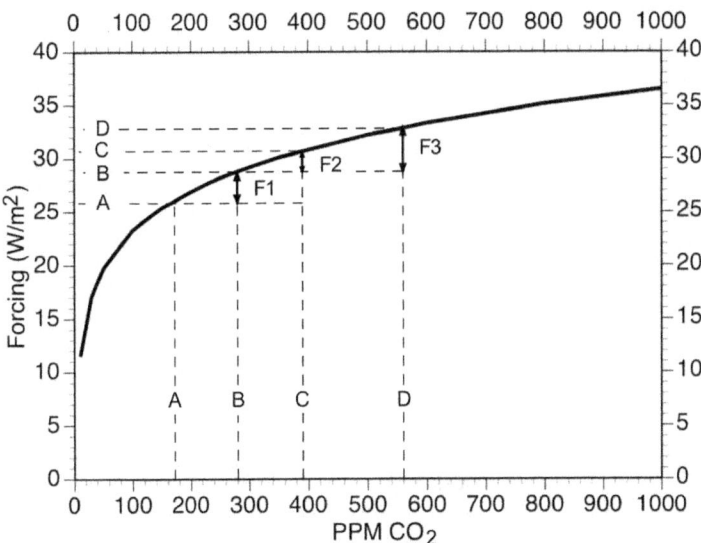

Figure 5.6. Estimated forcing of the climate due to changes in CO_2 concentration (Hansen, *et al.*, 2000).

However, if CO_2 rises from ~280 to ~560 ppm, various secondary consequences would undoubtedly occur. Some of these could change the heat balance of the Earth, producing additional temperature changes. These factors are called *feedbacks*. One feedback factor is that as the Earth warms, glaciers and ice sheets tend to retreat. Since these regions reflect incoming sunlight, the net effect will be additional solar energy absorbed by the Earth, resulting in additional warming. Another feedback factor is due to the fact that as the oceans warm, they will tend to evaporate more water vapor, and water vapor (like CO_2) is a greenhouse gas that can absorb IR radiation emitted by the Earth. Some climate modelers have assumed that as the Earth warms due to doubling of CO_2, the whole Earth will experience a uniform increase in water vapor content in the atmosphere, resulting in a significant amplification of the original warming effect of CO_2. Estimates of the combined effect of warming due to a doubling of CO_2 plus the effects of feedbacks range from a temperature rise of 2°C to 9°C. However, there remains great uncertainty in these estimates. While many estimates have been made, the consensus value often used is ~3°C. Like the porridge in "The Three Bears", this value is just right – not so great as to lack credibility, and not so small as to seem benign. Unfortunately, all of the estimates made to date by various procedures lack adequate data and require considerable speculation. The various estimates of increased humidity suffer from the facts that much of the increase in water evaporation will occur in the tropics where (i) the air is typically heavily laden with humidity and the water absorption bands are already saturated, and (ii) much of the heat transfer from the surface to the atmosphere is via upward cumulus convection, not by radiation. The only regions where increased humidity would provide a strong positive feedback would be desert areas, but if the Earth warms, these areas might become drier, not wetter. Another important factor is the effect of clouds. Several climatologists of the alarmist persuasion have written papers claiming that the effect of rising temperature due to increased CO_2 will be to reduce cloudiness, producing even more warming of the Earth by allowing more solar energy to reach the surface. However, there are many good reasons to believe that increasing evaporation will produce more (not less) clouds and will act contrary to the warming due to CO_2. Our knowledge of the Earth's cloudiness is very limited and this issue remains not understood to any reasonable degree.

The bottom line is that climatologists of the alarmist persuasion are convinced that the full effect of doubling CO_2 from ~280 to ~560 ppm will be a global temperature rise of around 3°C or more, whereas an honest examination of their analyses shows that they are quite specious and unreliable. Some warming due to CO_2 from ~280 to ~560 ppm is expected but it is not yet clear whether the net feedback is positive or negative.

5.3 CO_2 and Climate

The connection between CO_2 concentration and climate is uncertain.

Characterizing the Earth's climate is not a simple matter. The common approach taken by contemporary climatologists is to characterize the complex climate of the Earth using a single global average temperature (T_G) of the Earth's surface. We can state with some confidence that the Earth has warmed on average, roughly 0.7°C over the past 120 years, albeit not continuously, and not uniformly over all regions. Most of the warming occurred in the higher northern latitudes. Warming has not been in lockstep with rising CO_2 levels.

Climatologists have also attempted to estimate T_G using proxies over geological time periods as long as hundreds of millions of years. Proxies are indirect indicators of past temperature based on some natural process that occurred in the past that was dependent on temperature. Many proxies have been proposed and utilized. The proxies that are greatest value in estimating global temperatures over tens or hundred of millions of years are oxygen isotope ratios in benthic ocean sediments in which the ^{18}O concentration is an inverse measure of T_G. While the conversion of the direct signal $\delta^{18}O$ to T_G is only approximate so that absolute values are uncertain, the $\delta^{18}O$ measurements appear to be reliable and we have fairly good relative measures of how T_G varied over the past 500 million years. These data indicate that the Earth has undergone major climate changes over geologic time from a hothouse Earth where all polar and mountain glaciers were melted and the Earth basked in warmth at all latitudes, to an ice house Earth where polar glaciation extended down to low latitudes, possibly including equatorial zones.

Believing that every effect has a cause or causes, climatologists have searched for possible causes of long-term climate changes and inevitably, after eliminating all other candidates, they have assumed that variability of CO_2 concentration was the major factor that caused long-term climate changes over many millions of years:

> "The major transitions between climatic icehouse and greenhouse conditions are ultimately most probably driven by the deep Earth processes of plate tectonics, as a function of the long-term balance between CO_2 degassing at spreading centers and the conversion of atmospheric CO_2 to mineral carbon through long-term silicate weathering and oceanic carbonate formation." (NAS, 2011)

As Pierrehumbert (2009) said: "... the Urey reaction removes CO_2 from the atmosphere. When CO_2 dissolves in water, it forms a weak acid (carbonic acid), which reacts with silicate minerals (e.g. $CaSiO_3$) to form carbonate minerals (e.g. $CaCO_3$, or 'limestone'). The reaction takes place only in the presence of liquid water".

The argument goes (more or less): "If it wasn't CO_2, what else could it have been?" This argument has some merit. We can estimate from first principles the heating effect that rising CO_2 will produce in the atmosphere. If that was the only thing that occurred – that is, only the CO_2 concentration changed and no secondary effects took place– we would be able to predict the effect of changing CO_2 with some precision. The problem is that other effects take place as a consequence of the climate changes induced by changing CO_2, such as changes in humidity, cloudiness, winds, ocean currents, land cover, ice sheets and glaciers, etc., and these secondary changes may be of greater magnitude than the original stimulus of CO_2 change, and they are very difficult to predict.

It has occurred to a number of climatologists that perhaps by studying the past (tens of thousands of years ago to hundreds of millions of years ago) and finding relationships between CO_2 and climate during those periods, we might be able to obtain real world data on how climate and CO_2 are connected. This real world data will presumably have built into it all the secondary processes that take place. For example, we have very good data on CO_2 concentration during the Last Glacial Maximum, some 20,000 years ago, so that is one important historical point for further study. In addition, there are also a variety of estimates of CO_2 concentration that go back as far as 500 million years, but unfortunately such data are very scattered and do not appear to be very reliable. But climatologists are usually willing to derive a dollar's worth of conclusions from a penny's worth of data.

What we seek is a relationship between CO_2 concentration and the Earth's climate over long geological periods during which the CO_2 concentration varied over a wide range. There is evidence that the CO_2 concentration may have been well over 20,000 ppm in the distant past, and it has been as low as ~180 ppm only 20,000 years ago. It would be very nice if there were a single curve relating T_G to CO_2 concentration such as that shown in Figure 5.7. In that case, if we could find several points on the curve, we could attempt to map out a good portion of the curve.

However, over long time periods, the variation of T_G with CO_2 concentration depends on various factors such as the placement of the continents on Earth, the functionality of ocean currents, the past history of the climate, the orientation of the Earth's orbit relative to the Sun, the luminosity of the Sun, the presence of aerosols in the atmosphere, volcanic action, land clearing, biological evolution, etc. Hence, there is probably no single curve relating T_G to CO_2 concentration, but rather, a set of curves that depend on the above factors (see Figure 5.8). The holy grail of climatology is to seek an estimate of how T_G varies with CO_2 concentration over the range 280 – 560 ppm. To put this in perspective, we show this range in Figure 5.9. Not only do we not know *a priori* which curve applies to our present situation, but the vertical slice of greatest interest is a very narrow one in the total scheme of things.

Climatologists have mainly concentrated on the realm of CO_2 concentration between 280 ppm and 560 ppm, with some concern for higher concentrations up to ~ 900 ppm. In this regard, we can magnify the grey slice from Figure 5.9, and combine this with known values of T_G over the past ~120 years, as shown in Figure 5.10. Curves 1 to 4 show various estimates of the temperature rise that will be induced by further increases in CO_2 concentration. In this figure, the pre-industrial climate pertains to that prevailing prior to the mid 19th century. (Of course, this climate varied with time and therefore it is not quite right to represent it as a single point). The current climate is also shown. Curve 1 represents a rough estimate by climate models of how the climate would change if there were no secondary feedbacks. Curve 3 represents the consensus view of alarmist climatologists based on climate models with feedbacks. Curves 2 and 4 are arbitrarily drawn.

Over the relatively small range of CO_2 concentrations in Figure 5.10, climatologists tend to assume that the relationship between T_G and CO_2 concentration can be approximated as linear. The holy grail is how much T_G increases when CO_2 goes from 280 to 560 ppm.

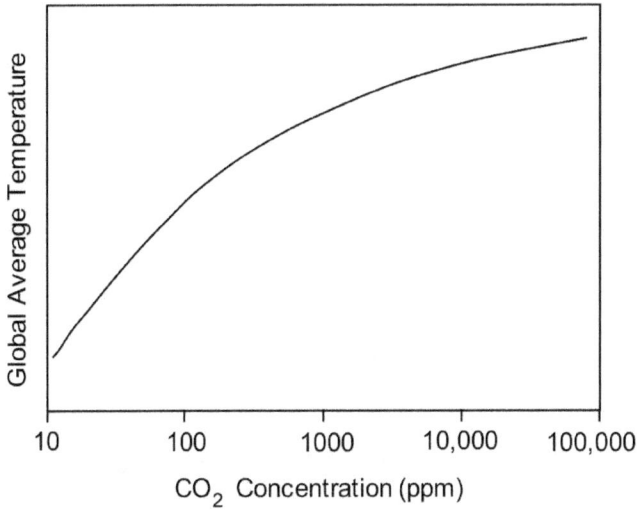

Figure 5.7. Hypothetical single curve relating T_G to CO_2 concentration.

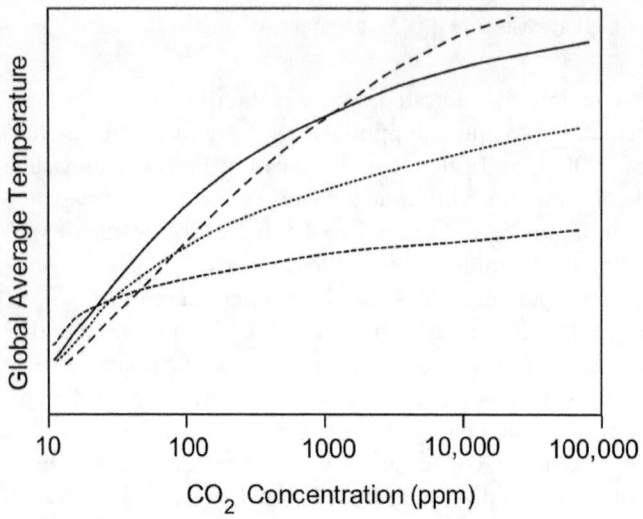

Figure 5.8. Hypothetical curves relating T_G to CO_2 concentration.

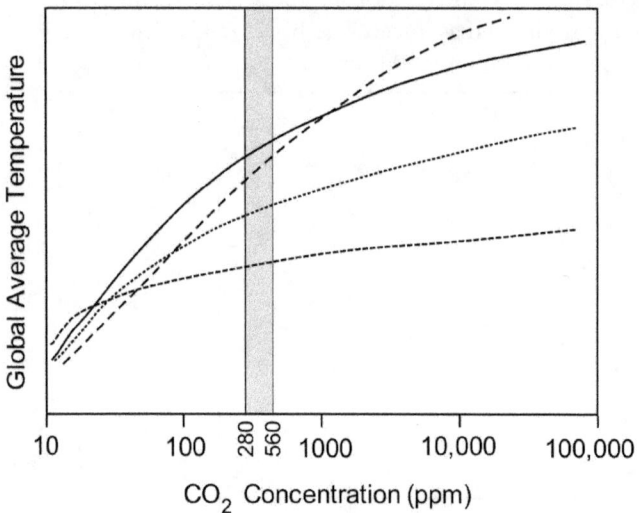

Figure 5.9. Range of CO_2 concentration (280 – 560 ppm) for 21st century climate change (grey slice).

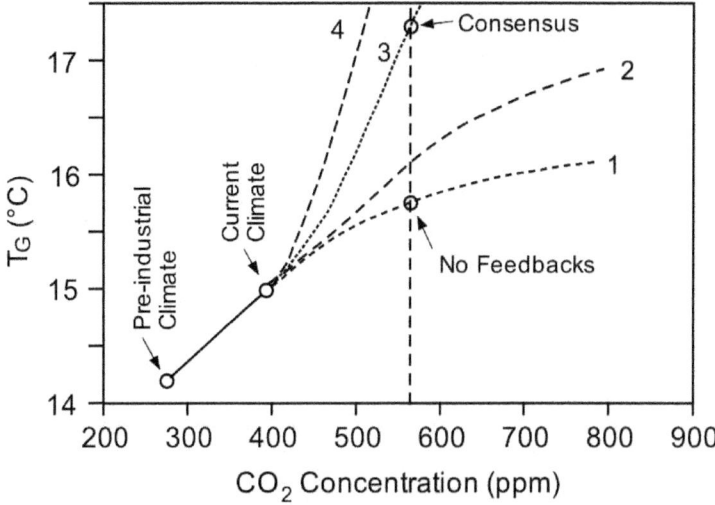

Figure 5.10. Hypothetical variation of T_G with CO_2 concentration.

5.4 CO₂ and Ancient Climates

While our principal interest is in the climate change induced in the 21st century by increasing the CO_2 concentration from 280 to 560 ppm, multiple efforts by many climatologists cannot seem to overcome the uncertainties inherent in performing this analysis, and their models still lack credibility and consistency. It has therefore occurred to a number of climatologists that perhaps by studying the past (tens of thousands of years ago to hundreds of millions of years ago) and finding relationships between CO_2 and climate during those periods, we might be able to obtain real world data on how climate and CO_2 are connected. This real world data will presumably have built into it all the secondary processes that take place. For example, we have data on CO_2 concentration during the Last Glacial Maximum, some 20,000 years ago, so that is one important historical point for further study. In addition, there are also a variety of estimates of CO_2 concentration that go back as far as 500 million years, but unfortunately such data are very scattered and do not appear to be very reliable.

What we seek is a relationship between CO_2 concentration and the Earth's climate over long geological periods during which the CO_2 concentration varied over a wide range. There is evidence that the CO_2 concentration may have been well over 20,000 ppm in the distant past, and it has been as low as ~180 ppm only 20,000 years ago. It would be very nice if there were a single curve relating T_G to CO_2 concentration such as that shown in Figure 5.7. In that case, if we could find several points on the curve, we could attempt to map out a good portion of the curve.

However, over long time periods, the variation of T_G with CO_2 concentration depends on various factors such as the placement of the continents on Earth, the functionality of ocean currents, the past history of the climate, the orientation of the Earth's orbit relative to the Sun, the luminosity of the Sun, the presence of aerosols in the atmosphere, volcanic action, land clearing, biological evolution, etc. Hence, there is probably no single curve relating T_G to CO_2 concentration, but rather, a set of curves that depend on the above factors (see Figure 5.8).

About 20,000 years ago, the most recent ice age was at its maximum extent with gigantic ice sheets in the higher latitudes of the Northern Hemisphere. There is reliable evidence from ice cores that the CO_2 concentration at that time was roughly 170 – 180 ppm. The first question is what was T_G at the LGM? A number of estimates have been made by various investigators. Taylor *et al.* (2001) carried out an analysis in which they took into account the reduced CO_2 concentration and the extended ice sheets of the LGM in climate models to estimate the amount of cooling at the LGM compared to pre-industrial times. Using six different climate models, they obtained values of 3.5, 3.7, 3.8, 4.4, 5.2 and 5.9, for an average of 4.4°C. Crucifix (2006) provided a less optimistic view of the precision to which this is known: "The global temperature change is therefore is estimated to be comprised between 3°C and 9°C with 95% confidence". He also estimated that the tropical ocean sea-surface temperature decreased between 1.7 and 2.7°C, and Antarctic surface air temperature decreased by 7 to 11°C at the LGM compared to pre-industrial times. Hansen and Sato (2011) also estimated ΔT_G relying heavily on the paper by Zachos *et al.* (2001). Hansen and Sato concluded that the global average temperature change from the LGM to recent pre-industrial times was 5°C. However it should be noted that Hansen and Sato's Figure 1c indicates a deep ocean temperature change from the last glacial maximum (LGM) to recent pre-industrial times of ~3°C, which, after applying their "2/3 rule" would indicate a global average temperature change from the last glacial maximum to recent pre-industrial times of $3/2 \times 3 = 4.5$°C, rather than 5°C. (They estimated that the ratio of deep ocean temperature change to global average temperature change was 2/3). Shakun and Carlson (2010) carried out an extensive review of the LGM–Interglacial transition. They found, as expected, that the ΔT in this transition varied with latitude as shown in Figure 5.11. Their estimate of ΔT_G (the temperature at the LGM minus the pre-industrial temperature) was – 4.5°C. If we couple the temperature during the LGM (14.3°C – 4.5°C = 9.8°C) with an estimated CO_2 concentration of 170-180 ppm, we can plot a point representing the LGM, as shown in Figure 5.12.

Figure 5.12 shows the following:

A = Conditions at the LGM

B = Conditions in pre-industrial times
C = Consensus projection temperature change for doubling CO_2
D = Computer model projection for doubling CO_2 with no feedbacks
E = Present conditions
F = Range of climate model predictions including feedback
Curve 1 = Path we seem to be on
Curve 2 – Transition from LGM to pre-industrial times

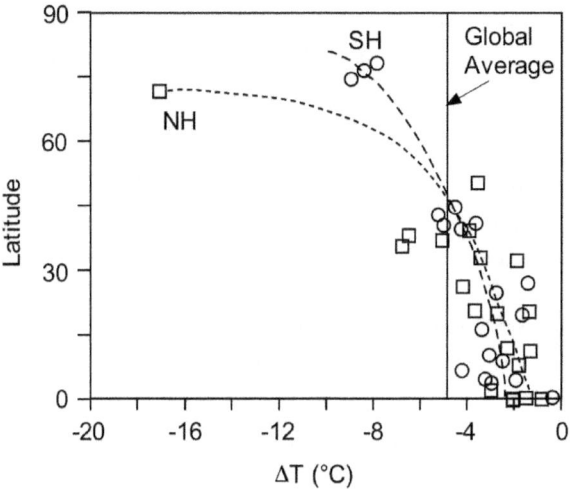

Figure 5.11. Variation of ΔT_G with latitude. (Shakun and Carlson, 2010).

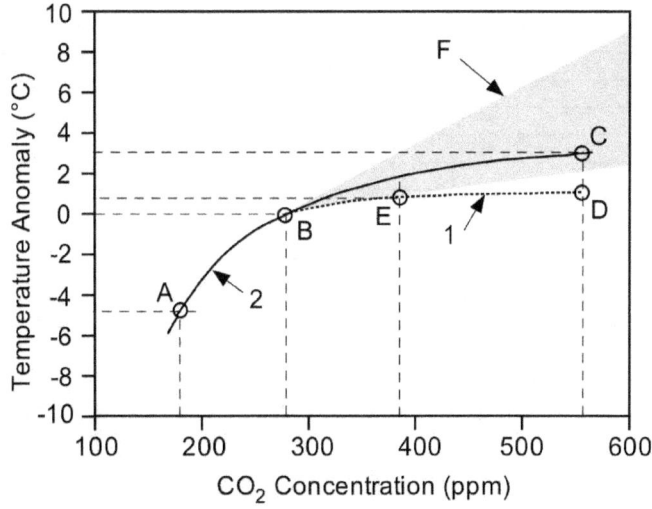

Figure 5.12. Inclusion of LGM point in relationship between CO_2 concentration and T_G.

The reason why the LGM point lies so low is because important changes took place on the surface of the Earth as the ice sheets expanded. These changes go beyond the purely spectroscopic effect of less absorption of IR by CO_2 in the atmosphere as the CO_2 concentration was lowered to below 200 ppm. The growth of large ice sheets from recent pre-industrial times to the LGM resulted in an increase in the Earth's albedo across the ice sheets, as well as for mountain glaciers. In addition, the drop in sea level moved shorelines outward, converting ocean to land, thereby further increasing the Earth's albedo. As the climate got colder, biomass and vegetation grew less abundantly, increasing the Earth's albedo still further. Undoubtedly, there were other effects as well (humidity, cloudiness, dust, ...). Thus, lowered CO_2 can only have contributed a fraction of the temperature reduction at the LGM. Kohler *et al.* (2009) said: "Although water vapor is the most important GHG, the following compilation does not consider any changes in water vapor in the past due to missing constraints on its variability". In other words, they more or less said: *Water vapor may be the biggest factor, but since we have no data on it, we will neglect it!*

Table 5.1. Parameters for analyzing LGM – pre-industrial transitions. Forcings are in W/m^2. Blank elements are not available. Elements with dashes represent items that were not included.

	Chylek & Lohmann (2008)	Kohler et al. (2009)	Hansen & Sato (2011)	Hansen & Sato (modified)
CO_2 forcing	2.4	2.1	2.25	3.0
CH_4 forcing	0.27	0.4	0.43	
N_2O forcing	–	0.3	0.32	
Total GHG forcing	2.67	2.8	3.0	3.7
Land Cryosphere		4.54		
Land ice		3.17		
Sea ice		0.55		
Snow cover		0.82		
Sea Ice		2.13		
Sea ice - north		0.42		
Sea ice - south		1.71		
Vegetation		1.09		
Total Albedo	3.5	7.76	3.5	3.5
Dust/Aerosols	3.2	1.88		1.0
Water vapor, lapse rate and clouds	–	–	–	
Total Forcing	9.4	12.43	6.5	8.2
Climate Sensitivity	0.48	0.36	0.69	0.55
Implied ΔT due to doubling CO_2 from 280 to 560 ppm (°C)	1.8	1.3	2.6	2.0

Several groups have attempted to estimate the sensitivity of the climate to a net forcing by comparing the LGM with recent pre-industrial times, estimating the forcings due to various climatic changes, and attributing the

4.5°C global temperature change to the estimated forcing. Table 5.1 provides the estimates by three groups. The Hansen and Sato (2011) estimate was modified by increasing the forcing due to CO_2 to 3.0 W/m² and the forcing due to dust by 1 W/m². If the estimates of total forcings are compared to the temperature change of 4.5°C, the bottom row provides the climate sensitivity (°C per W/m²).

Rapp (2010) provided a review of estimates of CO_2 concentration and climate for various historical time periods (3-5 million years ago, the past 20 million years, initiation of Antarctic glaciation 34-33 million years ago, peak warming around 40 million years ago, 40-60 million years ago, 100 to 300 million years ago, and 300-500 million years ago). His conclusions were as follows:

The widely held view amongst geologists and climatologists alike, is that the primary cause of long-term climate changes was variability of CO_2 concentration due to long-term imbalances between CO_2 degassing at spreading centers and the conversion of atmospheric CO_2 to mineral carbon through long-term silicate weathering and oceanic carbonate formation. The argument goes (more or less): "If it wasn't CO_2, what else could it have been?" Foster *et al.* (2009) described this as the "accepted paradigm" that requires CO_2 to vary in unison with global temperature. Thus, paleoclimatologists have been trying for decades to establish a relationship between climate and CO_2 concentration over many millions of years. There is some evidence that over many millions of years, higher CO_2 concentrations are often, but not always associated with warmer climates. However, there is a great deal of scatter in the CO_2 proxy data, and this relationship is difficult to pin down quantitatively. Royer (2010) began his commentary with the statement:

"Global temperatures have covaried with atmospheric carbon dioxide (CO_2) over the last 450 million years of Earth's history. Critically, ancient greenhouse periods provide some of the most pertinent information for anticipating how the Earth will respond to the current anthropogenic loading of greenhouse gases. Paleo-CO_2 can be inferred either by proxy or by the modeling of the long-term carbon cycle. For much of the geologic past, estimates of CO_2 are consistent across methods."

This seems to be a rather optimistic view, considering the data from his paper show very large scatter. This argument seems to make some sense from 120 million years ago to ~90 million years ago. Yet, there are difficulties from 70 million years ago to 50 million years ago when the sea temperature remained high, yet the CO_2 concentration appears to have been much lower. In any event, the extreme scatter in the data do not convey confidence that any valid conclusions can be drawn. Despite the lack of precise data, the close relationship between CO_2 concentration and global average temperature is widely accepted. Such beliefs based on inadequate data are the hallmark of climatology.

The best chance to use paleoclimatic data to infer climate sensitivity is probably the last glacial maximum (LGM) some 20,000 years ago, when the total negative forcing produced a decreaes in global average temperature of roughly 4.5°C. We discussed these analyses in previous paragraphs. The contribution of the diminution of CO_2 at the LGM to the total cooling was estimated by these studies to be in the range 16% to 33%. While it seems likely that solar input to higher latitudes triggered the cycles, the variability of CO_2 concentration likely played a part in determining the extremity of the temperature cycle that resulted from this trigger. The changes in CO_2 concentration between glacial maxima and interglacials (~180 ppm to ~ 280 ppm) are well documented in ice core records, although no one seems to have a satisfactory explanation for why the CO_2 concentration changed this much (simple solubility in the oceans does not suffice). However, the estimates of forcings, particularly due to dust, vary considerably from investigator to investigator and it is difficult to pin down the climate sensitivity to CO_2 change. Moreover, changes in humidity and cloudiness are unknown and may be very large factors. There are good estimates available of the global average temperature and the CO_2 concentration at the Last Glacial Maximum (LGM) 20,000 years ago, and if these data are compared with values in the pre-industrial era (a few hundred years ago) one can thereby estimate the sensitivity of the climate to CO_2 concentration over the range ~ 180 ppm to ~ 280 ppm. Using this estimated climate sensitivity, one can then estimate the global average temperature rise in going from 280 ppm to 560 ppm. The various investigators have come up with a range of projections. It is noteworthy that this range of estimates for the real world ΔT due to doubling CO_2 from 280 ppm to 560 ppm is from ~1°C to ~3°C. However, as we stated above, these estimates do not take into account possible differences in humidity and cloudiness.

The data on temperature and CO_2 over hundreds of millions of years are far less reliable, and conclusions drawn from these time periods are dubious at best.

Our conclusion here is that CO_2 is probably one of several major factors in long-term climate change, but other factors such as the placement of the continents on Earth, the functionality of ocean currents, the past history of the climate, the orientation of the Earth's orbit relative to the Sun, the luminosity of the Sun, the presence of aerosols in the atmosphere, volcanic action, land clearing, biological evolution, internally generated changes, etc. Hence, there is probably no single curve relating global average temperature to CO_2 concentration, but rather, a set of curves that depend on the above factors.

5.5 Climate Models for the 21st Century

5.5.1 The Good, the Bad and the Ugly

Global climate models divide the atmosphere into many small three-dimensional cells with typically 10 to 20 vertical layers and 50,000 to 100,000 horizontal cells. Land areas are divided into surface cells, and oceans are also divided into multiple vertical layers combined with many horizontal cells. Within each atmospheric cell, the various parameters such as temperature, humidity, barometric pressure, and wind velocity are uniform, but are updated frequently, typically every 30 minutes. Each cell interacts with its neighbors according to physical laws that are expressed as mathematical equations. Solar energy impinges from above, and radiation is emitted outward. The rotation of the Earth is also taken into account. Atmospheric models calculate the state variables of the atmosphere, such as temperature, pressure, humidity, kinetic energy, etc., as a function of space and time. The set of model equations is formulated by using geophysical fluid dynamics theory and physical laws governing the exchanges of the mass and energy. The atmosphere is divided into discrete vertical layers, which are then overlaid with a two-dimensional horizontal grid, producing a three-dimensional mesh of grid elements. Models differ in spatial resolutions and configuration of model grids. Typical models have spatial resolution of 200 kilometers in the horizontal direction, and 20 vertical levels below the altitude of 15 km. Clouds reflect solar radiation to space, cooling the Earth–atmosphere system. Clouds also trap infrared radiation, keeping the Earth warm. The net effect depends on the height, location, microphysical and radiative properties of clouds, and their appearance in time with respect to the seasonal and diurnal cycles of incoming solar radiation. Cloud feedback refers to the changes in cloud amounts and properties that can either amplify or moderate a climate change. Uncertainties of cloud feedbacks in climate models have repeatedly been identified as the leading source of uncertainty in model-derived estimates of climate sensitivity. The fidelity of cloud feedbacks in climate models is therefore important to the reliability of their prediction of future climate change.

Global climate models have been used to extrapolate global temperatures backward in time as much as 1,000 years. Global climate models have also been used extensively to try to explain the underlying physical reasons for the global climate changes that we have experienced over the past century or so. Global climate models have been used to analyze the climatic effects of volcanoes. However, the main purpose of global climate models is typically to predict the future climate of the Earth and how it depends on future scenarios for greenhouse gas emissions.

The IPCC presented a rather positive view of GCMs:

"Climate models are based on well-established physical principles and have been demonstrated to reproduce observed features of recent climate and past climate changes. There is considerable confidence that Atmosphere–Ocean General Circulation Models (AOGCMs) provide credible quantitative estimates of future climate change, particularly at continental and larger scales. Confidence in these estimates is higher for some climate variables (e.g., temperature) than for others (e.g., precipitation)" (IPCC, 2001).

The IPCC Report tended to emphasize the improvements that have been made in models, and provided a mostly upbeat evaluation.

"There is considerable confidence that climate models provide credible quantitative estimates of future climate change, particularly at continental scales and above. This confidence comes from the foundation of the models in accepted physical principles and from their ability to reproduce observed features of current climate and past climate changes. Confidence in model estimates is higher for some climate variables (e.g., temperature) than for others (e.g., precipitation). Over several decades of development, models have consistently provided a robust and unambiguous picture of significant climate warming in response to increasing greenhouse gases" (IPCC, 2001).

This is rather a strange statement for several reasons. One is that it is difficult to see how one can conclude that models "provide credible quantitative estimates of future climate change" because the future is yet to occur and at best, one would have to extrapolate from past success. But have the models been successful in the past? The IPCC seems to think so. The IPCC Report claims:

"Models are routinely and extensively assessed by comparing their simulations with observations of the atmosphere, ocean, cryosphere and land surface."

The IPCC Report claims that models are able "to reproduce features of past climates and climate changes."

However, we must distinguish between use of models with adjustable parameters that can be tweaked to approximate trends discerned from proxies, as opposed to *a priori* predictions of trends without one eye on a known result. Thus, models tend to fall into the syndrome of models that can explain everything and predict nothing.

The fact that the models are based on "accepted physical principles" is a *necessary* requirement but not necessarily *sufficient* to assure that models deal adequately with the complexities of the Earth's climate.

On the positive side, GCMs provide a basic framework for modeling the Earth's climate, and although these models are still rather primitive in many respects (clouds, aerosols, rain, inadequate spatial resolution, land use, lack of consideration of regional variations in humidity, etc.) these are all issues that are amenable to improvement in the future, and such improvements can be incorporated into the frameworks of models that have been developed. Various climate models have produced a wide range of results. This lack of convergence suggests considerable uncertainty. Eventually, the models may begin to converge. However, even after the various models converge, will they converge to the right answer?

The majority of studies with climate models have addressed the question of how much the future global temperature will rise as a result of a putative future doubling of carbon dioxide concentration in the atmosphere (doubling compared with the pre industrial level of ~280 ppm). The equilibrium response, the response expected if one waits long enough (several hundred years) for the system to re-equilibrate, is the most commonly quoted measure. The range of equilibrium climate sensitivity to doubling of CO_2 predicted by various models is typically 1.5°C to 4.5°C, although some models predict much higher values. The difficulty in simulating the Earth's clouds and their response to climate change are given as the major reason it has proven difficult to reduce the range of uncertainty in model-generated climate sensitivity. Uncertainty still remains considerable and is not decreasing rapidly, due in part to the difficulty of cloud simulation but also to uncertainty in the rate of heat uptake by the oceans and a variety of other factors.

One major problem with GCMs is that it is difficult to test them against actual data. While some models claim to reproduce measured temperatures in the 20th century, these seem to be contrived after the fact, with forcings chosen to obtain agreement. Such methods explain everything and predict nothing. Hoyt (2006) pointed out that the global temperature rise of ~0.6°C during the 20th century was lower than would be predicted by GCMs based on the known increase in CO_2 concentration. However, other factors (aerosols, land use, etc.) may have affected 20th-century temperatures significantly.

"Uncertainties in the climatic effects of man-made aerosols (liquid and solid particles suspended in the atmosphere) are a major stumbling block in quantitative attribution studies and in attempts to use the observational record to constrain climate sensitivity. We do not know how much warming due to greenhouse gases has been cancelled by cooling due to aerosols. Uncertainties related to clouds increase the difficulty in simulating the climatic effects of aerosols, since these aerosols are known to interact with clouds and potentially change cloud radiative properties and cloud cover" (CCSP, 2007).

Haerter *et al.* (2009) analyzed the uncertainty in climate models due to aerosols and found that the uncertainties were large.

Atmospheric carbon dioxide has increased ~30% over the industrial period. Radiative forcing of climate change by increased concentrations of CO_2 and other long-lived greenhouse gases has been estimated to be 2.4 ± 0.2 W/m^2 relative to the pre-industrial era. However, simple models predict that a doubling of CO_2 would result in a forcing of about 3.7 W/m^2. The most recent IPCC Report estimated that this would produce a temperature rise in the range 1.5 to 4.5°C, based on a temperature sensitivity factor in the range 0.38 to 1.13 °C/(W/m^2). The factor-of-three uncertainty in present estimates is certainly unacceptably large for planning for mitigation or adaptation (Schwartz, 2003).

The increase in global mean temperature to date, relative to the pre-industrial era has been estimated to be 0.6 ± 0.2°C, suggesting a sensitivity of 0.25 ± 0.09 °C/(W/m^2), well below the low end of the IPCC range. But such an empirical estimate assumes that climate response is near equilibrium and that forcing is due entirely to long-lived greenhouse gases. The equilibrium assumption is claimed to be valid. However forcings other than by greenhouse gases, particularly the cooling influence of anthropogenic aerosols due to their scattering of solar radiation, are thought to offset much of the greenhouse gas forcing on a global basis, resulting in a much lower total forcing and consequently much greater sensitivity. Present uncertainty in total forcing is so great as to preclude a meaningful empirical estimate of climate sensitivity from the temperature record and forcing over the industrial period (Schwartz, 2003).

Schwartz (2004) concluded:

"The sensitivity of global mean temperature change to an increase in atmospheric carbon dioxide (CO_2) is not well established. The complexity of the climate system precludes calculation of the response of Earth's climate to a change in a radiative flux component (forcing) from well-established physical laws. Consequently, determination of global climate sensitivity is a subject of intense research. This work is reviewed from time to time by pertinent national and international bodies. One such landmark review was that of a 1979 National Research Council panel, which concluded: 'We estimate the most probable global warming for a doubling of CO_2 to be near 3°C, with a probable error of 1.5°.' More recently, the IPCC concluded that 'Climate sensitivity [to CO_2 doubling] is likely to be in the range 1.5–4.5°C.' These estimates must be considered somewhat subjective. They are based mainly on calculations with climate models constrained, especially for the IPCC estimate, by observation of the extent of warming over the industrial period and concurrence of modeled and observed warming. Neither the Charney panel nor the IPCC quantitatively specified the meaning of their uncertainty bounds, but in the case of the Charney estimate, a National Research Council panel three years later expressed its understanding that 'the Charney group meant to imply a 50% probability that the true value would lie within the stated range.' Remarkably, despite some two decades of intervening work, neither the central value nor the uncertainty range has changed. The large uncertainty range, a factor of 3, in present estimates of climate sensitivity renders such estimates not particularly useful from the perspective of developing policy regarding either reduction of greenhouse gas (GHG) emissions or adaptation to a new, increasingly warm climate."

Schwartz, Charlson, and Rodhe (2007) reviewed the latest (2007) IPCC assessments of climate change based on results from "an ensemble of 58 runs with 14 climate models" and concluded:

"[IPCC] estimates [of] total anthropogenic forcing [were] 0.6 to 2.4 W/m² (95% confidence range). This factor of four range greatly limits the ability to evaluate the 'skill' of climate models in reproducing past temperature changes and to infer climate sensitivity from observed change because a given temperature increase might result from a large forcing and low climate sensitivity or alternatively from a small forcing and high climate sensitivity."

By comparison, the range of predicted temperature rise (0.4°C to 0.8°C) from 1910 to 2000 was only a factor of 2. Schwartz, Charlson, and Rodhe (2007) raised questions whether the temperature estimates spanned the full range of potential forcing, and concluded:

"The narrow range of modeled temperatures gives a false sense of the certainty that has been achieved."

Schwartz, Charlson, and Rodhe (2007) never raised the question as to whether the purveyors of GCMs "fudged" their models with one eye on the known (or at least the believed) 0.6°C temperature rise from 1910 to 2000. However, this seems to be a likely possibility.

Meehl *et al.* (2007) provided an upbeat report on climate models, saying that projections of the future warming pattern are "robust." However, Stainforth *et al.* (2007) discussed the credibility of atmosphere/ocean global circulation models. They discussed two aspects of model credibility: uncertainty and inadequacy. Model uncertainty deals with uncertainty as to which parameterizations to use and which values of the parameters are best. "Model inadequacy captures the fact that we know *a priori*, there is no combination of parameterizations, parameter values and initial conditions which would accurately mimic all relevant aspects of the climate system." Finally, they concluded:

"Complex climate models, as predictive tools for many variables and scales, cannot be meaningfully calibrated because they are simulating a never before experienced state of the system; the problem is one of extrapolation. It is therefore inappropriate to apply any of the currently available generic techniques which utilize observations to calibrate or weight models to produce forecast probabilities for the real world. To do so is misleading to the users of climate science in wider society."

McWilliams (2007) concluded: "climate models are structurally unstable in various ways that are not yet well explored, and this implies a level of irreducible imprecision in their answers that is not yet well estimated."

Rind et al. (2008) reviewed the uncertainties in climate models with particular emphasis on the ability of models to predict relative climate sensitivities in the tropics and at high latitudes. They concluded that little progress has been made in resolving latitudinal variations of climate sensitivity, even after thirty years of model development.

Huybers (2010) provided a very insightful review of climate models. He pointed out that Schwartz et al. (2007) identified "an important interdependence between the radiative forcing and climate sensitivity across the CMIP3 models" in that, "while twentieth-century changes in radiative forcing differs by a factor of 4 across the models, the resulting temperature spread differs by only a factor of 2". He also pointed out that Kiehl presented evidence that "this narrow temperature range results from an anti-correlation between radiative forcing and climate sensitivity". He then suggested that:

"Inter-model compensation between climate sensitivity and radiative forcing underscores that the models are not based purely on theory but are also conditional upon observations and, possibly, expectations".

Huybers (2010) showed that the treatment of clouds was the "principal source of uncertainty in models". Indeed, his Table I shows that whereas the response of the climate system to clouds by various models varied from 0.04 to 0.37 (a wide spread), the variation of net feedback from clouds varied only from 0.49 to 0.73 (a much narrower relative range). He then examined several possible sources of compensation between climate sensitivity and radiative forcing. He concluded:

"Model conditioning need not be restricted to calibration of parameters against observations, but could also include more nebulous adjustment of parameters, for example, to fit expectations, maintain accepted conventions, or increase accord with other model results. These more nebulous adjustments are referred to as 'tuning'."

He suggested that one example of possible tuning is that "reported values of climate sensitivity are anchored near the 3±1.5°C range initially suggested by the ad hoc study group on carbon dioxide and climate (1979) and that these were not changed because of a lack of compelling reason to do so".

Huybers (2010) went on to say:

"More recently reported values of climate sensitivity have not deviated substantially. The implication is that the reported values of climate sensitivity are, in a sense, tuned to maintain accepted convention."

Additional quotations from this paper are given below:

"Although substantial changes to GCM cloud parameterizations have been implemented since 1990, it is not clear that a general increase in their accuracy is the sole explanation for the present trend toward convergence. It may be that current models are producing similar errors, while the earlier models produced different errors".

"Tuning climate sensitivity to lie within the observed spread across the CMIP3 models is a sufficient explanation for the origins of the compensation between [clouds] and the other feedbacks".

"The covariance between the CMIP3 model feedbacks may be symptomatic of the uneven treatment of outlying model results".

"The specter of tuning leading to a curtailment of the inter-model spread in climate sensitivity is difficult to dismiss".

"Convergence between model results, if not truly driven by a decrease in model uncertainty or clearly understood as a result of calibration, could have the unfortunate consequence of lulling us into too great a confidence in model predictions or inferences of too narrow a range of future climates. To the extent that it occurs, tuning the models based on expectation or convention renders the modeling process a partially subjective exercise from which it is very complicated to derive a statistical interpretation".

"Focusing on maximally inconsistent possibilities seems more likely to lead to scientific discoveries and to uncover climate surprises. A maximally inconsistent ensemble of state-of-the-art model realizations would also have the advantage of suggesting outer bounds upon the range of climate sensitivity and, therefore, be complimentary to existing estimates".

Translated into simple terms, the implication is that climate modelers have been heavily influenced by the early (1979) estimate that doubling of CO_2 from pre-industrial levels would raise global temperatures $3\pm1.5°C$. Modelers have chosen to compensate their widely varying estimates of climate sensitivity by adopting cloud feedback values countering the effect of climate sensitivity, thus keeping the final estimate of temperature rise due to doubling within limits preset in their minds. Had they not done this, the spread in estimates of temperature rise would be much greater. Thus, they have imposed their preconceived notions of the expected temperature rise on the models to make them come out "right". As we stated previously, this is like the Three Bears children's story where the pprridge was not too hot or too cold; the canonical 3°C temperature rise is large enough to be alarming, but small enough to be be credible.

A fundamental characteristic of global climate models is that they predict a positive feedback due to water vapor and clouds that adds to the predicted temperature rise due to increasing CO_2 concentration. Most of this positive feedback is due to trapping of outgoing long wave radiation, and this increases the climate sensitivity to rising CO_2 above that due solely to CO_2. Pinning down a good estimate of the feedback, and hence to the climate sensitivity is one of the most important things needed in climate science. As we discussed previously in Section 2.4.2, Lindzen and Choi (2011) found that feedbacks were primarily negative, resulting in relatively low climate sensitivity. This is contrary to the alarmist position that feedbacks are positive leading to higher climate sensitivity (and therefore a greater increase in global temperature as greenhouse gas concentrations increase).

Roe and Baker (2007) pointed out that if one plots a histogram of number of estimates vs. estimated equilibrium climate sensitivity (to doubling of CO_2), one obtains a function that rises sharply above 2°C, peaks around 3°C, and has a long tail that extends out beyond 10°C. They then posed the question: "Why is uncertainty not diminishing with time?" To answer this, they provided the following model. A reference climate system is perturbed by a forcing, ΔR. After equilibrium is achieved, the change in system temperature is ΔT (upper part of Figure 5.13). In the case of doubling CO_2, many models would agree (within a moderate range) that

$$\Delta T_o = K \, \Delta R$$

where R = the forcing due to doubling of CO_2 (R ~ 4 W/m²) and K is the climate sensitivity to CO_2 doubling without feedback $\{K \sim 0.3 \, [°C/(W/m^2)]\}$. Thus, they would calculate that $\Delta T_o \sim 1.2$°C without feedback.

Any number of feedbacks may operate in the system to impose additional forcings (positive or negative) that are proportional to ΔT (lower part of Figure 5.13). These feedbacks enter the equation as follows:

$$\Delta T = K \, (\Delta R + c_1 \, \Delta T + c_2 \, \Delta T + \ldots)$$

where ΔT is different from ΔT_o due to the feedback. This can be rearranged as

$$\Delta T = K \, (\Delta R)/(1 - c_1 \, \Delta T - c_2 \, \Delta T - \ldots)$$

and if we define the feedback factor as

$$f = c_1 \, \Delta T + c_2 \, \Delta T + \ldots$$

we obtain

$$\Delta T = \Delta T_o/(1 - f) = G \, \Delta T_o$$

where the "gain" is $G = 1/(1-f)$. The literature contains a number of estimates of f, and therefore, G. As Roe and Baker (2007) pointed out, if one plots ΔT vs. f, the curve rises steeply when f exceeds about 0.7 ($\Delta T \sim 3°C$) – see Figure 5.14. There are a number of estimates of f, each presumably with some uncertainty function that describes the unknowns in the model, as in function A in Figure 5.14 (drawn for a most probable estimate of $f \sim 0.65$). This corresponds to an uncertainty distribution in ΔT as shown as function B in Figure 5.14. There is a long tail extending to high temperatures. The nature of feedbacks was further elaborated by Roe (2009).

Roe and Baker (2007) thus pointed out that if feedback is small, one can estimate the ΔT that results from the feedback quite precisely, but if the feedback is great, the uncertainty in ΔT rises sharply. Many climate models predict feedback factors f in the range 0.6 to 0.7, whereas Lindzen and Choi (2009) found a much lower value.

Roe and Baker (2007) then went on to statistically analyze the probability that f lies within various bounds, based on the various climate model estimates that have been made. In doing this, they make an implicit assumption (not stated) that all the climate models are basically correct, but they approximate uncertain parameters in different ways. One can then arrive at probabilities that ΔT lies within certain bounds. However, if (as seems likely to this writer) none of the climate models are credible, then statistical correlation of their results is not meaningful. It is a simple case of "GIGO".

The 2007 paper was followed by Baker and Roe (2009) in which the authors defined *climate sensitivity* as "the equilibrium response of the global ... surface air temperature to a doubling of carbon dioxide over preindustrial values" and distinguished this from the "transient climate change" that is delayed and damped by ocean heat uptake. Figure 5.14 is implicitly based on an assumption that the CO_2 concentration is instantly doubled in a step function, and the change in temperature is the long-term equilibrium response to this change. The 2009 paper explored the transient temperature change along the way toward the new equilibrium. They concluded that the transient response is more predictable than the ultimate equilibrium response. However, they did not treat the realistic case of a gradually evolving rise in CO_2.

Reacting to doubt expressed by alarmists, Roe and Armour (2011) defended the asymmetrical nature of the dependence of ΔT on f.[41]

---

[41] Also note: "Comment on 'Another look at climate sensitivity' by Zaliapin and Ghil (2010)" *Nonlin. Processes Geophys.*, **17**, 1–3, 2010.

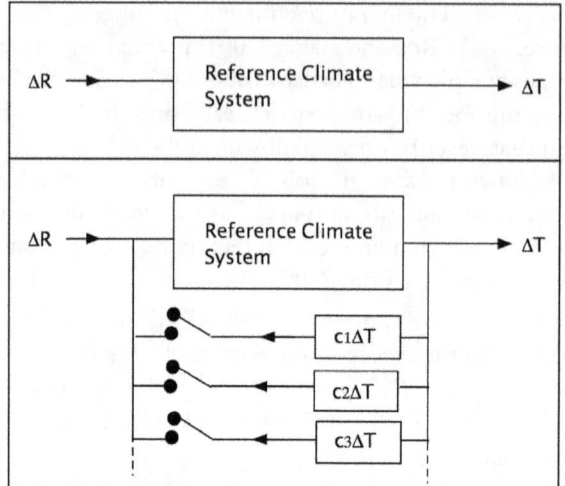

Figure 5.13. Model used by Roe and Baker (adapted from Roe and Baker, 2007).

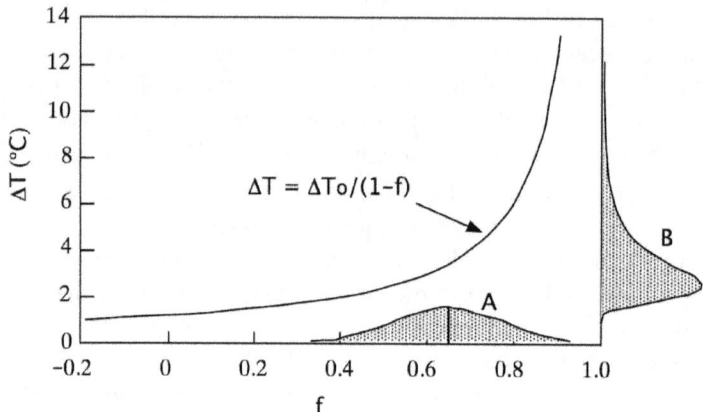

Figure 5.14. Dependence of ΔT on f. For any estimate of f with its probability distribution function (A), there is a corresponding distribution of ΔT shown as function B (adapted from Roe and Baker).

(yosemite.epa.gov/ee/epa/eerm.nsf/vwAN/EE-0564.../EE-0564-117.pdf)

Roe distinguished between *climate sensitivity, transient climate response* (see above) and *climate commitment* – which is "is a measure of the climate change we already face because of emissions that have already occurred" (the implicit climate change yet to come from past emissions as the transient response gradually approaches a new equilibrium)

Armour and Roe (2011) discussed climate commitment. These authors place great faith in the IPCC climate models. According to their viewpoint, the effect of greenhouse gases over the past century or so would have been much greater, had it not been for the aerosols generated along with greenhouse gas emissions, with the aerosols acting to mitigate the forcing due to greenhouse gases. Thus an immediate and total cessation of emissions would produce "an immediate and significant warming following the cessation of emissions as aerosols are quickly washed from the atmosphere, and the large uncertainty in current aerosol radiative forcing implies a large uncertainty in the climate commitment". The longevity of CO_2 in the atmosphere is estimated to be hundreds of years, so its' forcing will slowly diminish over many centuries. Hence Armour and Roe (2011) concluded that after cessation of emissions, temperatures will rapidly rise significantly and remain high for millennia. The amount of additional temperature rise built into the system beyond that already experienced over the past century was estimated using several assumptions. It was estimated that this temperature rise is most probably 0.6°C with a very wide uncertainty range of 0.3°C to 6.3°C.

5.5.2 Water Vapor and Cloud Feedbacks

We have shown that the prediction of simple models is that a doubling of CO_2 from 280 ppm to 560 ppm would produce a global average temperature rise of about 1.2°C if there were no other secondary changes associated with this temperature rise. A number of feedbacks can significantly alter this result. Of particular importance is the fact that as the oceans warm, evaporation increases, and this tends to increase atmospheric humidity. Since water vapor is itself a potent greenhouse gas, this can produce a positive feedback that amplifies the original warming effect due to CO_2. However, increased humidity in the tropics may not have much effect since they tend to be saturated, and much of the heat transfer from Earth to atmosphere in this region is via cumulus cloud formation rather than radiation. Warming in desert areas will not produce a positive feedback if they become drier as a consequence of global warming. As humidity and temperature increase, the cloud cover might also change, producing further feedback influences. Clouds can reflect incoming solar energy and trap outgoing IR emissions. Depending on the latitude, altitude and characteristics of the clouds, the feedback can be positive or negative. Understanding the feedback effects of humidity and cloud cover are the most crucial aspects of climate modeling, yet relatively little has been done, and the attempts that have been made to date tend to be feeble, inadequate, limited or pathetic. Humidity and cloud cover are two variables that vary wildly from region to region, and within any region they are chaotically variable with time. Therefore, to discern any possible trend, one must have synoptic data covering the whole Earth for lengthy time periods. Unfortunately, such data do not exist.

Water vapor is by far the most important greenhouse gas, in the sense that it absorbs more irradiance from the Earth than all other greenhouse gases combined. A number of skeptical blogs have criticized the global climate modelers and alarmists (often the same people) who typically do not list water vapor as a greenhouse gas. However, the reason for this is that climate modelers are not interested in the steady-state relatively unchanging role of water vapor as a greenhouse gas, but rather, they are interested in the changes induced by human activity.

Thus, according to the *realclimate.org* blog:

> "Whenever three or more contrarians are gathered together, one will inevitably claim that water vapor is being unjustly neglected by IPCC scientists. 'Why isn't water vapor acknowledged as a greenhouse gas?' 'Why does anyone even care about the other greenhouse gases since water vapor is 98% of the effect?' 'Why isn't water vapor included in climate models?' 'Why isn't water vapor included on the forcings bar charts?' etc. Any mainstream scientist present will trot out the standard response that water vapor is indeed an important greenhouse gas, it is included in all climate models, but it is a feedback and not a forcing." (*http://www.realclimate.org/index.php?p=142*)

The point here is that water vapor concentrations are not being considered by modelers to increase directly by injection from anthropogenic sources, but rather, as a consequence of indirect processes due to feedback from other greenhouse gases. If climate models properly include such feedbacks,[42] then water vapor would be properly accounted for by climate models. However, widespread changes in land use with irrigation appear to be changing water vapor inputs to the atmosphere over wide regions, so there may be some changes in water vapor aside from the feedback effect from greenhouse gas warming. Furthermore, water vapor is the predominant greenhouse gas and whether it increases directly, or as a result of warming produced by other greenhouse gases, its action is still important. Recent investigations into depletion of groundwater are also relevant here (see discuassion just before Section 8.1.4).

Realclimate.org admitted that water vapor is the single most important absorber but suggested that it accounts for "between 36% and 66% of the greenhouse effect," and "together with clouds makes up between 66% and 85% of the temperature change induced [by greenhouse gas emissions]. CO_2 alone makes up between 9 and 26%..." Realclimate.org went on to conclude:

> "The maximum supportable number for the importance of water vapor alone is about 60–70% and for water plus clouds 80–90% of the present day greenhouse effect. (Of course, using the same approach, the maximum supportable number for CO_2 is 20–30%, and since that adds up to more than 100%, there is a slight problem with such estimates!)."

---

[42] A consummation devoutly to be wished.

Estimates in the literature vary widely. According to Marsh (2002), estimates range from water vapor accounting for 64% of the greenhouse effect, carbon dioxide 21%, ozone 6%, and other trace gases 9%, to water vapor being responsible for 90% of the greenhouse effect, leaving 10% for carbon dioxide and the other greenhouse gases. Kiehl and Trenberth (1997) estimated 59% from water and 28% from CO_2. Van Dorland (1998) estimated that water vapor accounts for 70% of the greenhouse effect, clouds account for 10%, CO_2 accounts for 15%, and other gases account for 5%. Schmidt *et al.* (2010) now claim that water vapor accounts for 50%, clouds account for 25%, and CO_2 accounts for 20%. Note that Gavin Schmidt and *realclimate.org* appear to be essentially one and the same.

If *realclimate.org* is correct in its claim that changes in humidity and clouds produce between 66% and 85% of the temperature change induced by greenhouse gas emissions, then pinning down changes in humidity and clouds as a consequence of CO_2 emissions is *the* crucial issue in predicting future climate change. The question of how credible climate models are in making such predictions then revolves about how well they account for changes in humidity and cloudiness.

According to Hall and Manabe (1999):

"The latest IPCC assessment indicates that the likely equilibrium global-mean temperature response to a doubling of CO_2 ranges from 1.6°C to 4.6°C. The source of this uncertainty is our inability to quantify the role of feedback mechanisms in the climate system, including water vapor, cloud, lapse rate, and albedo feedback. Water vapor feedback has long been thought to be a positive feedback mechanism. This is due to the dependence of the saturation water vapor mixing ratio on temperature, as predicted by the Clausius–Clapeyron equation. Thus a CO_2-induced warming of the surface–troposphere system will lead to a water vapor increase in the atmosphere. Since water vapor is itself a greenhouse gas, this increase will...make the warming larger than it would be otherwise. While there is a consensus that water vapor feedback is positive in the context of global warming, it remains unclear exactly how strong the effect is. Large uncertainties also exist regarding the magnitude (and sign, in the case of cloud feedback) of the other feedback mechanisms."

Hall and Manabe (1999) carried out long-term (500-year) GCM analyses of a modeled Earth's climate in two ways: not allowing, or allowing water vapor to increase as greenhouse gases warm the Earth. In the former case, amplification by water vapor is eliminated, and in the latter case, it is included. They presented arguments in favor of a model that assumes that the relative humidity will not change as the Earth warms, resulting in a sizable increase in absolute humidity with temperature everywhere on Earth. The results of Hall and Manabe (1999) indicated that without amplification by water vapor, the global temperature rise due to a doubling of CO_2 was 1.1°C, but with amplification by water vapor, it rose to 3.4°C.

Held and Soden (2000) discussed water vapor feedback effects at some length. They derived an equation of the form:

$$dT/dlog[CO_2] = A/(1 - B - C)$$

in which the surface temperature (*T*) depends logarithmically on the CO_2 concentration in the troposphere with a coefficient *A*. To the extent that surface warming generates additional water vapor, and the water vapor contributes a feedback to more surface heating, the parameter *B* is inserted. In addition, surface heating could reduce the ice/snow albedo of the Earth, and the term *C* is included for this effect. Held and Soden (2000) pointed out that the combined effect of *B* and *C* could have much greater impact than either term alone. However, clouds will also form under these conditions, and the effects of clouds on radiation balance are not well understood. The feedback effects of water vapor in the tropics are complex. Held and Soden (2000) discussed aspects of this uncertainty but did not seem to reach any quantitative conclusions. Bauer *et al.* (2002) also studied water vapor feedback in this era.

Soden *et al.* (2002) studied the after-effects of the eruption of Mt. Pinatubo that produced a temporary cooling of the Earth. They also found that there was a reduction in the column water vapor during this period. Attributing this reduction in water vapor to the cooling event, they concluded that there is a positive water vapor feedback. By analogy, during a period of warming, the water vapor concentration should increase. Del Genio (2002) wrote a brief review of this work, saying:

> "A few cautionary notes are in order. Although the agreement shown by Soden *et al.* is good, volcanic eruptions are not perfect reverse proxies for greenhouse gas climate change. Volcanic aerosols affect incoming solar energy more than they do Earth's thermal radiation, whereas the reverse is true for greenhouse gases. Volcanic forcing decreases more from equator to pole than does greenhouse gas forcing. And both types of climate change reduce the rate at which temperature decreases with height from the surface to the upper troposphere, even though one is a global warming and the other a global cooling".

Soden *et al.* (2005) reported "We use satellite measurements to highlight a distinct radiative signature of upper tropospheric moistening over the period 1982 to 2004". The reliability of these measurements is uncertain. But more importantly, when one examines the data reported by this paper, one finds rather large yearly fluctuations in column integrated water vapor from year to year, presumably due in part to El Niño - La Niña cycles. As we have noted in regard to tropospheric temperatures (see Figures 4.12 and 4.13), the effect of the great El Niño of 1998 seems to have been long lasting, producing a step function change in humidity as well as temperature. Thus, Figure 1 of Soden *et al.* (2005) shows a pattern of humidity fluctuating about a mean prior to 1998, and fluctuations about a higher mean after 1998. Hence the change in water vapor over the period 1982-2004 seems to have been driven by El Niño - La Niña cycles rather than temperature change, and the entire thesis of El Niño - La Niña cycles relating humidity to greenhouse driven temperatures seems to be fragile. Santer *et al.* (2007) continued this line of argument. They began by asserting: "Data from the satellite-based Special Sensor Microwave Imager (SSM/I) show that the total atmospheric moisture content over oceans has increased by 0.41 kg/m² per decade since 1988". This would seem to imply a steady rise in humidity over two decades. However when their Figure 1 is examined, one finds essentially the same result as that of Soden *et al.* (2005); there are large oscillations in humidity with an apparent step function in the mean after the 1998 El Niño. As in the case of Soden *et al.* (2005), Santer *et al.* (2007) utilized a very short period of data to infer great conclusions that remain very uncertain.

Minschwaner and Dessler (2004) wrote:

"Observational studies have attempted to verify the positive water vapor feedback by examining the response of atmospheric humidity to changes in surface temperature caused by inter-annual variability, the annual cycle, volcanic eruptions, and the El Niño–Southern Oscillation; however results have been inconclusive, with some studies yielding a positive feedback and others indicating a negative response. There are several plausible mechanisms for creating a negative water vapor feedback in the upper troposphere. One oft-cited mechanism invokes the drying effects of deep cumulus convection, arguing that the mean detrainment altitude of deep convection will be both higher and cooler in a warmer climate compared to the present. Because the water vapor content of air pumped into the upper troposphere by convection is governed by the saturation vapor pressure at the temperature of cloud detrainment, this would imply a reduced supply of water at warmer surface temperatures, leading to drying and a negative feedback on climate."

Minschwaner and Dessler (2004) attempted to test the convective drying mechanism using a model specifically designed to examine the moisture content of the upper troposphere in the Tropics. Implications for water vapor feedback were also examined using measurements of relative and specific humidities in the tropical upper troposphere (UT) from microwave and infrared limb-viewing instruments in space. As the surface temperature rises, the saturation water vapor pressure increases as required by the Clausius–Clapeyron equation, which implies that air can hold more water before it becomes saturated (i.e., the maximum possible absolute humidity increases). The relative humidity (ratio of absolute humidity to maximum possible absolute humidity) may vary in ways that are difficult to predict. In the absence of any better knowledge, some investigators assumed that as the temperature changes, the relative humidity remains roughly constant, which then leads to an increase in absolute humidity because the maximum humidity increases with temperature. This led to significant amplification of the temperature rise due to CO_2 alone.

Minschwaner and Dessler (2004) found that the drying mechanism (mentioned previously) occurs weakly so that the relative humidity decreases slowly with increasing temperature. Nevertheless, the absolute humidity was claimed to increase somewhat with increasing surface temperature. At the bottom line, Minschwaner and Dessler (2004) found that with no water vapor feedback, a doubling of CO_2 would produce a tropical surface warming of 0.8°C, and when water vapor feedback is included, the increase in temperature is 1.2°C. Note the huge discrepancy with Hall and Manabe (1999) who concluded that with constant relative humidity, the change in temperature would be 3.4°C.

Minschwaner, Dessler, and Sawaengphokhai (2006), in an update to Minschwaner and Dessler (2004), presented an analysis of the water vapor feedback in the tropical upper troposphere as simulated by 17 coupled ocean–atmosphere climate models. The strength of the water vapor feedback in the IPCC models was inferred using methods described by Minschwaner and Dessler (2004). The models indicated less drying than was found by Minschwaner and Dessler (2004). This would seem to imply a change in temperature due to doubling CO_2, including the effect of water vapor, between 1.2°C and 1.6°C.

According to a NASA press release:

"A NASA-funded study found some climate models might be overestimating the amount of water vapor entering the atmosphere as the Earth warms. Since water vapor is the most important heat-trapping greenhouse gas in our atmosphere, some climate forecasts may be overestimating future temperature increases. In response to human emissions of greenhouse gases, like carbon dioxide, the Earth warms, more water evaporates from the ocean, and the amount of water vapor in the atmosphere increases. Since water vapor is also a greenhouse gas, this leads to a further increase in the surface temperature. This effect is known as 'positive water vapor feedback.' Its existence and size have been contentiously argued for several years."

The size of the positive water vapor feedback is a key debate within climate science circles. Some climate scientists have claimed atmospheric water vapor will not increase in response to global warming, and may even decrease. General circulation models, the primary tool scientists use to predict the future of our climate, forecast the atmosphere will experience a significant increase in water vapor.

"Using UARS data to actually quantify both specific humidity and relative humidity, the researchers found, while water vapor does increase with temperature in the upper troposphere, the feedback effect is not as strong as climate models have predicted. The increases in water vapor with warmer temperatures are not large enough to maintain a constant relative humidity." (NASA text release)

An important analysis of the effect of CO_2 on climate was made by Lindzen (1997). Many aspects are covered in his paper; only a brief report on some parts is given here. Lindzen emphasized:

"Water vapor, the atmosphere's main greenhouse gas, decreases in density rapidly with both height and latitude. Surface radiative cooling in the tropics, which has the highest concentration of water vapor, is negligible. Heat from the tropical surface is carried upward by cumulus convection and poleward by the Hadley circulation and planetary-scale eddies to points where radiation can more efficiently transport the heat to space. Where radiation can more efficiently carry the heat depends on the radiative opacity and the motions themselves. In point of fact, without knowing the dynamical heat fluxes, it is clear that one cannot even calculate the mean temperature of the Earth. It is interesting, in this regard, to look at model intercomparisons of meridional heat flux, and their comparison with observationally based estimates...Such differences [are] roughly equivalent to differences in vertical fluxes of about 25 W/m^2—much larger than the 4 W/m^2 change that a doubling of CO_2 is expected to produce."

There are two points here: (1) the tropics mainly lose heat by processes other than radiation, and (2) meridional heat transfer is much greater than putative CO_2 forcing.

As we discussed previously, the prevailing view amongst climatologists is that global warming due to increased CO_2 is amplified by increased water vapor content in the atmosphere. Lindzen provided a detailed discussion of several aspects of the regional distribution of water vapor in the atmosphere and its relationship to global warming induced by increased CO_2. Most climate models make the assumption that relative humidity does not change with global warming, and since warm air can hold more water vapor than cool air, a constant relative humidity implies an increase in absolute humidity as the Earth warms. The basis for the assumption that relative humidity does not change with global warming lies in some rather old radiosonde data that indicate that the average distribution of relative humidity (when plotted on altitude vs. latitude axes) does not change much from winter to summer. The argument then goes that over the smaller temperature change characteristic of global warming, relative humidity would also not change. However, Lindzen raised serious questions about the accuracy of the radiosonde data. Clearly, the assumption of constant relative humidity rests on a weak foundation, and that assumption is critical to the alarmist position that doubling CO_2 produces unacceptable global warming due to increased absolute humidity.

But Lindzen went further than this. He emphasized that the degree of water vapor feedback as a heating force in any region depends on absolute humidity. In desert regions with very low absolute humidity, an increase in humidity provides a significant heating force. However, in regions with high absolute humidity, an increase in humidity provides a very modest heating force (e.g., an increase in relative humidity from 10% to 20% produces a forcing of 1.5 W/m², whereas an increase in relative humidity from 50% to 60% produces a forcing of only 0.15 W/m²). Tropical regions that already have high humidity do not gain much additional heating from an increase in humidity. And as Lindzen pointed out:

> "Given the nonlinearity of the radiative effect of water vapor, the average radiative response to water vapor is not equal to the response of the average water vapor."

It has been estimated by climate modelers that a doubling of CO_2 implies a forcing at the tropopause of about 3.7 W/m². The question of climate sensitivity amounts to asking how much must the Earth's surface warm to compensate for this forcing. This requires estimation of the globally integrated total radiative flux at tropopause levels. A global change in the distribution of moist and dry regions can lead to a change in outgoing long-wavelength radiation (OLR) even in the absence of change in mean temperature. Changes in circulation and changes in temperature can both play a role in the moisture budget. Lindzen suggested:

> "...the interesting possibility that the primary feedback process might consist in the change in areal coverage of the very dry regions. Presumably, natural variations include a full range of such possibilities so that observed ratios of average temperature variations to variations in total OLR would show a significant scatter."

According to Lindzen, global climate models do not do a good job of estimating the coupling between tropical and extratropical regions and therefore do not allocate the global distribution of water vapor accurately; this has a profound effect on the putative heating effect of increased CO_2. It seems likely that global warming might decrease the humidity of air descending above desert areas of the Earth, and since these regions are by far the most sensitive to changes in humidity, they would counterbalance the smaller heating effect of increased humidity in regions where the humidity is higher. The regional effects of changes in humidity far outweigh the effects of changes in net global humidity. Drying of already dry regions is more important than net humidifying of the globe. Net moistening of the Earth could have a negative water vapor feedback if most of that moistening occurs in already moist regions. Climate models that take an average humidity for the whole Earth are overly simplistic.

The effect of water vapor feedback on amplifying global warming produced by increasing CO_2 concentration requires an understanding of the distribution of humidity changes resulting from warming; a global average of humidity change does not suffice. In addition, of course, even with a thorough understanding of the regional dependence of humidity change, one must still cope with the problem of changes in cloudiness. Lindzen and co-workers (2001, 2007) have studied this and made four major points: (1) The cloud and water vapor feedbacks are intimately connected. (2) Feedbacks are primarily associated with changing areas of moist and cloudy regions vs. regions that are dry and cloud-free (as opposed to mean humidity). (3) Models must have spatial and temporal scales (5–10 km and hours) characteristic of clouds in order to evaluate feedbacks. (4) The effect of cumulus activity must be included. A simplistic model that merely treats humidity as a global average that increases when surface temperatures rise, that ignores regional changes in humidity, and crudely treats clouds will always overestimate the temperature rise due to increased CO_2.

While most climate models deal with such elements as clear-sky humidity, average humidity, or differences between regions of high and low humidity, Lindzen and co-workers have studied feedback involving changes in the relative areas of high and low humidity and cloudiness. Their results suggest that cloudy moist regions contract when the surface warms and expand when the surface cools. In each case, the change acts to oppose the surface change, and thus presents a strong negative feedback to climate change as a sort of Le Chatelier's Principle. They concluded that the relevant feedbacks are negative rather than positive, and very large in magnitude. Spencer et al. (2007) studied the effect of changes in clouds to changes in temperature in tropical regions and found a negative feedback of 6 W/m² per degree of temperature rise. This provides some support for Lindzen's hypothesis.

Dessler et al. (GRL, 2008) attempted to derive water feedback sensitivity by comparing data on global temperature and humidity during the winter months of 2006–2007 and 2007– 2008. However, Christy demonstrated a strong correlation of global temperature with an El Niño index since 1978, and particularly for 2006–2008 (see Figures 4.12 and 4.13). Dessler et al. (GRL, 2008) also found a good correlation of global temperature with an ENSO index for 2006–2008. Hence it seems clear that global temperature changes for 2006–2008 were driven primarily by changes in the oceans, and changes in humidity during that period were not a cause of global temperature change, but an effect. The effect of changing CO_2 concentration and putative water vapor greenhouse effect are buried in the noise of a much stronger signal due to El Niño variability during these years. Therefore it is physically impossible to derive a water feedback sensitivity from data limited to these two winters. Yet, Dessler *et al.* claim that they have done so and quote a value in agreement with climate models. This seems impossible to this writer. They then reached the rather incredible conclusion:

> "The existence of a strong and positive water-vapor feedback means that projected business-as-usual greenhouse gas emissions over the next century are virtually guaranteed to produce warming of several degrees Celsius."

This conclusion is utterly unsupportable from the analysis of a mere two winters' data controlled by El Niño activity.

Dessler *et al.* (JGR, 2008) analyzed a mere one month's data in 2005 to infer clear-sky top-of-atmosphere outgoing long-wave radiation (OLR) and its relationship to humidity. It is not clear to this writer that this paper sheds any light on water feedback sensitivity.

Gettleman and Fu (2008) analyzed the changes in humidity produced by temperature changes from 2002 to 2007. As before, temperatures during this period appear to have been determined by El Niño variability, and changes in water vapor content appear to be effects of this temperature change. There is little or no connection to heating produced by CO_2 and water feedback sensitivity does not seem to be derivable from this work.

It is now becoming apparent that the high temperatures experienced by the Earth in the late 1990s, and particularly in 1998, were related to the prevalence of El Niño conditions, and with the advent of La Niña in 2007–2008 world temperatures dropped. While it is likely that growth in CO_2 contributed to global warming during the 20th century, it is clear that large chaotic fluctuations dictated by ocean conditions, aerosol emissions, humidity, cloudiness and unknown factors have masked the putative CO_2 effect, making it very difficult to unravel the contribution of rising CO_2. The work by Soden *et al.*, Santer *et al.* and Dessler *et al.* and others has attempted to ferret out information from very limited amounts of data, some of which is of uncertain reliability. But ultimately, the credibility of their results is limited by the scarcity of good long-term data. Parameters such as humidity and cloudiness vary widely from day-to-day and year-to-year even in the absence of any forcing. In attempting to determine how these parameters respond to a forcing, one must have data over very long periods to overcome the low signal-to-noise ratios inherent in them. The same problem occurs in sea level measurements. However, whereas climatologists studying sea level have emphasized the need for very long-term data, those who infer feedbacks from humidity and cloudiness seem to be content with very short-term data.

Looking back on the work of Soden *et al.*, Santer *et al.*, Dessler *et al.*, and Gettleman and Fu, we must conclude that these papers should never have been accepted for publication because the data upon which they depended were too limited in duration, and were too obfuscated by other climatic effects than greenhouse gas impacts.

Cloud cover is a very important determinant of the Earth's climate. The heat fluxes in the Earth–atmosphere system were estimated by Kiehl and Trenberth (1997) and updated by Trenberth, Fasullo and Kiehl (2009). According to their analysis, the global average albedo decreased from 31.3% for 1985-1989 to 29.8% in 2000-2004, suggesting that solar power absorbed in 2000-2004 was 5 W/m^2 greater in 2000-2004 than it was in 1985-1989. This is a much greater forcing than that attributed to greenhouse gases. But it seems far more likely that it represents either a fluctuation independent of greenhouse gases; or perhaps it was just noisy data – as opposed to a diminution of cloudiness due to global warming (as for example Dessler insists).

Clouds play an important role in the heat balance of the Earth. Climate models have included the effects of clouds in various ways. However, most climate modelers agree that uncertainties in regard to the effects of clouds is a principal cause of disagreement between models. According to Bony *et al.* (2006):

"Clouds strongly modulate the earth's radiation budget, and a change in their radiative effect in response to a global temperature change may produce a substantial feedback on the earth's temperature. But the sign and the magnitude of the global mean cloud feedback depends on so many factors that it remains very uncertain. Cloud feedbacks have long been identified as the largest internal source of uncertainty in climate change predictions, even without considering the interaction between clouds and aerosols."

Bony *et al.* (2006) said that climate models exhibit "systematic biases" in simulating clouds that "restrict their ability to predict the magnitude of cloud feedbacks." They therefore concluded:

"Defining strategies for evaluation of cloud feedback processes in climate models is thus of primary importance to better understand the range of model sensitivity estimates and to make climate predictions from models more reliable."

Norris and Slingo (2009) reviewed measurements and models for variability of the Earth's cloudiness and it effect on the Earth's radiation budget (ERB) and the Earth's climate. Their overall assessment was that our knowledge of past variability of cloudiness is very poor, our present capabilities for monitoring global cloudiness are weak, and the prospects for future measurements are worse. Global cloudiness is a very important factor in determining the Earth's climate. As Norris and Slingo noted, it is difficult to resolve any putative effects of greenhouse gases on climate when we don't understand how cloudiness varies, because variability of cloudiness has a much greater forcing effect than changes in greenhouse gas concentration.

> "Small cloud changes are important because they can exert more leverage over ERB than equivalent changes in greenhouse gases."

While the effect of greenhouse gases is expected to be spatially coherent, changes in cloudiness will vary widely from region to region and it is necessary to make measurements at many sites globally to obtain a global average. In addition, Norris and Slingo said:

> "Since various cloud types have strikingly different radiative effects, it is not a simple matter to determine the overall global impact of cloud changes; each cloud type and climate regime must be examined in particular. Moreover, alterations of cloud albedo, cloud emissivity, and cloud height can affect ERB even when cloud amount remains the same."

As Norris and Slingo described in some detail, measurements of clouds, whether from the ground or satellite, appear to suffer from various artifacts and are not trustworthy. Global climate models (GCMs) suffer from an inability to represent clouds. As Norris and Slingo said:

> "Many studies comparing simulated clouds with observed clouds have found that GCMs poorly represent clouds when evaluated on terms for which they were not explicitly tuned."

> "Some GCMs suggest that the horizontal extent of low-level clouds over low-latitude oceans will increase with higher global temperature, whereas other GCMs suggest that low-level cloud amount will decrease."

Andronova *et al.* (2009) analyzed evolution of the tropical mean radiation budget at the top of the atmosphere since 1985 for the latitude range 20°S to 20°N. They found that since 1985, the Earth became less reflective to incoming short-wave radiation and more absorbent of outgoing long-wave radiation. Both effects produce heating of the atmosphere. Upon comparing with climate models, they found that "none of the models simulates the overall 'net radiative heating' signature of the Earth's radiative budget over the time period from 1985-2000." These changes occurred during a period when the Earth warmed. It is not at all clear whether these changes were induced primarily by rising CO_2 levels, or whether these changes were due to other factors such as El Niños, and it is difficult to assess the role of rising CO_2 levels.

The effect of possible changes in cloud cover on surface forcing when CO_2 is doubled could be significant. Hunt (1981) suggested that changes in cloud cover "raise the question whether general circulation models with self-predicting cloud properties would be able to attain sufficient accuracy to realistically evaluate such subtleties in an assessment of the CO_2 climate problem". Eastman *et al.* (2011) concluded that long-term secular changes in cloud cover over the oceans may have occurred during the period 1954-2008. Norris (2005) reported decreases in cloud cover over the oceans from 1952 to 1997. However, Evan *et al.* (2007) pointed out problems with cloud data. They said: "We have demonstrated that the long-term global trends in cloudiness from the ISCCP record are influenced by artifacts associated with satellite viewing geometry. Results from earlier studies based on these trends may be influenced by these non-physical artifacts, and we therefore suggest that development of a correction for the data is warranted." They suggested that the *International Satellite Cloud Climatology Project* (ISCCP) data set of cloud amounts "may not be appropriate". Wang *et al.* (2012) claimed that "decadal variations of surface incident solar radiation ... can be accurately estimated using globally available measurements of Sunshine Duration (SunDu)". They claim that since the late 1980's, Europe brightened but China dimmed. They concluded that averaged over the entire NH, changes in aerosols and cloud cover brightened the NH by about 0.9 W/m² per decade from 1982 to 2008. If this is correct, it would swamp putative forcings from rising greenhouse gas concentrations.

5.5.3 Model of Spencer and Braswell

Spencer and Braswell (2008, 2010) developed an analysis of feedback effects in response to a forcing of the Earth's climate. In general, they were concerned with the Earth being subjected to a forcing F (W/m²) at the top of the atmosphere causing temperature departures from equilibrium. Specifically, they were concerned with a positive forcing due to rising greenhouse gas concentrations. The deviation from normal global average surface temperature (T) rises with time in response to this forcing. If no feedback changes occurred, the equation representing the changing T would be:

$$C\, dT/dt = F$$

where C represents the effective heat capacity of the Earth (essentially the oceans), and the temperature would continue to rise under the forcing.

However, as the temperature rises, other changes occur. The Earth emits more radiation as it warms, and changes occur in the hydrological cycle (evaporation, cloud formation, water vapor distribution, etc.) Some of these factors act in opposition to the forcing to reduce the temperature rise resulting from the forcing (e.g. increased long-wave emission by the Earth), and some may either resist further temperature rise or enhance it (increased water vapor concentration and some types of clouds would provide an enhanced greenhouse effect by absorbing radiant emission from the Earth, while some types of clouds would reflect incident sunlight, thus providing a negative feedback). These feedback factors will become stronger as T

increases; hence as a first approximation, it is assumed that they are proportional to the temperature. Thus a feedback factor (β) is defined:

$$\beta = \beta_{SW} + \beta_{LW}$$

where SW = short wave and LW = long wave, and the equation for temperature change is now:

$$C \, dT/dt = F - \beta T$$

The minus sign in front of β in this equation is arbitrary since β can be positive or negative. The increase in long-wave emission by the Earth as it warms provides a positive contribution to β_{LW} that we can designate as β_{SB} (where SB represents the Stefan-Boltzmann emission increase). If there were no other feedbacks except for increased radiant emission by the Earth, β would equal β_{SB} and T would rise until $F - \beta T = 0$, at which point $dT/dt = 0$ and a new equilibrium would be established at a higher temperature. At this point,

$$T = F/\beta_{SB} = \lambda_{SB} F$$

where $\lambda = (1/\beta)$ is the climate sensitivity parameter (change in temperature per unit forcing).

Most climate models estimate that a doubling of CO_2 concentration would produce a forcing or roughly 4 W/m². However, there is a wide range of estimates of the climate sensitivity parameter. If $\lambda_{SB} \sim 0.3°C/(W/m^2)$, corresponding to $\beta_{SB} \sim 3.3$ (W/m²)/°C, it would be estimated that $T \sim 1.2°C$ for a doubling of CO_2 in the absence of other feedback effects. However, estimates for λ vary.

Some clouds have a principal effect of reflecting incident sunlight, thus increasing β_{SW}. Other clouds have a principal effect of making β_{LW} negative by absorption of long wave radiation. In general, increased water vapor concentrations produce greater negative values of β_{LW}. Most global climate models estimate that due to increased water vapor concentration and increased clouds, the net value of β is reduced to a less positive value than β_{SB} and therefore λ is increased to a value greater than λ_{SB}. Typical estimated values of λ by alarmists are around 0.8°C/(W/m²), thus leading to a value of T for doubling CO_2 of roughly 3°C.

Spencer and Braswell (2008, 2010, 2011) pointed out that in addition to the aforementioned feedback factors (Stefan-Boltzmann emission, increased clouds and water vapor due to rising temperature) "internal factors" operating near the Earth's surface may affect global temperatures and heat flows. These might include random variations in cloud cover in the Earth's climate system, "brought about through circulation–induced changes in tropospheric wind shear, frontal system behavior, precipitation efficiency, trade wind inversion strength, or any other of the myriad processes that can potentially affect cloud formation other than feedback upon temperature" and non-radiative forcing of temperature change such as tropical intra-seasonal oscillations in the rate of heat transfer from the ocean to the atmosphere".

According to Spencer and Braswell, if these internal factors are relatively small, then measurements of the Earth's heat budget at the top of the atmosphere should provide radiative fluxes proportional to surface temperature. Spencer and Braswell (2008, 2010, 2011) analyzed heat budget data and found little correlation of heat flows with surface temperature, suggesting that internal factors were obfuscating the other heat flows, making it impossible to experimentally determine feedback factors.

The gospel of the orthodoxy, for example as preached by Dessler (2011), is that CO_2 is the driver of climate change. In this view, the climate would remain quite constant as long as the CO_2 concentration remains constant. The degree of cloudiness would remain within narrow confines dictated by the CO_2 concentration. If the CO_2 concentration were to rise, cloudiness would change as a response to the increase in CO_2 (via a change in global average temperature). Thus, according to this viewpoint, one should be able to plot a measure of cloudiness vs. a measure of global average temperature, and find a direct correlation within limited scatter. However Dessler did not examine how cloudiness varies with CO_2 concentration or how cloudiness varies with increased surface temperature due to rising CO_2. Instead, he relied on very short-term data from volcanic eruptions and El Niño events to examine how cloudiness varied with temperature during these events. He found a very large amount of scatter, which is not surprising. However, he persevered by finding the best straight-line fit to the widely scattered data, and concluded that cloudiness decreased slightly as the temperature rose. It seems more likely that over this short time period, temperature and cloudiness are uncorrelated, and the data are the result of stochastic variations, rather than a cause-effect relationship. Spencer and Braswell (2011) put it very succinctly:

> "The sensitivity of the climate system to an imposed radiative imbalance remains the largest source of uncertainty in projections of future anthropogenic climate change. Here we present further evidence that this uncertainty from an observational perspective is largely due to the masking of the radiative feedback signal by internal radiative forcing, probably due to natural cloud variations.... It is concluded that atmospheric feedback diagnosis of the climate system remains an unsolved problem, due primarily to the inability to distinguish between radiative forcing and radiative feedback in satellite radiative budget observations."

What this means, is that there is too much scatter to determine whether changes in cloudiness are directly due to changes in surface temperature vs. innate changes in cloudiness in the internal climate system.

Aside from the science and pseudoscience involved in these analyses, there are social issues as well. The alarmists refer to their interpretation of climate science as simply "climate science". It is not one interpretation. It is **THE** CLIMATE SCIENCE – in their view. We see evidence of this in many publications and press releases. In particular, in regard to the effect of clouds, Dessler said: "In recent papers, Lindzen and Choi (2011), and Spencer and Braswell (2011) have argued that ... clouds are the cause of, and not a feedback on, changes in surface temperature. If this claim is correct, then significant revisions to *climate science* may be required". In other words, he regards "climate science" as that which the orthodoxy subscribes to. It is not his interpretation of climate science – IT **IS** CLIMATE SCIENCE!

Another bizarre aspect of Dessler's publication was discussed by Pielke, Sr. He said: "Dessler's paper was received 11 August 2011 and accepted 29 August 2011. This is some type of record ... and indicates that the paper was fast-tracked. This is certainly unusual" – to say the least.

"It is not clear whether the Editor of *GRL* included Roy Spencer as one of the referees, [and if they did not] they were derelict in their responsibilities". Dessler's paper should have been submitted to *Remote Sensing* as a Comment [on Spencer's paper]. Then Roy Spencer would submit a Reply." (*http://pielkeclimatesci.wordpress.com/2011/09/06/comments-on-the-dessler-2011-grl-paper-cloud-variations-and-the-earths-energy-budget/*)

We are now witnessing a phenomenon in climatology publications that is occurring repeatedly. The climatology orthodoxy has united into an informal association dedicated to (1) prevent contrary analyses and interpretations from being published, and (2) to quickly respond to those few contrarian publications that slip through their net with vitriolic attacks on the paper on orthodoxy blogs, and in the literature via rapid rebuttal publications such as that of Dessler (2011). It is evident that many Editors are in cahoots with the orthodoxy; certainly the Editor of *GRL* is, and the Editor of *Remote Sensing* who let Spencer and Bradwell's paper through the net, suddenly resigned for unclear reasons.

Ban-Weiss *et al.* (2011) added another viewpoint on cloudiness change due to global warming. They pointed out that increased evaporation would produce a significant surface cooling effect that would be counterbalanced by an equal heating effect as moisture condensed in the atmosphere. Thus there is no net heating effect. However, the increased evaporation produces an increase in low elevation cloudiness that increases the Earth's albedo, producing a net cooling effect. However, is not clear how reliable their estimates of changes in cloudiness are.

5.6 The Climate Debate Revisited

Table 5.2. Evidence for the climate debate.

Aspect	Evidence
Rising CO_2 was the cause of rising temperatures in 20th century	Higher CO_2 levels are generally associated with higher global temperatures although the correspondence is not clear cut. Rising CO_2 undoubtedly contributed somewhat to rising temperatures in the 20th century. There are too many divergences between CO_2 and temperature in the 20th century to conclude that rising CO_2 was the major contributor to temperature rise. The contribution of CO_2 remains uncertain.
Climate models provide reasonable estimates of future warming in business-as-usual scenario	Climate models do not have adequate data to estimate secondary effects due to water vapor, clouds and aerosols, and therefore cannot predict future temperatures with any reliability.

In the process of consuming fossil fuels, cement production, and other industrial activities, the world produces large amounts of CO_2. The CO_2 concentration has increased from about 280 ppm for at least several centuries prior to the 20th century, and rose steadily after about 1880 to its present value of ~395 ppm. If the world continues in a business-as-usual scenario, CO_2 production will continue to rise in the 21st century, leading to higher CO_2 concentrations in the atmosphere. According to Figure 5.4, the CO_2 concentration may rise by year 2100 to roughly 750 ppm in a business-as-usual scenario, 570 ppm if emissions are frozen at the 2010 level after 2010, and 450 ppm if emissions are rapidly reduced after 2010. An increase in CO_2 concentration is expected to produce some global temperature rise, but how much? Most studies have concentrated on analyzing the rise in temperature if CO_2 doubles from the pre-industrial level of 280 ppm to 560 ppm. Studies of CO_2 in ancient climates are not sufficiently precise to answer this question. Climate models do not have adequate data to estimate secondary effects due to water vapor, clouds and aerosols, and therefore cannot predict future temperatures with any reliability. Hence we cannot answer this question with any reliability. It seems likely that the global average temperature will rise between 1°C and 3°C from the pre-industrial level if CO_2 reaches 560 ppm. About 0.7°C of this warming has already been observed at CO_2 = 395 ppm. See Table 5.2.

6 THE EARTH'S HEAT BALANCE AND THE GREENHOUSE EFFECT

6.1 Estimates of Average Heat Balance and the Effect of Rising CO_2

The surface of the Earth (land, oceans and atmosphere) is warmed by irradiance from the Sun. The Earth warms up and emits infrared radiation until a balance is achieved between incoming solar power and outgoing radiant power. The solar power approaching the Earth is about 1,367 W/m² passing through a plane above the Earth, facing the Sun. Since the area of this plane projected by the Earth is πR^2, where R is the radius of the Earth, and the surface area of the Earth is $4\pi R^2$, the solar power falling on the Earth would be 1,362/4 ~ 340.2 W/m² per unit surface area of the Earth if it impinged uniformly on all areas of the Earth (which of course, it doesn't).[43] This is illustrated in Figure 6.1. In reality, a high percentage of this solar radiation impinges on the tropics and a lesser percentage falls on higher latitudes. Nevertheless, this unrealistic picture in which the Earth is treated as if it were uniformly the same for all latitudes is often used in discussions of the Earth's energy balance.

The Earth's albedo (reflection coefficient for solar irradiance) is estimated to be roughly 30%. If ~30% of the solar irradiance on the Earth is reflected back into space, the net solar irradiance absorbed by the Earth (per unit surface area) in this simplistic model of the Earth is roughly 0.7 x 340.2 ≈ 240 W/m².

---

[43] After 2008, the estimate of average solar power at the Earth distance from the Sun was revised downward by about 4 W/m² from the previous value of 1366 W/m².

The solar power input to the Earth in the simplistic model is uniformly distributed about the Earth at the level of about 340 W/m². An approximate analysis of the fate of this solar power is given in Figure 6.2.

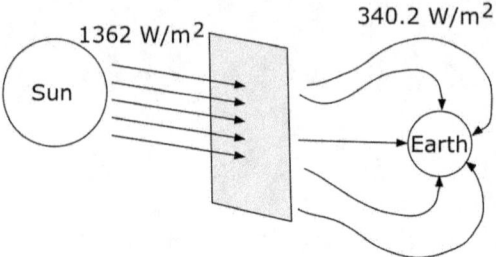

Figure 6.1. Solar power distributed uniformly around the Earth.

A total of ~ 240 W/m² is absorbed by the Earth and atmosphere, and about 100 W/m² is reflected back into space. These figures are approximate, and represent a hypothetical average for the whole Earth. In reality, the power balance varies very widely from latitude to latitude. Furthermore, cloud cover varies constantly producing rapid changes in the power balance. The figures given here are rough averages.

As we see in Figure 6.2, on average the Sun deposits about 240 W/m² into the Earth-atmosphere system and about 100 W/m² is refelected into space. Note that although Stephens *et al.* (2012) claim that the reflected solar flux is 100.0±0.2 W/m², the sum of the three components in Figure 6.2 adds up to 97.5 W/m². Furthermore, the uncertainties in these components are much larger than the claimed uncertainty in the reflected flux. It seems unlikely that any of these energy fluxes can be known to within ±0.2 W/m².

The solar flux input causes the Earth and the atmosphere to warm, thus inducing heat flows between the Earth, the atmosphere and space. These heat flows have been estimated in the simple average model of the Earth, as shown in Figure 6.3. The warming Earth sends rising warm air currents upward to heat the atmosphere (sensible heat transfer). In addition, evaporation of water from the oceans, fresh water and the soil sends water vapor upward where it condenses into clouds, releasing the heat of condensation (latent heat transfer). The Earth and the atmosphere exchange long wave radiation. The upward flux is 398±5 W/m² and the downward flux is 345.6±9 W/m², so the net result is a "back radiation" from the surface to the atmosphere of ~52.4 W/m².

Whether the Earth is warming or cooling depends on an energy flux balance at the top of the atmosphere (TOA). Hansen and co-workers have claimed that the Earth is out of balance by about 1 W/m²; that is the outgoing long wave radiant flux is about 1 W/m² less than the incoming solar flux. Lyman *et al.* (2010) and Stephens *et al.* (2012) assumed that the net warming imbalance is about 0.6 W/m² based on estimates of ocean warming.

Knox and Douglass (2010) argued that one can obtain very different estimates of the imbalance ranging down to zero, depending on the time period chosen for analysis. The theory being that any imbalance in the Earth energy budget at the TOA must end up as ocean warming. The Earth's imbalance cannot be calculated from the measured energy fluxes. The uncertainties in the various energy fluxes are large compared to the claimed imbalances. The imbalance would be calculated as the difference between large numbers. Thus the estimate of the imbalance is like weighing the captain by weighing the ship plus the captain and sutracting the weight of the ship. In actuality, the Earth's heat balance depends heavily on cloud cover. The heat balance is constantly shifting as the Earth's albedo meanders. The Earth's heat balance can only be assessed over a period several decades to average out these fluctuations. Yet, the data reported in Figures 6.2 and 6.3 are only for the decade 2000-2010.

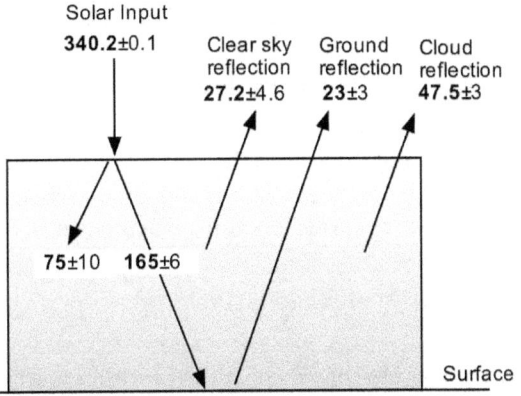

Figure 6.2. Solar Power Input to the Earth (W/m²) (Stephens, *et al.* 2012)

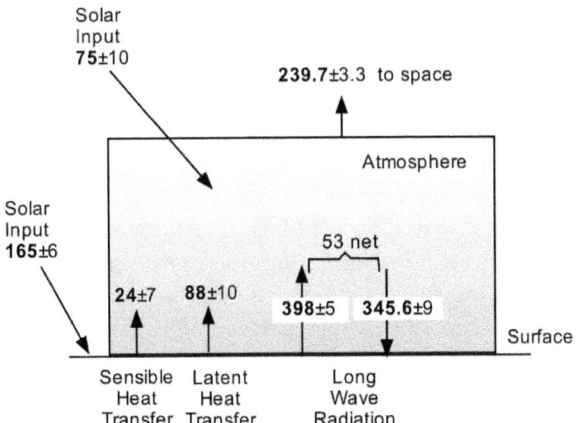

Figure 6.3. Heat flow between Earth, the atmosphere and space (W/m²). (Stephens, *et al.* 2012)

Figure 6.4 shows a comparison of spectral distribution of upward irradiance from the surface with that at the top of the atmosphere showing CO_2 and O_3 bands for assumed global cloudy conditions.

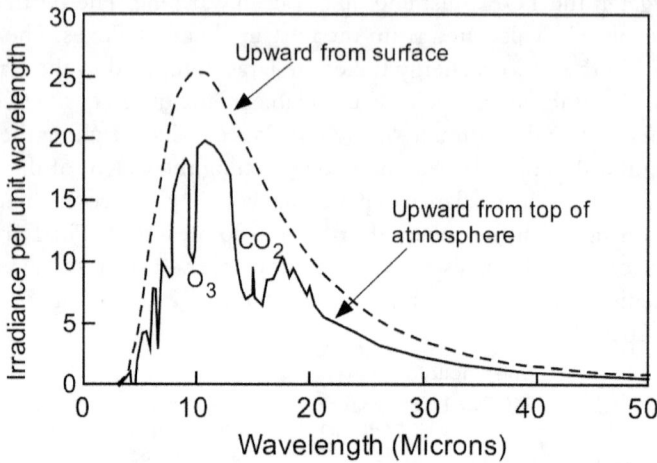

Figure 6.4. Comparison of spectral distribution of upward irradiance from the surface with that at the top of the atmosphere showing CO_2 and O_3 bands for assumed global cloudy conditions. Water absorption occurs throughout the spectrum in various amounts. Adapted from Schimel *et al.* (2001).

Lindzen and Choi (2011) provided a succinct description of the greenhouse effect:

> "This simply refers to the fact that the Earth balances the heat received from the Sun (mostly in the visible spectrum) by radiating in the infrared portion of the spectrum back to space. Gases that are relatively transparent to visible light but strongly absorbent in the infrared (greenhouse gases) will interfere with the cooling of the planet, thus forcing it to become warmer in order to emit sufficient infrared radiation to balance the net incoming sunlight. By the net incoming sunlight, we mean that portion of the Sun's radiation that is not reflected back to space by clouds and the Earth's surface. The issue then focuses on a particular greenhouse gas, carbon dioxide. Although carbon dioxide is a relatively minor greenhouse gas, it has increased significantly since the beginning of the industrial age from about 280 ppm to about 390 ppm, and it is widely accepted that this increase is primarily due to man's emissions."

A variety of climate models have been developed with the primary purpose of predicting the effect of future greenhouse gas emissions on the climate in the 21st century. Most of this work has been carried out by climate alarmists, and it is not clear to what extent the findings influenced the scientists, or the extent to which scientists influenced the findings. Just as earthquake specialists always seem to predict that "the big one" is coming, so one must wonder about the climatologists who predict a dire future for the planet. Any climate model that attempts to estimate the future temperature rise due to doubling CO_2 from $\sim$ 280 ppm must make assumptions regarding the issues of changes in water vapor concentration and changes in cloud cover resulting from warming induced by increased CO_2. Uncertainty in how to treat these factors leads to a wide swath of estimates from various climate models. In Section 5.5.2 we discussed the evidence for changes in water vapor concentration and changes in cloud cover resulting from warming induced by increased CO_2. Several climatologists of the alarmist persuasion have published papers in this regard (Soden, Santer, Dessler, Gettleman, ...). These studies have several characteristics in common:

- They all assume there is a unique water vapor content and cloud cover in the atmosphere determined by the current Earth surface temperature, whereas the truth is that the water vapor content and cloud cover vary chaotically and no relationship exists between these variables and Earth surface temperature over the short term.
- They all rely on short-term data (a few months to a few years) whereas it would take many decades, perhaps even centuries to separate the signal from noise for a putative relationship between water vapor content and cloud cover and Earth surface temperature.
- They all rely on short-term atmospheric phenomena, such as aftermath of volcanic eruptions or El Niño – La Niña transitions, neither of which have much to do with long-term evolution of the atmosphere due to rising CO_2.

Hence none of these studies can be validated with data, and we simply do not know how secondary processes will alter the fundamental temperature rise due to doubling CO_2.

6.2 Heat Imbalance of Earth from 1950 to 2005

Initiated by various publications by James Hansen, a number of investigators have estimated that due to the rise in greenhouse gases, principally CO_2, the Earth is out of heat balance at the top of the atmosphere (TOA) and has been so for a number of decades. In these models, the steadily rising greenhouse gas concentrations have gradually reduced heat loss from the Earth. This is treated as if it were a downward "forcing" of heat flux. Over this same time period a series of volcanic eruptions partly mitigated this forcing by reducing the solar heat input to the Earth for a period of a few years after each event. The net forcing is estimated as the difference between the greenhouse forcing and volcanic effects.

Murphy *et al.* (2009) estimated the annual heat balance of the Earth from 1950 to 2005. They estimated the forcing due to greenhouse gases over this period. They found that about 60% of total forcing was due to CO_2, about 20% was due to CH_4, and the remainder was divided amongst halocarbons, ozone and N_2O. They also estimated the mitigating effect of volcanic eruptions. In addition, they used satellite measurements to estimate how radiant heat loss from the TOA varied for the recent part of this time period. In doing this, they used ERBE and CERES data over short time periods to infer dependences of short wave and long wave energy fluxes at the top of the atmosphere (TOA) on global average temperature. Then they summed these to get the dependence of net radiant energy flux at the TOA on global average temperature. Using known values of the global average temperature each year from 1950 onward, they inferred the change in net radiant energy flux at the TOA for each year. Their estimates of the three terms are shown in Figure 6.5. The forcing due to greenhouse gases is the dashed line at the top. The curve marked "volcanic aersols" represents the mitigating effect of volcanic eruptions, and the sum of greenhouse gas forcing plus the mitigating effect of volcanic eruptions is shown as the solid curve with large downward spikes due to volcanoes. The change in downward thermal radiation at the TOA is also shown.

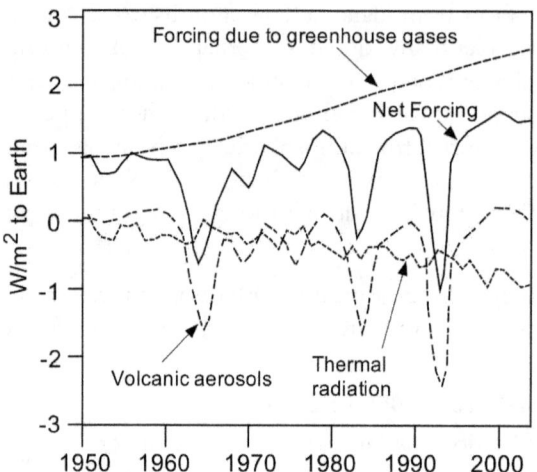

Figure 6.5. Factors contributing to heat balance of the Earth. Upward fluxes are negative in sign and downward fluxes are positive in sign. (adapted from Murphy *et al.*, 2009).

There are problematic aspects of this analysis. One factor is this. If the net radiative flux at the TOA increased negatively (increased outward) since 1950, why did the Earth warm at all? Somehow, Murphy *et al.* are treating the forcing as a real energy flux in the downward (positive) direction that is only partly offset by increased outward radiant flux. But what happens down below is irrelevant if the outward radiant flux at the TOA increases. In my view, the forcing is not really a downward flux, but rather, a decrease in upward flux from Earth to the TOA. Yet, if the radiant outward flux from the TOA increases, does the forcing down below matter?

There is no reason to believe that the short wave radiation budget depends on global average temperature, except indirectly, insofar that it might depend on cloud cover and cloud cover might depend on global average temperature. However, the time periods over which the authors correlated the short wave radiation budget with global average temperature was too short to arrive at a credible conclusion.

There are additional forcings that affect the Earth's heat balance. These revolve about water vapor in some form (aerosols, water vapor, clouds, ...). Furthermore, Murphy *et al.* did not seem to consider changing humidity or cloudiness over the time period 1950-2005, but they did allow for anthropogenic aerosols. Murphy *et al.* did not attempt to estimate this *a priori*. Instead, they separately estimated the yearly change in ocean heat content of the Earth (the oceans hold most of the heat content of the Earth – land and atmosphere heat content are minor). This result was presented as an 8-year moving average (see Figure 6.6). Then they took an 8-year moving average of the "net forcing" in Figure 6.5 (also shown in Figure 6.6) and noted that it had a similar shape to the ocean heat content curve except it was considerably higher than the ocean heat content curve. If the Earth is being heated by greenhouse gases and the calculated net forcing does not appear in the oceans, there is a considerable amount of heat that is not accounted for. Murphy *et al.* (2009) assumed that anthropogenic aerosols reflected enough incident sunlight that the difference between the 8-year smoothed net forcing and the ocean heat content in Figure 6.6 was due to this effect.

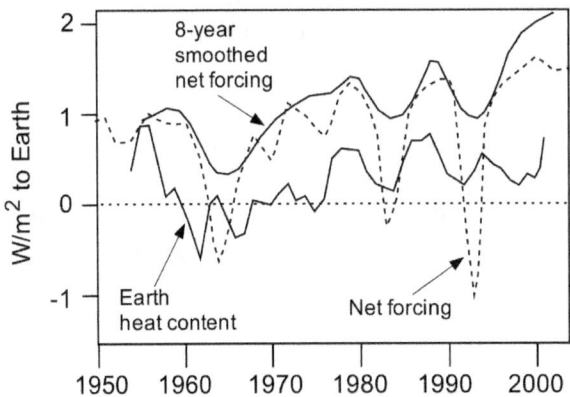

Figure 6.6. Comparison of net forcing due to greenhouse gases with increase in ocean heat content (adapted from Murphy *et al.*, 2009).

However, it is unclear why this was not accounted for in the radiation budget at the TOA. The difference between the calculated net forcing and the observed heat gain of the coeans was huge. Heat gain by the oceans only accounted for 10% of the calculated forcing. According to Murphy *et al.*, only 10% of the greenhouse gas forcing resulted in heating of the Earth, and the other 90% of heating induced by greenhouse gases was counterbalanced by the following factors:

- increased radiation to space from a warming Earth: 20%
- rejection of incident sunlight by volcanic aerosol emissions: 20%
- rejection of incident sunlight by anthropogenic tropospheric aerosols: 50%

According to Murphy *et al.* (2009), had there been no volcanic activity or anthropogenic aerosols over the period 1950-2005, the ultimate net forcing would have been considerably higher. This would have warmed the Earth a good deal more than the actual temperature rise of 0.55°C, resulting in further increases in radiation to space from a warming Earth. (However, as pointed out previously, if the change in radiation at the TOA was outward, it is not clear why the Earth warmed over this period.)

It is not clear from the work of Murphy *et al.* (2009) what the roles of changing humidity and cloud cover might be. Solomon *et al.* (2010) reported:

"Stratospheric water vapor concentrations decreased by about 10% after the year 2000. Here we show that this acted to slow the rate of increase in global surface temperature over 2000–2009 by about 25% compared to that which would have occurred due only to carbon dioxide and other greenhouse gases. More limited data suggest that stratospheric water vapor probably increased between 1980 and 2000, which would have enhanced the decadal rate of surface warming during the 1990s by about 30% as compared to estimates neglecting this change. These findings show that stratospheric water vapor is an important driver of decadal global surface climate change".

6.3 Solar Power Input to Earth

The Sun is the powerhouse that drives the Earth's climate. No understanding of climate change is possible without an understanding of the behavior of the Sun. There are four aspects of the solar input to the Earth that are most important: (1) the total solar irradiance (TSI), (2) the spectral distribution of irradiance, (3) the magnetic fields emanating from the Sun and interacting with the Earth's magnetic field, and (4) particle emissions and flares from the Sun.

6.3.1 Total Solar Irradiance Measurements

The TSI has received the most attention since it represents the principal input to the Earth's heat budget. Unfortunately, the TSI can only be measured in space above the atmosphere, and we only have three decades of data. Even this data is not entirely reliable. Absolute calibration of an instrument to measure the solar irradiance is not easy to achieve. However, it is more important to know how the TSI varies with time than it is to know the absolute value of the TSI. A single self-consistent instrument should be able to make such measurements. Although ten such instruments have been flown on satellites, unfortunately, their longevity has ranged from about 5 to about 12 years.

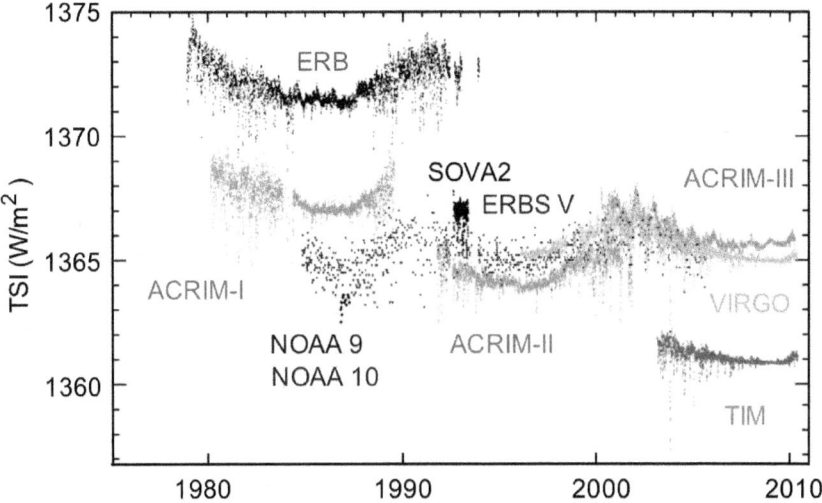

Figure 6.7. Result of the series of redundant, overlapping satellite TSI monitoring experiments that have provided a continuous record since late 1978. Adapted from Kopp. *(http://spot.colorado.edu/~koppg/TSI/).*

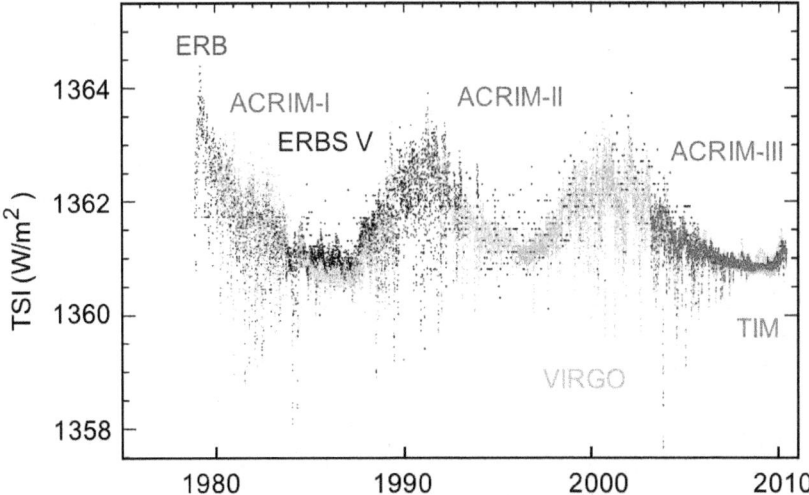

Figure 6.8. Composite of TSI measurements as developed by Kopp. *(http://spot.colorado.edu/~koppg/TSI/).*

Several investigators have scaled these various measurements so as to join them in a continuous record from 1980 to 2010, although this process involves some subjectivity. The raw data are shown in Figure 6.7 and the scaled continuous record is shown in Figure 6.8. It is evident that over the past three decades, the TSI has undulated over an ~11-year cycle with an amplitude of roughly ±1 W/m² about the mean.

6.3.2 Appearance of the Sun

Observations of the appearance of the Sun have been made over the past 400 years. The most conspicuous feature is the occurrence of sunspots. Sunspots are dark, planet-sized regions that appear on the surface of the Sun. Sunspots appear dark because they are colder than the areas around them. A large sunspot might have a temperature of about 4,000 K. This is much lower than the 5,800 K temperature of the bright photosphere that surrounds the sunspots. However, sunspots are only dark in contrast to the bright face of the Sun. If you could cut an average sunspot out of the Sun and place it in the night sky, it would be about as bright as a full moon. Sunspots have a lighter outer section called the penumbra, and a darker middle region named the umbra. Sunspots are caused by the Sun's magnetic field welling up to the photosphere, the Sun's visible surface. The powerful magnetic fields around sunspots produce active regions on the Sun, which often lead to solar flares and Coronal Mass Ejections (CMEs). Sunspots form over periods lasting from days to weeks, and can last for weeks or even months. The average number of spots that can be seen on the face of the Sun is not always the same, but goes up and down in a cycle. Historical records of sunspot counts show that this sunspot cycle has an average period of roughly 11 years although this period has varied widely over the past century. Our Sun isn't the only star with spots. Just recently, astronomers have been able to detect "starspots" (sunspots on other stars). Some sunspots expand with time and evolve into groups.

Faculae are hot structures in active regions of the Sun. Faculae = plural of facula, Latin for "small torch". They were originally discovered near the limb in full-disk images of the Sun. Faculae are the small, bright, patterns around dark sunspots and in the photospheric network.

Observations of sunspots and faculae on the surface of the Sun date back to the early 17th century although the most complete and reliable data are after about 1800. These observations showed the existence of a solar cycle long before space measurements of TSI were initiated around 1980. The incidence of sunspots varied according to a cycle of period around 11 ± 2 years. The peak sunspot incidence varied from cycle to cycle but in recent years numbered well over 100 at Solar Maximum (SMAX) and bottomed out at less than 10 at Solar Minimum (SMIN). A historical summary of group sunspot numbers is given in Figure 6.9. A historical record of sunspot numbers is given in Figure 6.10.

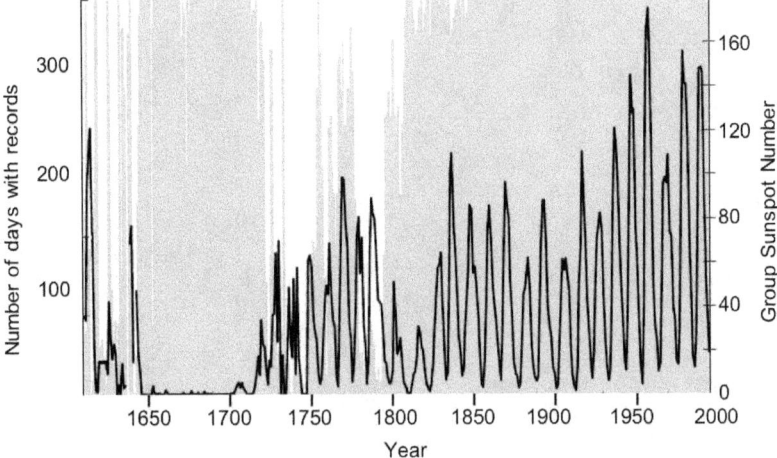

Figure 6.9. Group sunspot numbers (right scale) since 1600. Gray bars are number of days with observation records. Adapted from Vaquero (2007).

The various solar cycles are numbered, with the cycle that reached solar maximum just after year 2000 designated as Cycle 23, and Cycle 1 being the cycle that reached solar maximum around 1760.

Eddy (1976) wrote an authoritative paper on the Maunder Minimum (MM) that has greatly influenced thinking about global climate variations ever since. In this paper, Eddy traced out the history of accumulation of data on sunspot observations and suggested that the reliability of the data may be graded into four epochs.

- Reliable from 1848 to the present
- Good from 1818 through 1847
- Questionable from 1749 through 1817
- Poor from 1700 through 1748.

In addition, he traced out reports from various sources that indicated that the number of sunspots was minimal (or almost zero) for a 70-year period from 1645 to about 1715, and a more speculative indication that sunspot numbers may have been low for many years prior to 1645, which in turn might suggest that the era of higher sunspot numbers we have witnessed since 1715 may be an aberration from "normalcy". These sources were initially accumulated by E. W. Maunder around the turn of the 20th century, which is why the period of low sunspot numbers in the late 17th century is referred to as the "Maunder Minimum" (MM).

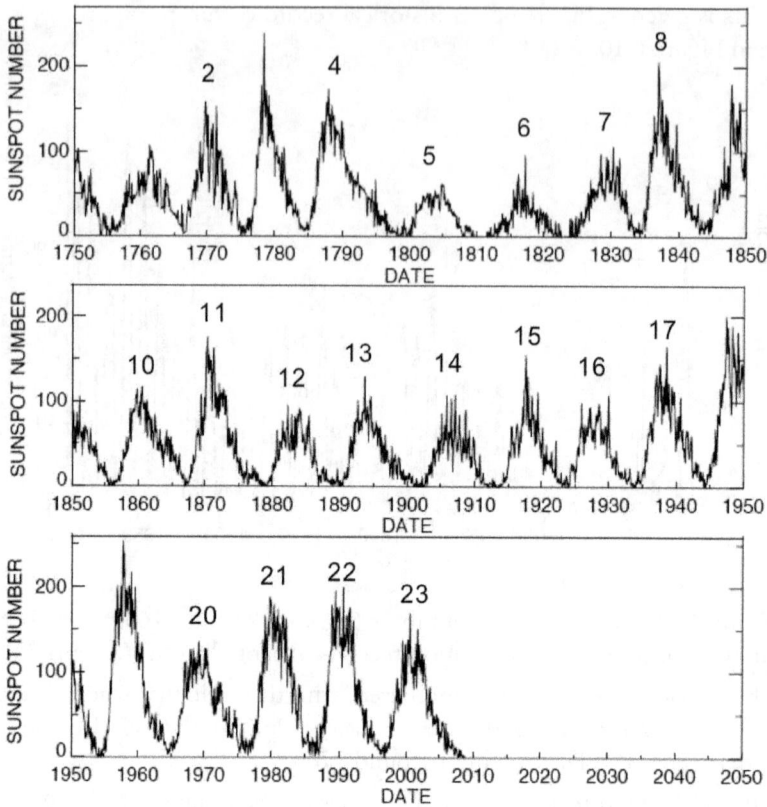

Figure 6.10. Historical record of sunspot activity (from a NASA website).

Eddy's paper is lengthy and detailed, and should be required reading for anyone interested in global climate change. Only a very brief summary is given here. Eddy provides the arguments that follow. Observation of sunspots was well within the capability of observers with simple optical telescopes in the 17th century, and probably the latter half of the 16th century. Many reports of sighting an occasional sunspot as a rarity during the MM indicate their prevalent absence during that period even though specific counts may not have been available. Eddy showed that "the period between 1645 and 1715 was characterized by a marked absence of aurorae. Far fewer were recorded than in either the 70 years preceding or following". He also presented evidence that the ^{14}C history had a marked and prolonged increase that reached its maximum between about 1650 and 1700, in remarkable agreement in sense and date with the Maunder Minimum. The flux of incoming galactic cosmic rays produces ^{14}C and trajectories of such charged particles are affected by the Sun's magnetic activity. According to Eddy (1976), historical accounts of the solar corona at total eclipse offer another possible check on anomalies in past solar behavior. He pointed out that the shape of the corona seen at eclipse varies with solar activity: when the Sun has many spots, the corona is made up of numerous long, tapered streamers that extend outward like the petals of a flower. As activity wanes, the corona dims and fewer and fewer streamers are seen. At a normal minimum in the solar cycle the corona seen by the naked eye is highly compressed and blank. Coronal streamers are rooted in concentrated magnetic fields on the surface of the Sun, which are associated with solar activity and sunspots. As sunspots fade, so do concentrated surface fields and associated coronal structures. Eddy said that first-hand descriptions of total solar eclipses during the Maunder Minimum seem entirely consistent with an absence of the modern structured corona.

Since the appearance of the Sun has changed significantly over the past four centuries, we may now ask whether this implies changes in the TSI over that period? Beckman and Mahoney (1998) commented extensively on Eddy's work. Eddy (1976) succeeded in convincing many researchers that there was real evidence for the sunspot absence in the Maunder Minimum period. He also showed that the solar corona at eclipse during the period was strongly suppressed compared with its present exhibition of major streamers. In addition, he also looked at the tree ring ^{14}C record, and confirmed a high incidence of ^{14}C during that period. While Eddy's arguments for a lull in solar magnetic activity during the Maunder Minimum seems to be widely accepted, some controversy has arisen out of his claim that the Maunder Minimum coincided in time with an era of colder weather, and that by implication the absence of magnetic activity was accompanied by a net fall in TSI. An implicit corollary is that TSI in later years has been increasing, with a consequent warming of the Earth. However, the connection of variable TSI to climate change since the Maunder Minimum (if any) is difficult to resolve.

In addition to the ~70-year period of solar quiet during the Maunder Minimum (1645–1715) Eddy discerned a period of prolonged solar quiet between about 1460 and 1550 (which he called the Sporer Minimum) and a prolonged sunspot maximum between about 1100 and 1250 that is sometimes referred to as the Medieval Warm Period (although some other studies have placed the MWP a century or two earlier). He speculated that if the prolonged maxima of the 12th and 13th centuries and the prolonged minima of the 16th and 17th centuries are extrema of a repetitive cycle of solar change, the cycle has a full period of roughly 800–1,000 years. If this change is periodic, he further speculated that the Sun might now be progressing toward a grand maximum that might be reached in the 21st to 23rd centuries. He pointed out that the overall envelope of solar activity has been steadily increasing since the end of the Maunder Minimum, giving some credence to this view. The coincidence of the Maunder Minimum with the coldest excursion of the LIA suggests possible relations between the Sun and the terrestrial climate. These coincidences suggest a possible relationship between the overall envelope of the curve of solar activity and terrestrial climate in which the 11-year solar cycle may be effectively filtered out or simply unrelated to the problem. The mechanism of this solar effect on climate may be the simple one of ponderous long-term changes in the total radiative output of the Sun, or "solar constant". These long-term drifts in solar radiation may modulate the envelope of the solar cycle through the solar dynamo to produce the observed long-term trends in solar activity. The continuity, or phase, of the 11-year cycle would be independent of this slow, radiative change, but the amplitude might be controlled by it. According to this interpretation, the cyclic coming and going of sunspots would have little effect on the output of solar radiation, or presumably on weather, but the long-term envelope of sunspot activity carries the indelible signature of slow changes in solar radiation which surely affect our climate. However, all of this is hypothetical and unproven.

Eddy made a strong case that during the Maunder Minimum, solar magnetic activity was at a minimum and may have ceased altogether for periods of years. This appears to have caused a reduction in TSI during that period, but it is difficult to evaluate the magnitude of this decrease.

6.3.3 Reconstructing Total Solar Irradiance (TSI) in the Past

A number of attempts have been made to reconstruct the TSI over the past few hundred years using various approaches and assumptions. The assumptions made by the various models are typically subjective and somewhat arbitrary. Unfortunately, there is no way to test the veracity of these models. They all seem dubious to this writer.

One of the most common approaches is to assume that changes in TSI are related to visual changes in the surface of the Sun; hence they assume that variations in TSI can be correlated to variations in sunspot number. The observed TSI record was shown in Figure 6.8. The variation in TSI between solar maximum (SMAX) and solar minimum (SMIN) is roughly 0.14%. Over the time interval covered by Figure 6.8, the sunspot numbers ranged from well over 100 at sunspot maximum to less than 10 at sunspot minimum. Any model that relates TSI merely to sunspot number based on these data will inevitably conclude that the lowest value that the TSI can ever reach (or has ever reached in the past) is when the sunspot number goes to zero, in which case the TSI would be close to what it presently is at SMIN, or about 0.14% less than it presently is at SMAX. That being the case, it would be concluded that at no time in history has the TSI ever been less than it presently is at SMIN, about 1,365.3 W/m². These models are referred to by Hoyt and Schatten (1993) as "constant quiet Sun models" (CQSMs) because the TSI for a quiet Sun is taken as constant for all time. Several models have been developed on this basis.

Note that this was written prior to more recent analyses that suggest that TSI at SMIN is closer to 1361 W/m² as shown in Figure 6.8. **Many graphs in the remainder of this chapter may need to be scaled to the lower estimate of TSI**.

On the timescale of decades to centuries, four classes of models were described by Hoyt and Schatten (1993) that postulate different variations of the Sun's output. These models can be described as:

Constant quiet Sun models. Constant quiet Sun models (CQSM) postulate that the solar irradiance has only an 11-year cycle and all radiation changes can be explained by the presence or absence of active features. Since all solar minima are essentially the same in these models, it is called the constant quiet Sun model. In this model, the historical TSI when solar activity was at a minimum (as, for example, in the Maunder Minimum) would be set equal to the contemporary TSI at solar minimum. Examples of such models include Tapping *et al.* (2006), Foukal and Lean (1990), and Vaquero *et al.* (2006).

The *solar diameter model*. The solar diameter model uses the solar diameter or its time rate of change as a proxy for solar irradiance variations. However, some controversy still exists about the history of solar diameter variations (Foukal *et al.*, 2006).

The *activity envelope model*. The activity envelope model postulates that long-term solar irradiance variations follow the envelope of solar magnetic activity so that solar minima irradiances vary over time according to such a cycle.

The *umbra/penumbra (U/P) variations model*. The U/P models are so-called because early models of this class used sunspot structure expressed as the ratio of umbral areas to penumbral areas as a proxy measure of solar irradiance.

Subsequent studies have used solar equatorial rotation rate and sunspot cycle duration to derive similar models.

In addition to these models mentioned by Hoyt and Schatten in 1993, several models were developed in ensuing years. These include the following (named by this writer):

The *MM temperature model*. The MM temperature model is based on (i) an estimate of the temperature lowering during the MM compared with today's temperatures, (ii) an estimate of the reduction in TSI needed during the MM to produce that lowering of temperature, and (iii) a linear scaling of the change in temperature from the MM to current times, with the change in sunspot number at SMAX from the MM (zero) to today (over 100).

The *stellar Ca HK index model*. The stellar Ca HK index model is based on (i) an estimate of the enhanced level of TSI vs. the solar Ca HK index model from recent measurements, (ii) observation of the Ca HK index for non-cycling Sun-like stars, (iii) the assumption that non-cycling Sun-like stars are representative of the Sun during the MM, (iv) linear extrapolation of the dependence of TSI from current solar measurements to the expected Ca HK index for non-cycling Sun-like stars in order to estimate TSI during the MM.

The *solar cycle duration model*. The solar cycle duration model utilizes the duration of the solar cycle from peak to peak rather than sunspot numbers as a measure of TSI.

The *coronal source flux model* is based on correlations of coronal source flux with TSI. The coronal source flux, F_s, is the total magnetic flux leaving the Sun, and thereby entering the heliosphere.

6.3.3.1 Constant Quiet Sun Models

The period between 1700 and the present is important because there is a continuous record of sunspot number, which is a directly measured index of solar activity, of known pedigree with established relationships with other activity indices, and which antedates the rapid increase in anthropogenic greenhouse gases that began with the industrial revolution. A vital aspect of climatology studies is the issue of how the TSI varied over the past several centuries, particularly back to the Maunder Minimum (1645–1715).

Group Sunspot Numbers of variable quality are available since 1610. Sunspot areas and white light facular areas are available since 1874. Various models use one or more of these proxies in order to reconstruct historical irradiance. One parameter that has been used by a number of investigators to estimate TSI as far back as the 17th century is the Group Sunspot Number. The essential assumption made in these models is that the relationship between sunspot number and TSI is embodied in Figure 6.8; the lowest TSI in this figure is associated with very low sunspot numbers and the highest TSI in this figure is associated with very high sunspot numbers. The underlying hypothesis is that over time periods of hundreds of years, the Sun's output changes by only small amounts. These changes are due to the fact that faculae are associated with sunspots, and the faculae produce more irradiance than the sunspots reduce irradiance. Nenvertheless, the sunspots provide an indicator for the small variations that take place about the basically quiet Sun. With this assumption, the TSI can never drop below about 1361 W/^{2} or exceed about 1363 W/m^2. (As we pointed out, these revised values are roughly 4 W/m^2 lower than the canonical values accepted prior to about 2008). It is rather unfortunate that many of these papers have provided the end result: plots of TSI vs. year for hundreds of years, but they have not usually provided the detailed algorithms from which these data were derived. Since some of these are essentially one-parameter models; there must exist a functionality:

$TSI =$ function of (N)

where N = Group Sunspot Number. This functionality does not seem to appear in most of the relevant papers. In some cases, the models use long-term sunspot data and other proxies, in which case the functionality would take the form:

TSI = function of (N, a, b, ...)

where a, b, ... are parameters associated with proxies, but once again these functionalities are not typically stated very clearly.

Krivova *et al.* (2009) reported on their SATIRE model – a CQSM that has been under development for a number of years. As they pointed out: "The assumption underlying SATIRE is that all changes in the solar irradiance on time scales longer than hours are solely due to changes in the solar surface magnetic flux as traced through surface features, such as spots and faculae". This has been shown to work quite well for the time period of the last few decades where TSI data are available. It has been estimated for example that in solar cycle 23, the TSI variation from SMIN to SMAX was estimated to be a darkening of ~ 0.8 W/m² due to sunspots and a brightening of ~1.7 W/m² due to faculae. Their model is pegged to sunspot data, and the underlying assumptions lead to a CQSM. Their results indicate that the TSI reached a minimum of 1364.8 W/m² during the Maunder Minimum, and then rose over the following centuries to about 1366 W/m². (As we pointed out, these revised values are roughly 4 W/m² lower than the canonical values accepted prior to about 2008).

Tapping *et al.* (2006) developed a model for long-term TSI based on group sunspot number.

> "... since total irradiance is highly correlated with sunspot number, it seems logical to plot irradiance against sunspot number and extrapolate back to zero sunspot number, and then conclude that the corresponding value of irradiance is the value that would be reached if solar activity remained low for an extended period. *This is almost certainly not the case.* Sunspots do not cause increases in irradiance; it is the accompanying active region structures, such as faculae and elements of the active network that do this. Although there might not be any sunspots, there are signs of activity during every observed minimum of the solar activity cycle ... Sunspots might be a good indicator of magnetic activity when present, but they are not useful when activity is low. When examining solar activity during a sustained change in the solar activity cycle, or even a temporary cessation, one needs to examine two issues: firstly, does the nature of the process by which magnetic flux is processed change, and secondly, what is the solar activity machine below the photosphere doing? ... In the case of this investigation, the input to the model has to be sunspot number, which is the only direct index of solar activity available. In this paper we develop a model for the processing of solar

magnetic flux and use it to model the historical record of total irradiance" (Tapping *et al.*, 2006).

Tapping *et al.* (2006) estimated the TSI indirectly from the historical record of sunspot numbers. Although the functional relationship of TSI to sunspot number was not explicitly stated in their paper, we can surmise what that relationship must have been by comparing their estimate for TSI with the sunspot record. By associating the maxima and minima in the TSI curve of Tapping *et al.* (2006) with sunspot number, we can derive the results shown in Figure 6.11.

Using this relationship, they obtained their estimate of TSI over the past four centuries as shown in Figure 6.12. However this model rests on the assumption that over four centuries, the TSI is directly related to the number of sunspots, and the relationship between TSI and sunspot number over that time period can be taken as that observed over the past three decades.

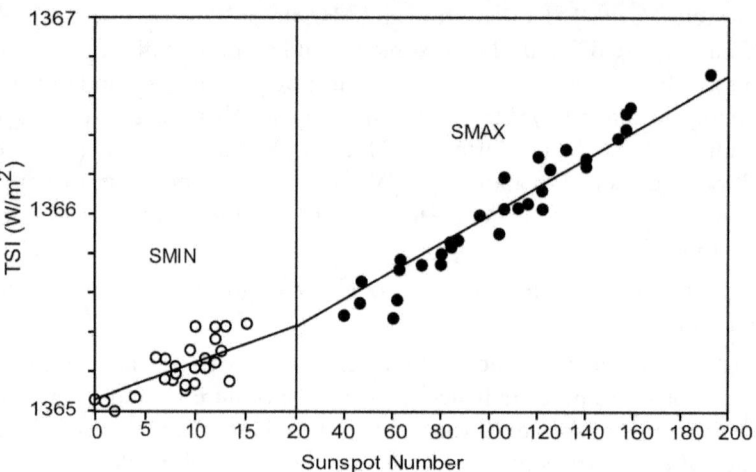

Figure 6.11. Relation between TSI and sunspot number in a CQSM. (As we pointed out, these revised values are roughly 4 W/m² lower than the canonical values accepted prior to about 2008).

According to Figure 6.12, the best estimate for the TSI during the Maunder Minimum was about 1,364.7 W/m², compared with a current average of about 1,366 W/m², for a decrease of about 1.3 W/m² (about 0.1%) during the Maunder Minimum. (As we pointed out, these revised values are roughly 4 W/m² lower than the canonical values accepted prior to about 2008). The interesting thing about this model is that although Tapping *et al.* (2006) went to great pains to emphasize

"... since total irradiance is highly correlated with sunspot number, it seems logical to plot irradiance against sunspot number and extrapolate back to zero sunspot number, and then conclude that the corresponding value of irradiance is the value that would be reached if solar activity remained low for an extended period. *This is almost certainly not the case.*"

Nevertheless they went ahead and carried out a CQSM model anyway, leading inevitably to the result that TSI during the MM was a mere 0.1% lower than it is today—*which is almost certainly not the case.*

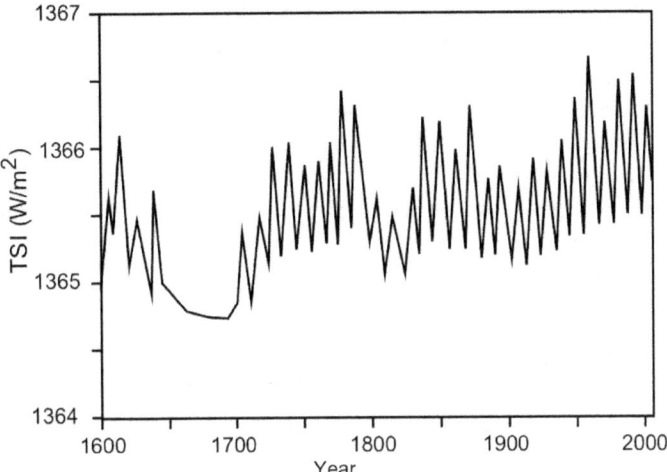

Figure 6.12. Modeled TSI through the MM up to the present. Adapted from Tapping *et al.* (2006).

A number of other CQSM models based on sunspots have been developed and they all lead to the same general result that TSI never wavers by more than about ±1 W/m² over several centuries. This would not be sufficient to account for observed temperature changes.

Several investigators have attempted to correlate TSI with solar cycle length. Archibald asserted that there is a correlation between solar cycle length and surface air temperatures during the solar cycle that follows. This is attributed to the theory as stated by Archibald:

> "... weak solar activity causes a weak solar wind, which in turn increases the number of galactic cosmic rays penetrating the Earth's atmosphere. This increases low-level cloud formation and the Earth's albedo. The Earth cools as a consequence."

The evidence for this claim is based on data that are not very convincing to this writer. It is noteworthy that solar cycle length can be determined in a number of different ways, based on maxima or minima, and some subjectivity is always involved. Thejll and Lassen (2000) provided data that appeared to suggest a correlation between solar cycle length and NH temperature. Agee *et al.* (2010) showed that the length of the solar cycle is closely related to the length of the "quiet period" defined as "the period from the first monthly sunspot number average of 10 to the last monthly sunspot number average of

10". However, Thejll (2009) provided data on cycle length that does not correlate at all with temperature.

6.3.3.2 Solar Cycles 23 and 24

During the MM in the late 1600s, there was an extended period with essentially no sunspots. Sunspot cycles 5 and 6 had unusually low sunspot numbers (see Figure 6.10). Sunspot numbers rose after about 1950 and temperatures also rose during that period. For those who would seek a relationship between climate and sunspot number (typically via TSI models) these dates are particulaly interesting. Is there a connection between rising sunspot counts after 1950 and global warming in the latter part of the 20th century? Were the climates of the MM and sunspot cycles 5 and 6 exceptionally cool? These questions are difficult to answer because the data are limited. The eruption of Tambora in 1815 cooled the Earth during solar cycle 6 but it is not possible to resolve any putative effects of low sunspot number.

A very interesting situation arose in the early 21st century. As solar cycle 23 waned, the tail of its extension through SMIN was longer than expected. SMAX for solar cycle 23 occurred around 2001, and the sunspot count continued to drop after that for eight years to 2009. As of the end of September 2008, the Sun had been blank (i.e., no visible sunspots) on 200 days of the year, representing a 50-year low. A dearth of sunspots in 2008 and 2009 was evident. Based on recent solar cycles, one would have expected the sunspot count to begin rising with the advent of solar cycle 24 at least two to three years earlier than it did. This stimulated a number of studies, commentaries and predictions for solar cycle 24. Several of these predicted that solar cycle 24 would have an unusually low number of sunspots. The effect of this on climate was subject to some speculation. Archibald (2009) showed that the sunspot vs. time plots for cycles 22 and 23 were very similar to cycles 3 and 4 and therefore suggested that cycles 24 and 25 might possibly resemble cycles 5 and 6 with their low sunspot counts. Feulner and Rahmstorf (2010) pointed out:

> "The current exceptionally long minimum of solar activity has led to the suggestion that the Sun might experience a new grand minimum in the next decades, a prolonged period of low activity similar to the Maunder minimum in the late 17th century."

A number of other predictions were made for solar cycle 24 and beyond. Svalgaard et al. (2005) predicted that cycle 24 would be the "smallest in 100 years." Mackey (2007) reviewed a number of studies of solar cycles and predictions for cycle 24. He suggested that as a consequence of the low solar activity, there might be "thirty years of global cooling commencing in 2008 that would have adverse consequences for humanity." Schatten and Presnel (2007) predicted that cycle 24 would peak in 2013 at about 130 sunspots—

about 65% of the maxima of cycles 21 and 22. Choudhuri *et al.* (2007) predicted a peak for solar cycle less than half of that of cycles 21 and 22, in the 2013 time frame. Bhatt (2009) predicted a slightly weaker cycle 24 with a peak of 111 sunspots, compared with 120 for Solar Cycle 23. Others have predicted peaks as low as 50 sunspots (e.g., Badalyan *et al.*, 2004). By contrast, Dikpati *et al.* (2006) predicted: "that cycle 24 will have a 30–50% higher peak than cycle 23, in contrast to recent predictions by Svalgaard *et al.* and Schatten." Kane (2008) indicated that depending on which method of correlation was used, the prediction for cycle 24 could vary over a wide range, but he seemed to predict that cycle 24 would be stronger than cycle 23, which does not appear to be occurring in reality.

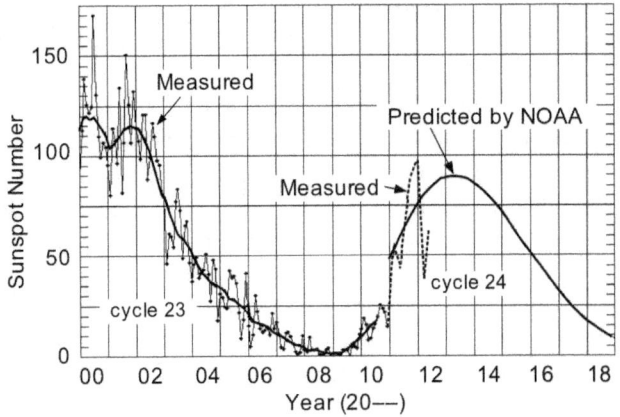

Figure 6.13. Measured sunspot numbers for solar cycle 23 and predicted sunspot numbers for solar cycle 24 as of January 2011 (NOAA). Dashed line shows initial measurements of cycle 24. Note sharp upward jump at beginning of cycle.

In December 2006, Hathaway and Wilson were quoted with an untenable prediction as follows:

"Evidence is mounting: the next solar cycle is going to be a big one. Solar cycle 24, due to peak in 2010 or 2011 'looks like its going to be one of the most intense cycles since record-keeping began almost 400 years ago, ' says solar physicist David Hathaway of the Marshall Space Flight Center. He and colleague Robert Wilson presented this conclusion last week at the American Geophysical Union meeting in San Francisco." (*http://www.physorg.com/news86010302.html*)

First of all, cycle 24 is clearly not going to a very intense one. Secondly, it appears that cycle 24 will likely peak around 2013, not 2010-2011. Callebaut (2008) predicted a deep minimum for cycle 26 with a climate similar to that of the MM. He suggested that global temperatures during cycle 26 might drop by 1°C. The actual data are shown in Figure 6.13. It appears that despite the

Donald Rapp

rather long extended tail of cycle 23 with low sunspot numbers persisting for some time, cycle 24 seems to be evolving in normal fashion. The problem with many areas of climatology is that they depend on a short period of limited data from which they extrapolate assuming that apparent short-term relationships persist over long periods. Usually they are wrong.

6.3.3.3 Other Reconstructions

Vaquero *et al.* (2006) developed a correlation of TSI with yearly average sunspot area based on measurements of sunspot area made since 1832. However, the units of sunspot area were not specified. Vaquero *et al.* (2006) based their paper on previous work by Solanki and Fligge (1998, 1999).

The fundamental assumption made in Vaquero *et al.* (2006) and Solanki and Fligge (1998, 1999) is that the variations in TSI that we currently observe during the solar cycle are due to changes in sunspots and faculae in active regions of the Sun. Their combined effect on TSI is embodied in a term $\Delta(AS)$ representing the change in TSI due to such solar activity. Based on observations of the Sun since 1978, this term is typically quite small. The network (and possible changes in solar convection) provide the main contribution to TSI variations on timescales longer than the solar cycle, and secular changes in the network are denoted as variations in the quiet Sun $\Delta(QS)$. However, these long-term variations are presumably much larger than those due to solar activity, and therefore use of the term "quiet Sun" is misleading. These models are similar in some ways to the CQSM, except the "constant quiet Sun" is allowed to vary slowly over time.

Thus, according to this concept, TSI at any epoch is the sum of three terms:

$$TSI = TSI(0) + \Delta(QS) + \Delta(AS)$$

where:

$TSI(0)$ = a constant to produce the correct absolute TSI

$\Delta(QS)$ = additive term to account for long-term secular variations of the so-called "quiet Sun"

$\Delta(AS)$ = additive term to account for solar activity via the solar cycle.

Since $\Delta(QS)$ varies slowly with time, measurements of TSI made in space since ~1980 over two decades can be presumed to include a current value of $\Delta(QS)$ corresponding to 1980–2000. The active Sun term $\Delta(AS)$ can be correlated with a parameter such as sunspot number or sunspot area by comparing the two decades of TSI measurements with variations in such parameters. Solanki and Fligge (1999) made such a comparison with the daily sunspot number. Although the data were very noisy, they were able to arrive at a correlation by binning the data. The result was:

$$TSI(0) = 1{,}365.4 \text{ W/m}^2$$

264

$$\Delta(AS) = 0.0161 \; (SN) - 0.000055 \; (SN)^2$$

where SN = Zurich sunspot number. As we pointed out, based on recent recalibration, TSI values should be reduced by roughly 4 W/m².

This function increases with SN until about $SN \sim 150$, and decreases for higher SN. For low SN, as SN increases, the facular area increases faster than the sunspot area. However, for high SN, the reverse occurs.

Vaquero *et al.* (2006) performed a similar correlation based on sunspot area. They obtained:

$$TSI(0) = 1,365.4 \; W/m^2$$

$$\Delta(AS) = 6.8 \times 10^{-4} \; (SA) - 1.0 \times 10^{-7} \; (SA)^2$$

where SA = sunspot area (units not given).

In either case, whether based on sunspot number or area, $\Delta(AS)$ contributes a relatively small oscillatory term to TSI that varies with the solar cycle.

The key to evaluating TSI is estimation of the term $\Delta(QS)$. However, as Solanki and Fligge (1999) said:

> "We stress, however, that determining the quantitative long-term variations of the quiet Sun is highly speculative and subject to large uncertainties."

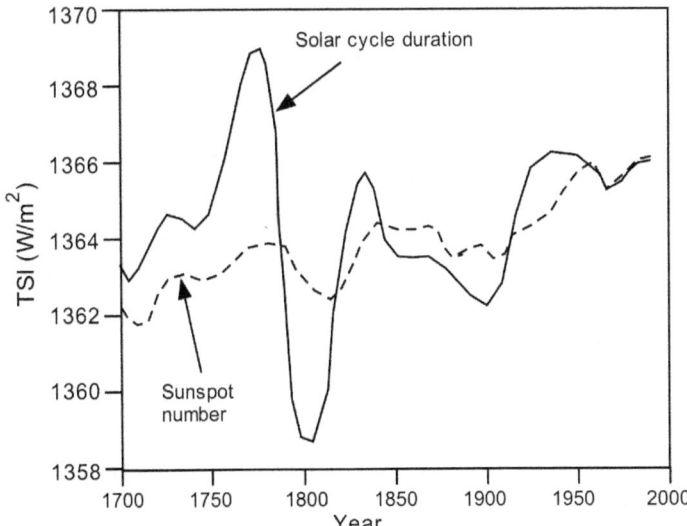

Figure 6.14. Reconstructed TSI based on sunspot number or cycle duration. Adapted from Solanki and Fligge (1999).

Solanki and Fligge (1998) generated two estimates for $\Delta(QS)$. Unfortunately, the descriptions of the procedures for doing this are rather murky to this writer. One procedure was based on an assumed linear relationship between chromospheric emission in the core of the Ca H and K lines and photospheric brightness. The other is a linear correlation between the length of solar cycles and observed brightness of solar-type stars. Each of these correlations was apparently used independently to estimate past values of $\Delta(QS)$ from observations of either the Sun's cycle duration or HK measurements. Since such measurements are limited to post-1880, Solanki and Fligge (1998) only reconstructed TSI back to 1880. Solanki and Fligge (1999) followed their 1998 paper with the intent to reconstruct TSI as far back in time as 1700. As before, Solanki and Fligge (1999) are equally obscure on the details of reconstructing $\Delta(QS)$. They mention Sun-like stars but it is not clear how stars relate to the procedure, which seems to involve either sunspot numbers or solar cycle duration. The procedure remains murky to this writer. Their results are shown in Figure 6.14. Figure 6.10 shows that the cycle length of cycle 4 was exceptionally long, while the length of cycle 5 was very short. This appears to have led to wild gyrations in the predicted TSI around 1800 based on cycle length.

6.3.4 TSI Reconstructions Based on Cosmogenic Isotope Proxies

Galactic cosmic rays are continually impinging on the Earth's atmosphere, producing nuclear reactions that generate radioactive isotopes. These isotopes gradually settle through the atmosphere and may be incorporated into biota, ice deposits, and other media that preserve the "cosmogenic isotopes" for long periods of time. Of particular interest is the occurrence of ^{14}C in tree rings and ^{10}Be in high-latitude ice cores. The interpretation of such proxies has mainly been in reference to historical surface temperatures, but proxies have also been used in a few instances to infer aspects of historical variation of TSI.

As Muscheler *et al.* (2005) pointed out:

> "The Sun influences the production rate of ^{14}C in the Earth's atmosphere by modulating the galactic cosmic-ray flux through its magnetic field. Increased magnetic field in the solar wind causes a stronger deflection of galactic cosmic rays and lower radionuclide production rates in the atmosphere, and vice versa."

Thus, the historical record of annual isotope production would have been influenced by the historical variation in the activity of the Sun. However, Muscheler *et al.* (2005) also pointed out:

> "The atmospheric ^{14}C concentration also depends nonlinearly on the geomagnetic field intensity and the global carbon cycle. These factors and their uncertainties need to be carefully included in the reconstruction of solar activity."

Similar considerations apply to other proxies such as [10]Be in ice cores. Thus, the historical isotope record contains information about past solar activity, but unraveling that information is difficult due to a number of confusing factors.

The use of cosmogenic radioisotopes as proxies for TSI variation introduces uncertainties. The radioisotopes are not produced in the Earth's atmosphere directly by the Sun's flux of energetic particles. They are produced mainly by high-energy galactic cosmic rays from outside the Solar System. Their modulation with the solar cycle is due to changes in the way these cosmic rays are shielded by the heliosphere (the solar wind). The efficiency of this shielding depends on solar plasma outflows from open magnetic fields in quiet regions and individual events such as flares and coronal mass ejections. Although this shielding increases roughly with the general level of solar activity, it is only very loosely proportional to the areas of the dark and bright magnetic structures that drive TSI. In view of this, it is unrealistic to expect a fixed relation between variations in [10]Be or ^{14}C production rate and TSI over the past millennium. The relation is complicated further by possible climate influences on the [10]Be and ^{14}C deposition rates, causing errors in the inferred [10]Be and ^{14}C formation rates (Foukal *et al.*, 2006).

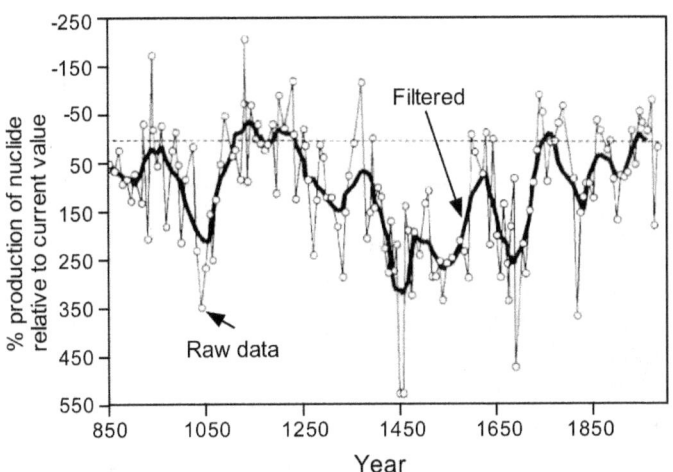

Figure 6.15. Cosmo-nuclide production as percent of present production based on [10]Be in polar ice and ^{14}C in tree rings. Adapted from Bard *et al.* (2000).

Bard *et al.* (2000) provided a profile of variation of [10]Be and ^{14}C during the past millennium. Their result is shown in Figure 6.15. The TSI is inversely related to cosmo-nuclide production, so the vertical scale is plotted inversely. The curve in this figure is claimed to be representative of TSI, but the quantitative relationship of nuclide production to TSI is not obvious.

Bard *et al.* (2000) scaled the nuclide data by setting TSI equal to the known value (1,367 W/m²) in the 1990s, and then using a parameterized decrease in TSI during the MM to generate a family of curves for TSI over the past millennium for each assumed minimum during the MM (see Figure 6.16). (As we pointed out, based on recent recalibration, these values should be reduced by roughly 4 W/m²). However, in doing this, apparently, Bard *et al.* (2000) only scaled the nuclide production rate during periods of low TSI (high nuclide production) and assumed that during periods of high TSI (low nuclide production) no scaling was needed. That is one possibility; it assumes that prior to (and after) the MM, during periods of relatively high TSI, the Sun acted as it does today. This model is the antithesis of the CQSM because in this model, the maximum TSI is fixed and the minimum TSI is varied and unbounded.

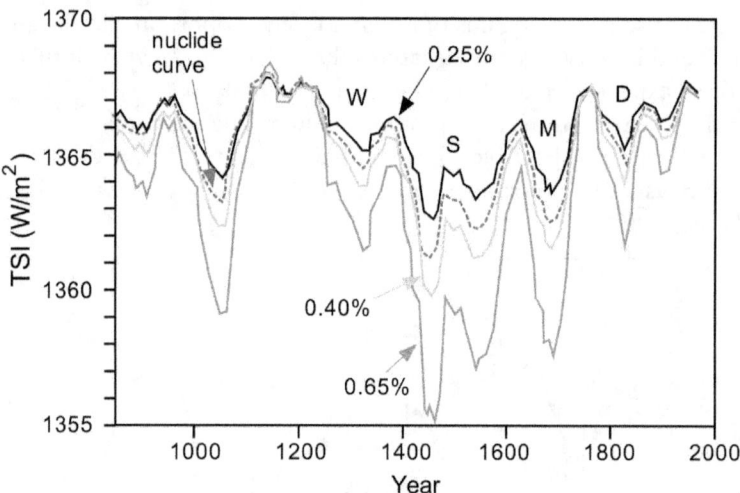

Figure 6.16. Modeled TSI during the period 1850 to the present. The blue curve is the raw cosmo-nuclide data taken from Figure 6.15. The various curves are scaled to produce 0.25%, 0.40% and 0.65% reductions in TSI during the MM as compared to today. Current TSI was set at 1367 W/m². Adapted from Bard *et al.* (2000).

Solanki *et al.* (2004) carried out a reconstruction of the sunspot number over 11,000 years back in time, based primarily on archival concentration of cosmogenic isotope ^{14}C activity in the atmosphere obtained from high-precision ^{14}C analyses on decadal samples of mid-latitude tree ring chronologies, but also including some comparison with ^{10}Be in ice cores from Antarctica and Greenland since 850. Their data for ^{14}C over 11,000 years show a slowly changing long-term decrease, with superimposed small "wiggles" of a few percent about the long-term trend. The long-term decline (indicated by the smooth curve) is caused by a reduction in ^{14}C production rate due mainly to an increase in the geomagnetic shielding of the cosmic ray flux and does not necessarily indicate any changes in TSI. The short-term fluctuations (duration one to two centuries) are assumed to reflect changes of the production rate due to solar variability. (See Figures 4.39 to 4.41 of Rapp (2010).

Thus, in order to estimate variability of TSI one would have to flatten out the smooth long-term curve to horizontal and deal only with variations about this horizontal line. Since the accuracy of the basic $\Delta^{14}C$ measurement is several percent, and the signal ("wiggles" about the trend line) amounts to only a few percent of the $\Delta^{14}C$ measurement, the accuracy of the signal will be limited. As Anon. (E) said:

> "... the extent to which cosmogenic isotope variations really indicate terrestrially relevant variations in solar energy outputs, either radiative or particle, and the scaling of the relationship over long times is poorly known; the paleo-climate record is similarly somewhat uncertain."

It is not clear to this writer how Solanki *et al.* (2004) processed their data. Their results suggest that the sunspot number is higher today than at any previous time in the past 8,000 years. These results must be considered to be highly specualtive.

An extension of the work of Solanki *et al.* (2004) was carried out by Muscheler *et al.* (2005). However, the entire procedure for data reduction and modeling remains obscure to this writer.

Steinhilber *et al.* (2009) carried out a reconstruction of TSI over the past 9,300 years. They estimated the historical open solar magnetic field from the cosmogenic radionuclide ^{10}Be measured in ice cores. The conversion of solar magnetic field to TSI was accomplished by the method of Frolich (2009) in which the recent TSI data over the past thirty years were compared with the solar magnetic field. However, this time period is too short to be certain that such a relationship existed over the past 9,300 years.

Vieira *et al.* (2011) carried out an estimate of TSI during the Holocene based on ^{10}Be measurements in ice cores. By some kind of "black magic" that is unintelligible to this writer, this led to estimates of TSI for the Holocene. Shapiro *et al.* (2011) presented their approach for modeling historical values of TSI. The methods used are also obscure to this writer.

Benestad (2005) provided an extended discussion of a theory that cosmic rays, controlled by the Sun's magnetic field, produce changes in cloud formation that affect the Earth's climate. He provides many references. Only a brief report is given here.

The theory here is that as variations in solar activity take place, the solar magnetic field and the solar wind change, and they control the amount of galactic cosmic rays from deep space that enter our solar system and penetrate the Earth's atmosphere. The solar wind thus acts like a control grid on an old-fashioned triode vacuum tube where the cosmic rays provide the "current to the anode." The theory then claims that cosmic rays enhance cloud formation by producing charged atmospheric aerosols that act as nuclei for cloud formation. Thus, according to this model, an increased flux of cosmic rays due to lower solar activity produces a cooling effect on the Earth. So, it is claimed that a putative correlation of solar activity with climate is an indicator of solar wind effects that in turn affect cosmic ray penetration, which affects cloud formation, which in turn produces cooling. Several versions of this concept have been proposed.

Patterson (2007) asserted all of this rather pompously as if it were self-evident and a proven fact.

Svensmark and Friis-Christensen (1997) compared the variation in low- to mid-latitude total cloudiness between 1984 and 1990 with the cosmic ray flux (which is inversely dependent on solar activity). During the period of minimum solar activity in 1986 total cloudiness was 3-4% higher than near solar maximum in 1990. From this they suggested that cosmic rays might enhance cloudiness possibly through a mechanism involving an increase in atmospheric ionization and formation of cloud condensation nuclei. Such an increase in cloudiness would produce a cooling effect. Over a sunspot cycle, the cosmic rays varied by 15–20%, and this correlated strongly with a 3% (absolute) variation in cloud cover over that same period. Since total cloud cover is roughly 63%, this is about a 5% relative change in cloud cover.

Kernthaler, Toumi and Haigh (1999) disputed the results of Svensmark and Friis-Christensen (1997) on the grounds that if higher latitude data are included, the correlation between cosmic rays and cloudiness is weakened. But a greater concern is that the short period involved in the study is statistically inadequate to draw firm conclusions.

Svensmark (2000) extended previous work. He showed that the production of radiocarbon-14 in the Earth's atmosphere was inversely related to the pattern of Earth temperature over the past 1,000 years, with low production of ^{14}C during the MWP and high production during the LIA. The production of ^{14}C decreased sharply in the 20th century along with global warming. Svensmark said:

> "In 1900 the cosmic rays were generally more intense than now and most of the warming during the 20th Century can be explained by a reduction in low cloud cover. Going back to 1700 and the even higher intensities of cosmic rays, the world must have seemed quite gloomy as well as chilly, with all the extra low-level clouds."

Lockwood and Fröhlich (2007) published a rebuttal to Svensmark's theory. They admitted that over the 20th century the trend in ^{10}Be has been downward as the temperature trend was upward which supports the Svensmark theory. However they claimed:

> "Over the past 20 years, all the trends in the Sun that could have had an influence on the Earth's climate have been in the opposite direction to that required to explain the observed rise in global mean temperatures."

Svensmark and Friis-Christensen (2007) developed a response to Lockwood and Fröhlich (2007). In this rebuttal they pointed out that the use of running means of global temperature data over about ten years obfuscated the fact that temperatures stopped rising after 1998. In addition, discrepancies between tropospheric temperature trends and surface temperature trends lead to different conclusions on temperature variations over the past few decades. Using tropospheric temperatures without averaging, and allowing for effects of El Niños and volcanic eruptions, Svensmark and Friis-Christensen (2007) found good anti-correlation between cosmic ray levels and global temperatures over the past few decades. It is also noteworthy that the bias of observers toward (or against) the alarmist position on global warming produced by CO_2 may have crept into the arguments. Lockwood and Fröhlich (2007) proclaimed the alarmist position with apparent satisfaction:

> "Our results show that the observed rapid rise in global mean temperatures seen after 1985 cannot be ascribed to solar variability, whichever of the mechanisms is invoked and no matter how much the solar variation is amplified."

Svensmark and Friis-Christensen (2007) took the opposite position:

> "The continuing rapid increase in carbon dioxide concentrations during the past 10-15 years has apparently been unable to overrule the flattening of the temperature trend as a result of the Sun settling at a high, but no longer increasing, level of magnetic activity. Contrary to the argument of Lockwood and Fröhlich, the Sun still appears to be the main forcing agent in global climate change."

Kniveton and Todd (2001) found a close correspondence between the cosmic ray flux and global precipitation efficiency.

Bond *et al.* (2001) found close correlations of the extent of ice-rafted debris in the North Atlantic with fluxes of nuclides produced by galactic cosmic rays over the past 12,000 years. Figure 6.17 illustrates their results. Higher levels of ice–rafted debris are expected to reflect warmer temperatures. According to Bond *et al.* (2001), high levels of [10]Be reflect lower solar activity and therefore lower TSI. However, this author is unable to find evidence supporting this assumption. It does seem likely that high levels of [10]Be may be indicative of greater cloud formation. In either case, higher levels of [10]Be would be associated with lower temperatures, not higher temperatures. Therefore the increase in ice rafted debris seems to be associated with lower temperatures, not higher temperatures contrary to the assertions of Bond *et al.* (2001).

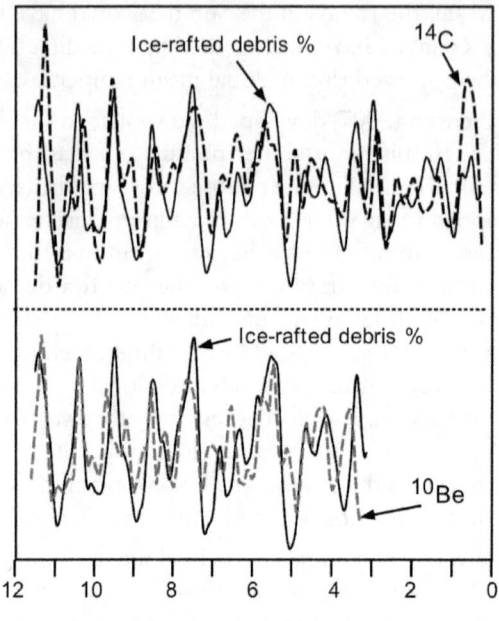

Figure 6.17. Comparison of radionuclide fluxes with relative amount of ice-rafted debris over the past 12,000 years (Bond *et al.* (2001)).

Kirkby (2008) reviewed the status of the cosmic ray theory over several time periods from the past few thousand years to hundreds of millions of years. While he concluded that "numerous paleoclimatic observations, covering a wide range of time scales, suggest that galactic cosmic ray variability is associated with climate change", he also admitted that there is considerable uncertainty in the mechanisms and the significance of the effect.

This topic should alert us to the possibility that complex processes may be at work in the Earth's climate that depend on factors seemingly unrelated to our climate. Most recently, Agee *et al.* (2011) reviewed the proposed hypothesis that "galactic cosmic rays (GCRs) are positively correlated with lower troposphere global cloudiness." They emphasized that Marsh and Svensmark (2000) and Svensmark (2007) utilized "lower troposphere cloud cover" rather than total cloud cover, and this appears to be more appropriate to the theory. However, several published papers have questioned the validity of the cloud data. Agee *et al.* (2011) examined the recent period between solar cycles 23 and 24 during which solar activity was very low, leading to "record high levels of GCRs" by correlating data on GCR levels with measurements of lower troposphere cloud cover. The found very poor correlation between cosmic rays and clouds, which seems to cast doubt on the GCR theory.

Shapiro *et al.* (2011) developed a model for historic TSI based on long-term proxies of solar activity, (^{10}Be isotope concentrations in ice cores and 22-year smoothed neutron monitor data). Instead of assuming that the current TSI at SMIN is the lowest that TSI can ever be, they assumed that in the past, the TSI at SMIN was proportional to the solar activity as recorded in proxies. During quiet periods such as the MM, the estimated TSI would be considerably lower than it is today. There were periods prior to the MM when they estimated that the TSI was even lower than it was in the MM. Over the interval from years 1600 to 2000, they found that the TSI varied from a low of 1359 W/m^2 to a current high of 1365 W/m^2, for a range of about 6 W/m^2. As they said: "The difference is remarkably larger than other estimations published in the recent literature". According to their results, TSI rose from about 1362 W/m^2 in 1900 to about 1365 W/m^2 in 2000; a forcing greater than that due to increased CO$_2$. They concluded:

> "We note that our conclusions can not be tested on the basis of the last 30 years of solar observations because, according to the proxy data, the Sun was in a maximum plato state in its long-term evolution. All recently published reconstructions agree well during the satellite observational period and diverge only in the past. This implies that observational data do not allow [one] to select and favor one of the proposed reconstructions. Therefore, until new evidence become available, we are in a situation that different approaches and hypotheses yield different solar forcing values...."

6.4 The Albedo of the Earth

The effective albedo (reflection coefficient) of the Earth for incident solar irradiance is an important factor in the global heat balance. In the simplistic picture of a solar irradiance of 342 W/m^2 spread uniformly over the area of the Earth, a 1% difference in albedo would produce a sizable forcing of 3.4 W/m^2. For lower latitudes where the solar irradiance is much higher than the average, the effect of albedo is significantly greater. Ice and snow have high albedos (0.5 to 0.9) while land (0.3) and oceans (0.1) have lower albedos. Clouds have an important effect on albedo and the presence of clouds tends to reduce differences in net albedo above land and oceans. Human-induced changes in aerosols and clouds can cause an enhanced albedo and hence cooling ("negative forcing"). Changes in land use/land clearing (LULC) also have a significant effect on albedo.

> "Many methods have been used to estimate albedo, which cannot be measured directly. These methods differ in their scattering geometries, calibration accuracy, and in spectral, space, and time coverage. The different modes of observation include measurements of earthshine

reflected from the Moon, broadband radiometer data from low orbits around Earth, geostationary cloud-cover observations, deep space radiometry, and surface radiometry. All of these methods require a theoretical model for relating the measured parameters to albedo, and they all rely on different assumptions... To date, the results from different measurement and modeling approaches are inconsistent among themselves and with each other. The magnitudes of the inconsistencies exhibited by both measurements and models of albedo changes and effects are as large as, or larger than, the entire enhanced greenhouse gas effect when compared in terms of the albedo change equivalent of climate forcing" (Charlson *et al.*, 2005).

Figure 6.18 shows a comparison of several estimates of Earth albedo. The differences between various estimates amount to over 3%, which translates into a difference in radiative forcing of over 10 W/m². Hence the uncertain overall average albedo of the Earth is a major factor in our lack of understanding of heat flows in the Earth's climate.

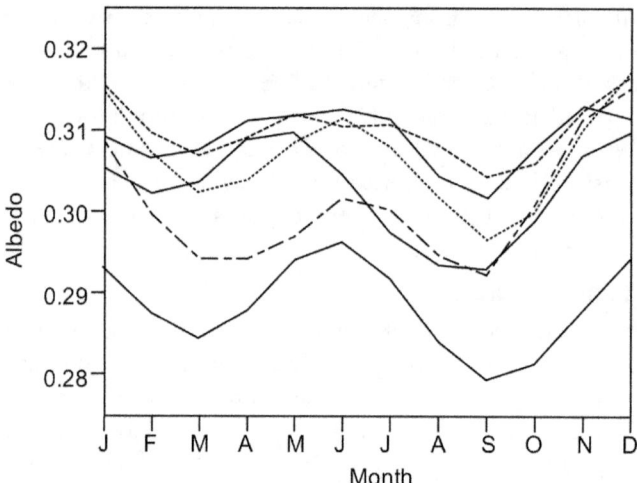

Figure 6.18. Comparison of various models for the albedo of the Earth. Adapted from Charlson et al. (2005).

Measurements of the Earth's heat budget by satellite indicate an average albedo of 34% to 36% over land, 28% to 30% over water, and 30% to 31% globally. Presumably, the presence of clouds reduces the innate difference between land and oceans.

A vital parameter for global climate analysis is the effective albedo at the top of the atmosphere (TOA) which determines the heat input to the Earth. High latitudes are typically very cloudy, especially in maritime areas, and it is expected that cloud cover would increase the TOA albedo over the open ocean, while making little difference over sea ice or snow. A distinction is therefore drawn between surface albedo and TOA albedo. The radiative effectiveness (RE) of snow or ice is defined as the change in TOA albedo for an ice/snow concentration change between 0% and 100%. It is diminished by the presence of clouds and altered by ice/snow properties. Based on data, Gorodetskaya *et al.* (2006) found RE values of about 0.2 for sea ice or land snow cover. This implies, for example, that if a 400 W/m^2 solar flux impinges on a polar area above the atmosphere, the flux reflected back to space at the top of the atmosphere from that area would increase by roughly 80 W/m^2 as the Earth's surface in the area changed from 0% to 100% ice/snow cover. Thus, even at RE ~ 20%, snow/ice cover provides a powerful feedback mechanism for climate change.

Ramanathan (1988) claimed that the albedo of the Earth would be about 10% if there were no clouds, but clouds bring it up to about 30%. An increase in the planetary albedo of 1% would zero out the predicted temperature rise for a doubling of CO_2 based on GCMs.

Based on ERBE, the Earth's albedo was estimated to be 31.3% for the 1985-1989 period. Based on CERES data the global average albedo for 2000-2004 was found to be 29.8%. The data for 2000-2004 differ somewhat from that of 1985-1989 and it is not clear how much of this was real and how much was due to differences in instrumentation. If the data are taken at face value, the solar power absorbed in 2000-2004 would have been 5 W/m^2 greater in 2000-2004 than it was in 1985-1989. This is a much greater forcing than that attributed to greenhouse gases. It seems evident from the foregoing discussion that clouds have a major impact on the Earth's energy balance. Does this represent a diminution of cloudiness due to global warming (as for example Dessler insists); is it a fluctuation independent of greenhouse gases; or is it just noisy data?

Large-scale changes in land use have undoubtedly had an effect on the Earth's climate through changes in the average albedo of the Earth. According to Bauer, Claussen, and Brovkin (2003), global forest cover diminished over the period 1000 to 1992 by 30%, from 57 × 10^6 km^2 to 41.5 × 10^6 km^2. Before 1900, forest was mainly removed in the northern subtropical and temperate regions. In the second half of the 20th century, agriculture in these regions stopped expanding and even reversed while tropical deforestation was intensified. Goosse *et al.* (2006) estimated deforestation reached 25% in the 20th century compared to year 1000.

Annual world land clearing amounts to over 100,000 km^2. A comparison of global maps showing land use (crop land, grazing land, evergreen forest, savannah, grassland, steppe, open shrub land, deciduous forest, and hot desert) reveals major changes worldwide from 1700 to 1900, and further changes from 1900 to 1990 (Pielke *et al.*, 2007b). Global cropland area increased from 2.6 million km² in 1700 to 15 million km² in 1980. Over the same period, the global extent of forest and woodland decreased by 17%, from 61 million km² to 51 million km². A significant portion of the natural vegetation of the world has been cleared to grow crops. In much of India, eastern China, the forests and shrub lands of Europe, the steppes of Asia, and the Great Plains of North America, over 75% of the land is now cropland. Bonan (2002) wrote:

"Changes in land cover alter net radiation at the surface, the partitioning of this available energy between sensible and latent heat, and the partitioning of precipitation into runoff and evapo-transpiration. This arises from differences among vegetation in albedo, roughness, leaf area index, root distribution, and stomatal conductance and changes in soil texture. For example, vegetation generally has a lower albedo than bare soil, and forests have a lower albedo than pastures or croplands. As a result, overgrazing grasslands and clearing forests for croplands increases surface albedo... Clearing forests for cropland or pastureland also reduces surface roughness. This reduces mechanical turbulence and sensible and latent heat fluxes. The partitioning of net radiation into sensible and latent heat is influenced by leaf area, stomatal conductance, root distribution, and soil texture. It is also affected by the overall hydrologic cycle. Evapo-transpiration is an important regulator of global climate.

The impact of historical land cover changes on climate is noticeable on a global scale, is comparable in magnitude to other climate forcings, and is important to reconstructing historical climate change. In particular, land cover changes in Europe and North America have likely cooled the climate of the Northern Hemisphere. If so, this land use change would have dampened the warming from increasing CO_2 and other greenhouse gases. Cooling from deforestation of Europe and North America may have contributed to the anomalously cold temperatures during the *Little Ice Age* from 1550 to 1850." (Bonan, 2002).

A number of studies have been conducted in which a land surface model is used to define the properties of the surface for various forms and distributions of vegetation on the surface. The properties relevant to climate modeling include albedo, solar absorptance, roughness, sensible heat/latent heat ratio, and heat capacity of the soil. Typically, these properties are derived from sophisticated vegetation models. The results of these surface models can be used as boundary conditions for GCMs, and the temperature at the surface can thereby be estimated for varying types of vegetation. By comparing the estimated temperature for the period prior to human intervention (land clearing, deforestation, establishment of croplands, etc.) with that after such activities, one can estimate the change in climate for any region that was induced by such human intervention. Several studies claim that deforestation produced a cooling effect of about 0.2°C from year 1000 to the 20th century.

6.5 Effect of Black Carbon Deposition on Ice and Snow

According to Jacobsen (2007) quoting the IPCC:

"Soot is an amorphous-shaped particle emitted into the air during fossil-fuel combustion, biofuel combustion, and biomass burning. Soot particles contain black carbon, organic carbon, and smaller amounts of sulfur and other chemicals.... Soot particles that fall to snow and sea ice surfaces, either on their own or within ice crystals or snow flakes, darken those surfaces, contributing to the melting of snow and ice and the warming of air above both. When soot particles age in the atmosphere, they become coated by relatively transparent or translucent chemicals, increasing their size and the probability that sunlight will hit and be absorbed by the particles. As such, aged, coated soot particles heat the air more than do new, uncoated soot particles.... Calculations suggest a strong net global warming by fossil-fuel plus biofuel soot. Soot particles containing black carbon, from fossil-fuel and biofuel burning sources, have a strong probability of being the second-leading cause of global warming after carbon dioxide and ahead of methane. Because of the short lifetime of soot relative to greenhouse gases, control of soot emissions, particularly from fossil-fuel sources, is very likely to be the fastest method of slowing global warming for a specific period."

It is well known that carbon particles resulting from incomplete combustion that deposit on high latitude snow and ice will increase the solar absorptivity of the snow and ice, producing a surface-warming effect. As the high latitude region warms, the extent and depth of snow and ice begins to diminish. This, in turn, provides a positive feedback via reduced solar reflection, producing further warming. The effect of various levels of carbon deposition on absorptivity was estimated by Grenfell *et al.* Hadley *et al.* (2010) measured the reduction in albedo due to deposition of BC on snow in the California and compared their results with other studies and models. They concluded that BC deposition is a significant factor in reduction of the snowpack in the California mountains.

BC is a mix of elemental and organic carbon emitted by fossil fuel combustion, biomass burning, and bio-fuel cooking (wood fires and cow dung) as soot. In the atmosphere, BC aerosols are mixed with sulfates and organics. Bond (2007) estimated BC emission rates by region since 1850, as shown in Figure 6.19. There was a sharp rise in BC emissions from 1900 to 1925, followed by a meandering period from 1925 to about 1970, and then a subsequent rise after 1970. The rise prior to 1925 derived mainly from the U.S. and Europe. After 1925, BC emissions from the U.S. dropped significantly and European emissions slowly diminished. Starting around

1970, emissions grew mainly from developing nations, particularly from China and other Asian nations. More recently, Bond (2010) estimated that in 2000, BC energy-related emissions (in ktons/year) were approximately as follows:

China	1,300
India	500
Other Asia	800
Latin America	300
North America	500
Europe	500
Former USSR	300
Middle East	100
Africa	500
Total	4,800

These figures do not include open burning which would add considerably to the above.

The BC emissions prior to about 1940–1950 were primarily from the U.S., Europe, and the former U.S.S.R. These locations are proximate to high-latitude regions of the NH. Beginning around 1970, BC emissions from the U.S., Europe and the former U.S.S.R. diminished and the increase in total BC emissions was due to Asian nations, further removed from the high-latitude regions of the NH. Hence, it is to be expected that BC deposition in high-latitude regions of the NH would have peaked around 1925. As Bond (2010) showed, BC emissions in the United States peaked at around 1,200 ktons/year in the late 1920s, and has now dropped to about 400 ktons/year despite growth in energy usage, due to regulation of power generation and industry.

McConnell *et al.* (2007) used measurements of central Greenland ice cores to assess the origin and climate forcing of BC in snow during the past 215 years. Air mass back-trajectory modeling suggested that the eastern and northern U.S. and Canada were likely source regions prior to about 1950. They concluded that conifer combustion was the major source of BC in Greenland before 1850 and it remained a significant source during summer throughout the 215-year record. They found that BC concentrations varied significantly during the past 215 years and were highly seasonal and erratic. They made rough estimates of surface radiative forcing which ranged from about 2 to 4 W/m^2 between 1900 and 1950.

It seems unlikely that BC alone would account for Arctic warming after 1976 because so much of the BC emissions of the past 30 years derive from Asian sources, further removed from the North Polar region.

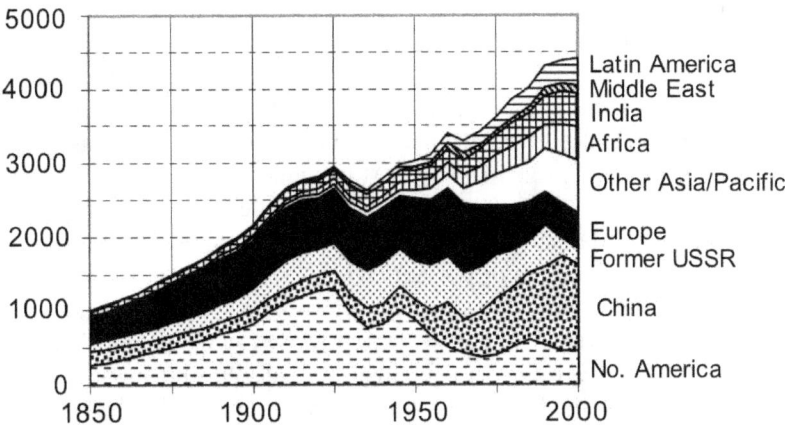

Figure 6.19. History of BC emissions by region since 1850 (Bond, 2007). Doherty et al. (2010) collected samples of Arctic snow

"... in Alaska, Canada, Greenland, Svalbard, Norway, Russia, and the Arctic Ocean during 2005–2009, on tundra, glaciers, ice caps, sea ice, frozen lakes, and in 10 boreal forests. Snow was collected mostly in spring, when the entire winter snowpack is accessible for sampling. Sampling was carried out in summer on the Greenland ice sheet and on the Arctic Ocean, of melting glacier snow and sea ice as well as cold snow".

They concluded: "the reduction of snow albedo is primarily due to BC, but other impurities, principally brown (organic) carbon, are typically responsible for 40% of the visible and ultraviolet absorption".

Several studies have been conducted using sophisticated climate models to estimate the contribution of black carbon deposition on snow and ice to global warming. Hansen and Nazarenko (2003) estimated that perhaps a quarter of the global warming of the 20th century was due to soot. These studies generally use models that were devised to estimate temperature rise due to increased greenhouse gas concentrations, particularly CO_2. The effect of BC can be taken into account by either adding the estimated forcing due to reduced albedo to the climate model, or by attributing the difference between the model and observed data to BC. In either case, questions remain as to how climate models utilized putative increased water vapor and clouds. Neither approach is likely to be very accurate. Although Jacobson (2002) continued to believe that greenhouse gases were the main cause of global warming, he concluded that "eliminating all fossil fuel BC + OM [organic matter] could eliminate 20–45% of net warming ... within 3–5 years if no other change occurred." He also concluded: "emission reduction of fossil-fuel particulate BC plus associated organic matter (OM) may slow global warming more than any emission reduction of CO_2 or CH_4."

Flanner *et al.* (2007, 2008) pointed out that "of 22 climate models contributing to the IPCC Fourth Assessment Report, 21 under-predict the rapid warming (0.64°C/ decade) observed over springtime Eurasia since 1979." They were therefore motivated to study the effects of BC + OM on climate change. To do this, they utilized sophisticated, comprehensive climate models to estimate the role of BC + OM in climate change. Their model requires a large number of parameters. This was the first global climate study that treated coupled snow aerosol heating and snow aging. The model required "plausible" ranges of present-day BC/snow forcing using combinations of BC emissions, BC optical properties, snow aging, meltwater scavenging of BC, and snow cover fraction. They estimated global annual mean BC/snow surface radiative forcings from all BC sources. While the authors clearly used the best values that could be estimated, it is difficult to resolve their reliability. There seems to be a considerable range of values, depending on the year chosen, as well as on the specific assumptions that were used. Nevertheless, the results are very interesting. These results should be read with the understanding that in most cases, they were presented as global averages, and not as specific effects for the 66.5°N–90°N Arctic region. But because all the effects of BC + OM are induced in the 66.5°N–90°N Arctic region, the estimated worldwide climate impact will depend on the model's ability to project regional forcings on global climate. The forcings produced in the Arctic region (W/m²) at the snow level due to various factors was estimated to be as follows:

BC emissions	0.54–2.00
Snow aging	0.58–1.58
Melt scavenging	0.69–1.08
Optical properties	0.88–1.12
Snow cover fraction	0.83–1.08

These are very significant radiative forcings. However, most of their data were scaled to worldwide forcings that reduced the magnitudes significantly. Whereas they claim that the effect of inclusion of BC in snow was about +0.10°C to 0.15°C, the effect on the 66.5°N–90°N Arctic region was estimated to be 0.5°C to 1.6°C. Because of large Asian emissions of BC, large forcings were estimated over the Tibetan Plateau (30– 40°N, 80–100°E), averaging 1.5 W/m² over all land. During some spring months, forcing over snowy areas exceeds 10 W/m² and 20 W/m² over parts of eastern China and the Tibetan Plateau, respectively. In this context, they examined the recent, rapid springtime warming observed over Asia, where BC emissions rose from roughly 1.6 Tg/yr to 2.6 Tg/yr during 1980–2000. For this region, global warming due to BC + OM was estimated to be comparable with that due to greenhouse gases. They concluded: "on the global scale, positive surface and atmosphere forcings from carbonaceous particles drive significant reductions in springtime snow cover."

Shindell and Faluvegi (2009) utilized sophisticated climate models to estimate the historical temperature of the Earth by region. Unfortunately, using the usual forcings, they were unable to come anywhere near the historical variation of temperature in the high NH latitudes. They postulated that deposition of BC on snow and ice in this region produced a much greater change in temperature than in other latitude regions. While they did not fully account for the dip from 1940 to 1970, they did point out that the rate of BC emissions slowed down after the 1920s and then turned up again in the 1970s. The total temperature rise in the 60°–90° latitude region of the NH in the 20th century was about 2.25°C. Shindell and Faluvegi (2009) estimated that somewhere between 0.5°C and 1.4°C of this amount was due to BC (22% to 62%). From 1976 to 2007, they estimated that the temperature increase was more than 50% due to BC. However, their method attributed the gap between climate theory and surface temperature observation to BC, but they did not invoke any direct measurements of BC levels or models to estimate forcing due to BC.

IPCC (2007) estimated the current forcing due to BC as +0.34 W/m² based on models of the global atmosphere. It can be compared with the forcing of carbon dioxide, which they estimated as +1.66 W/m². However Bond (2010) suggested that these results neglected other effects, and she suggested that the forcing due to BC is about 0.5 W/m². She also said:

"The emission rate of black carbon is another important factor in determining its forcing. Forcing is directly proportional to emission rate, so if emission estimates are doubled, the forcing estimate will double as well. Atmospheric measurements suggest that our current estimate of year 2000 emissions is too low in some regions. Forcing estimates as high as 1 W/m² have been published and are usually associated with models that assume more black carbon in the atmosphere than other models."

Wang (2002) modeled deposition of black carbon in Arctic areas. He was able to reproduce measured loadings in several regions. Surface warming was estimated in the range 0.6 to 0.9 W/m². However, black carbon aerosols suspended in the atmosphere can act in the opposite direction by reflecting incoming solar irradiance.

Hadley and Kirchstetter (2012) "developed processes for making both pristine and BC-laden snow and techniques for measuring the morphology, albedo and BC content of snow. These methods have allowed [them] to quantify the snow-albedo reduction associated with increasing amounts of BC and as a function of snow grain size. Their results are summarized in Figure 6.20. These results suggest that warming due to BC could have been a major factor in 20th century warming.

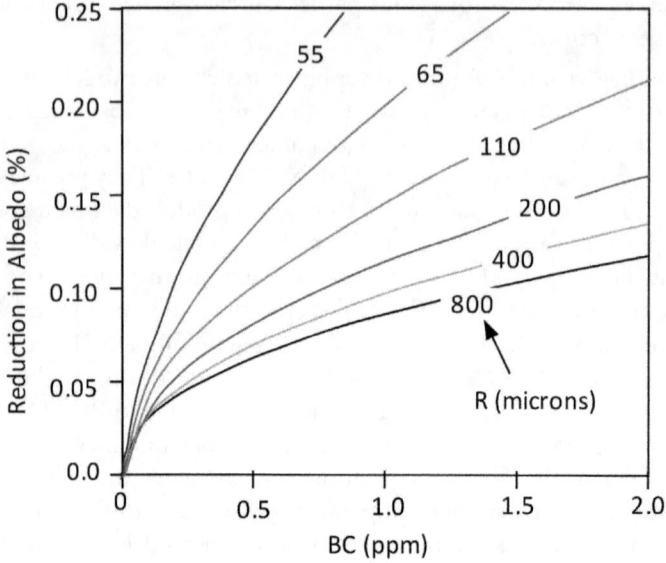

Figure 6.20. Snow-albedo reduction attributed to BC (Hadley and Kirchstetter, 2012).

In March 2012, the EPA published a 388-page report on BC including emissions, health effects, climate effects, and mitigation procedures. However, it seems possible that this report might have underestimated the effects of BC, considering that it depends heavily on papers written by Hansen *et al.*

6.6 The Dynamic Earth

The Earth is a dynamic system involving energy and mass transfers between solids, liquids and gases. Heat generation at the surface from the interior of the Earth is trivial compared to the solar input. The solar input is not completely steady. As we show in Section 6.3.1, there is a small variation in solar intensity across the ~ 11 year solar cycle, and there are small variations from solar cycle to solar cycle. There are larger variations in solar intensity during the course of a year due to the ellipticity of the Earth's orbit. At aphelion, the solar intensity at aphelion is about 6.6% lower than at perihelion. Since the global average solar intensity is about 341 W/m², this corresponds to an annual change in global average solar intensity of ±11.3 W/m². Furthermore, in the tropics, the typical solar intensity is over 1,200 W/m², so the annual variation in the tropics is more than ±40 W/m². These are huge variations compared to forcing from greenhouse gases. The peak intensity occurs around July 1, corresponding to midsummer in the Southern Hemisphere and midwinter in the Northern Hemisphere. Within the Earth system, there are many dynamic systems. In the atmosphere there occur air movements, storms, and weather. In the oceans, there are currents and surface variations such as El Niño – La Niña transitions. In addition, volcanic eruptions affect the Earth's heat balance. All of these factors (and more) result in the wide range of regional weather that we experience on Earth. Yet, averaged over longer time periods than a year, the Earth tends to establish a meandering equilibrium in which average incoming solar irradiance is balanced by average outgoing radiation to space. If no explicit external or atmospheric composition changes occur over an extended period of time (at least several decades) we might attribute the average of this meandering equilibrium as a "climate".

This meandering equilibrium can only be established because of the oceans. While most climate discussions focus on air temperatures, the oceans are the dominant factor in controlling variaitions in the Earth's climate. Oceans cover about 70% of the Earth. Most of the incident sunlight falling on the oceans is absorbed. The oceans also engage in an active exchange of IR radiation with the atmosphere. Because of the high heat capacity of the oceans and the relatively long time constant of the oceans for thermal change, the oceans stabilize the otherwise huge variations in the meandering equlibrium.

Schwartz (2007) estimated the heat capacity, time constant, and sensitivity of Earth's climate system. The following discussion of the Earth's heat capacity is excerpted from his paper.

The Earth's climate system involves a radiative balance between absorbed short-wave (solar) radiation Q and long-wave (thermal infrared) radiation emitted at the top of the atmosphere (TOA), E. At equilibrium,

$$Q \approx E$$

The global annual mean absorbed shortwave irradiance is:

$$Q = \gamma J$$

where γ is the mean planetary co-albedo (one minus albedo) and J is the mean solar irradiance at the top of the atmosphere ≈ 343 W/m². Satellite measurements indicate $Q \approx 237$ W/m², corresponding to $\gamma \approx 0.69$ (albedo ≈ 0.31). Schwartz sets the global annual mean emitted long-wave irradiance at

$$E = \varepsilon \sigma T^4$$

where ε is the effective planetary long-wave emissivity, σ is the Stefan-Boltzmann constant, and T is the global mean surface temperature. However, the heat loss at the TOA is depenedent on the temperature at the TOA, which is quite different than the temperature at the surface.

An energy imbalance $Q-E$ arising from a secular perturbation in Q or E results in a rate of change of the global heat content given by

$$dH/dt = Q - E$$

Donald Rapp

where dH/dt is the change in heat content of the climate system. But the definition of the heat capacity of the Earth is given by the equation:

$$dH/dt = C \, dT/dt$$

where C is an effective heat capacity that reflects only that portion of the global heat capacity that is coupled to the perturbation on the time scale of the perturbation. In the present context of global climate change induced by changes in atmospheric composition on the decade-to-century time scale, the pertinent heat capacity is that which is subject to a change in heat content on such time scales. Measurements of ocean heat content over the past 50 years indicate that this heat capacity is dominated by the heat capacity of the upper layers of the world oceans.

Combining previous equations, we have:

$$C \, dT/dt = \gamma \, J - \varepsilon \, \sigma \, T^4$$

This equation can be solved for a few simple special cases. For the case of a step-function forcing in which $F = (Q - E)$ makes a step change from one constant level to another constant level, the forcing F is defined as the difference between these two levels. In this case, the change in temperature (from the original temperature prior to the step-change) at a time (t) after the step-change is

$$\Delta T(t) = (\tau/C) \, F \, [1 - exp(-t/\tau)]$$

where τ is a constant of integration that characterizes the e-folding time over which the system readjusts to a new temperature.

For large t, the exponential becomes negligible, and

$$\Delta T(\infty) \rightarrow (\tau/C) \, F$$

But this is the expression for climate sensitivity in which the change in temperature is proportional to the forcing. Therefore we can identify the climate sensitivity parameter as:

$$\lambda = (\tau/C)$$
$$\Delta T(\infty) = \lambda \, F$$

and λ determines the ultimate temperature rise produced by a step-function forcing F. It turns out that:

$$\tau = C \, T_o/(4 \, J \, \gamma)$$

where T_o is the initial temperature before forcing, so that

$$\lambda = T_o/(4 \, J/\gamma)$$

As Schwartz (2007) showed, with $T_o = 288$ K, $J = 343$ W/m², and $\gamma = 0.69$, the estimated value for λ is 0.3°C per W/m². If there is an initial forcing F, that engenders additional forcings via feedback, we can multiply λ by a feedback parameter (f). Climate models tend to predict a climate sensitivity of

0.6 to 0.8°C per W/m², suggesting a feedback parameter of 2 to 2.5. This is based on assumptions regarding changes in humidity and cloud cover that attend surface warming, and these assumptions are fragile as we have discussed previously.

Schwartz (2007) also mentioned that if the forcing is not a step function, but a continuous ramp-up, as in $F = \beta\ t$, the solution of the time-dependent equation becomes

$$\Delta T(t) = (\beta\ \lambda)\ [(t - \tau) + \tau\ exp(-t/\tau)]$$

As before, for t $\gg$ τ the exponential is negligible, and we obtain

$$\Delta T(t) \rightarrow (\beta\ \lambda)\ (t - \tau)$$

showing that the temperature increases continuously in proportion to the forcing. The temperature follows the forcing with a time lag of τ.

The equation $\lambda = (\tau/C)$ defines a relation between C, λ and τ. If any two are known, we can calculate the third. Schwartz (2007) estimated τ by two methods. One was based on rates of decay of impacts of volcanic eruptions, and the other is an abstract technique based on the range of observed fluctuations of temperature over the past century, which is limited by the rate at which the system equilibrates to a perturbation. The result was $\tau \approx 5$ years. Both the values of τ and λ are lower than accepted values.

Based on $\tau \approx 5$ years and $\lambda \approx 0.3$°C per W/m², Schwartz (2007) estimated the heat capacity of the Earth to be ≈ 16 W-years/°C per m². About 85% of this is due to the oceans and about 15% to the land.

The rather short time constant of the climate system determined by this analysis implies that the climate system is in near-equilibrium with applied forcings.

Schwartz' paper suggested a lower value of λ (0.3°C/(W/m²)) than the values typically deduced by climate models (*e.g.* about 0.8°C/(W/m²)). This would suggest a lesser increase in global temperature (1.2°C) due to a doubling of CO_2 than climate models suggest (~ 3°C). As usual, the cabal sought damage control with a flurry of comments (e.g. Foster *et al.*, 2008) that led Schwartz to increase his estimate of λ from 0.3 to 0.5°C/(W/m²) which implies τ ~ 8 years and the estimated temperature rise due to the doubling CO_2 would be 1.8°C.

6.7 The Climate Debate Revisited

Table 6.1 provides a comparison of the results we have obtained in Chapter 6 with some of the questions raised in the climate debate.

Table 6.1. Dimensions of the climate debate.

Aspect	Results
Rising CO_2 was the cause of rising temperatures in 20th century	There is no credible estimate of the contribution of rising greenhouse gas concentrations to rising temperatures in the 20th century. It seems likely that greenhouse gases made some contribution but natural variability due to internal changes was far from negligible. The heat balance of the Earth depends on a delicate balance between opposing large numbers and is difficult to unravel accurately.
CO_2 concentration will rise further in 21st century in business-as-usual	Agreed
Climate models provide reasonable estimates of future warming in business-as-usual scenario	Climate models do not have adequate data on how clouds, humidity and aerosols vary as greenhouse gas concentrations rise. Nor are variations of the Sun understood at all despite many studies. Furthermore, these parameters may be controlled by factors other than greenhouse gas concentrations.

7 ENERGY AND CLIMATE IN THE 21ST CENTURY

7.1 World Energy Requirements

Hoffert *et al.* (1998) analyzed the requirements to provide the world with needed energy while keeping a lid on ultimate CO_2 levels. This study is particularly worthy of attention because it requires first and foremost that the world be provided with energy. Some proposals to cut CO_2 emissions do not require that the world be provided with enough energy to operate. In their formulation, the rate of emission of carbon to the atmosphere as CO_2 is the product of four terms:

(1) World population

(2) Gross domestic product (GDP) per person averaged for the world

(3) Energy required by the world per unit of GDP

(4) Mass of carbon emitted per unit energy consumed.

The first three terms define the amount of energy the world requires. The fourth term defines the amount of emissions for the energy consumed. Hoffert *et al.* (1998) utilized a baseline of the so-called "IS92a" projection made by the IPCC in 1992 as a "business as usual" scenario, and then departed from there to various hypothetical constraints on CO_2 production. The IS92a projection of world population is shown in Figure 7.1 with an estimated world population of 5.3 billion in 1990, ~6.6 billion in 2006, 7 billion in 2012, and an eventual plateau of 11.4 billion at the end of the 21st century. It should be noted that while IS92a allowed for significant improvements in energy efficiency and gradual expansion of renewable energy, it relied heavily on coal in the 21st century. It now seems likely that a significant fraction of the coal can be replaced by natural gas which would reduce future carbon emissions.

The GDP per person was estimated by IS92a to be U.S. $4,100 in 1990 and was projected to increase at around 1.6% per year through the rest of the 21st century. The energy per unit GDP was estimated by IS92a to be 0.49 Watt-years per U. S. $ in 1990 and was (optimistically) projected to decrease at the rate of 1% per year through the remainder of the 21st century.

The expected mix of energy sources for electric power is shown in Figure 7.2. In this scenario, the contribution made by renewables increases continuously, reaching about 30% by year 2100. The contribution made by nuclear slowly increases while coal remains the dominant source of electric power throughout the 21st century. However, as we stated previously, it seems likely that a significant portion of the coal in this figure could be replaced by natural gas.

The IS92a projection for mass of carbon emitted per unit energy consumed is shown in Figure 7.3. This model evidently builds in an implicit departure from fossil fuels, as well as increases in efficiency, leading to a steadily decreasing carbon factor with time.

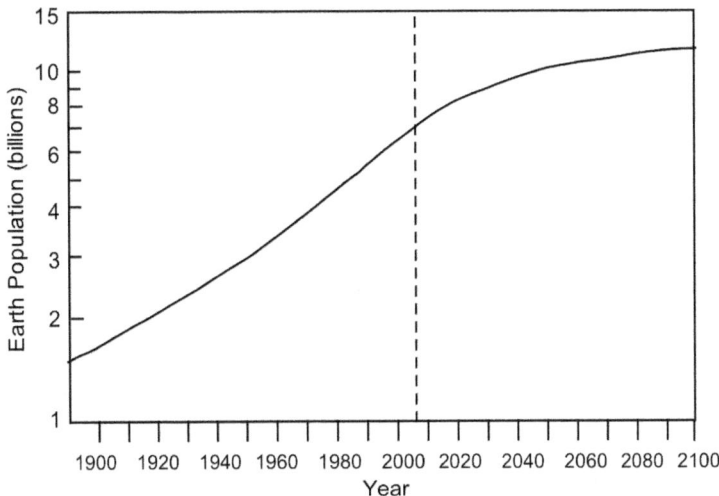

Figure 7.1. Projection of world population made by IS92a. Adapted from Hoffert *et al.* (1998).

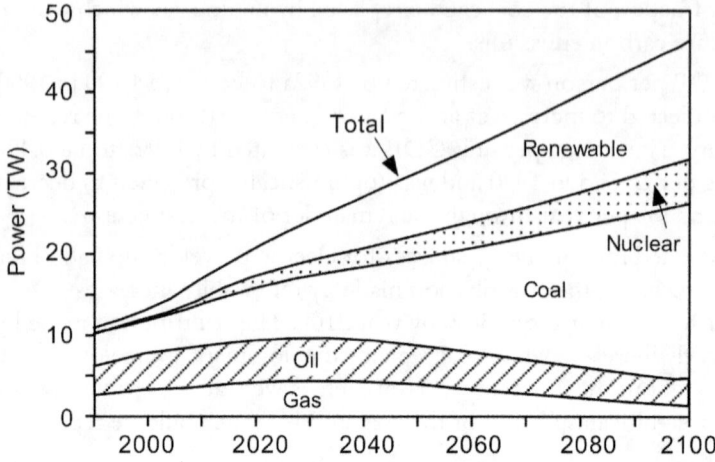

Figure 7.2. Energy mix for generation of electric power in the 21st century in the IS92a projection.

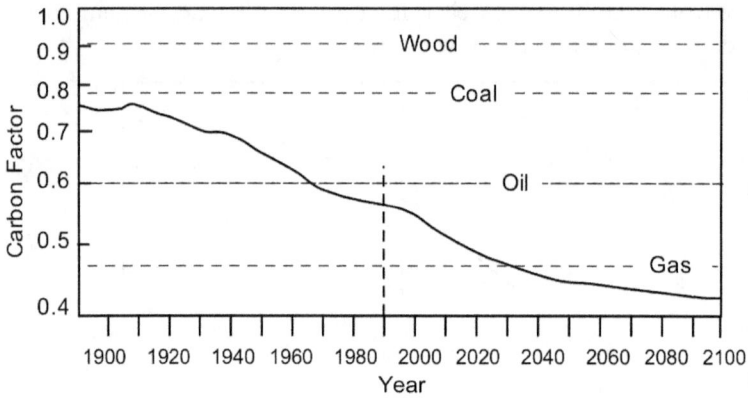

Figure 7.3. Projection by IS92a of "carbon factor" = kg of carbon emitted per Watt-year of energy generated in the 21st century. Adapted from Hoffert *et al.* (1998).

Hoffert *et al.* (1998) did not specify the ultimate rise in CO_2 concentration that results from the IS92a projection, but it appears from Figure 5.4 to be around 750 ppm by 2100. To achieve lower CO_2 concentrations, emissions must be reduced significantly.

> "Stabilizing atmospheric CO_2 at twice pre-industrial levels while meeting the economic assumptions of 'business as usual' implies a massive transition to carbon-free power, particularly in developing nations" (Hoffert *et al.*, 1998).

Hoffert *et al.* (1998) estimated the reductions in carbon emissions per year needed to stabilize the CO_2 concentration at various levels by 2100 as shown in Figure 7.4, based on the stabilization paths of Wigley, Richels, and Edmonds (1996). To achieve lower CO_2 concentrations, emissions must be significantly lower.

The increase in CO_2 concentration in the industrial era was shown in Figure 5.2. It may appear that the CO_2 concentration is headed ever upward in the 21st century. Climatologists tend to be extrapolators. Taking trends from the past, they project forward into the future, often without regard to the constraints of finite resources. Figure 7.4 shows the range of possible future CO_2 emissions that has been used by climatologists to predict future global warming. Each curve in this figure corresponds to a different stabilized CO_2 concentration around year 2100.

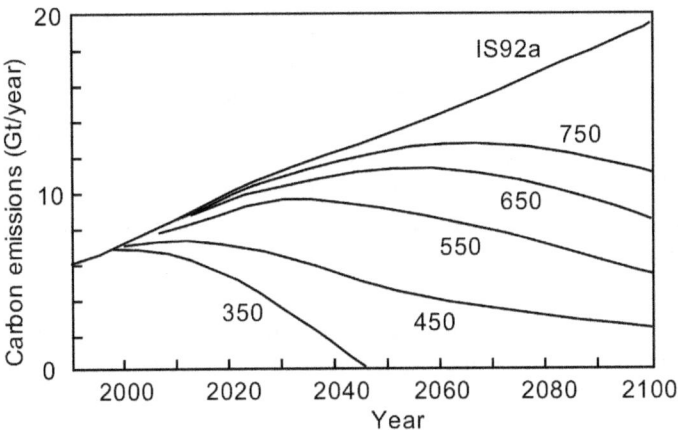

Figure 7.4. Carbon emissions per year needed to stabilize CO_2 concentration in the atmosphere at various levels of ppm. Adapted from Hoffert *et al.* (1998).

Lightfoot and Green (2002) estimated the required rate of world average annual energy intensity decline required to stabilize the level of CO_2 in the atmosphere at some level, such as 550 ppm in 2100 (about double the pre-industrial level). However, their projections of energy mix and energy efficiency for 2100 seem very optimistic.

The present world population of about 7 billion may grow to as much as 11 billion by 2100. As prosperity expands, this burgeoning world population will seek a better life by consuming more energy per capita. Most of these people have a major goal in life to emulate the lifestyle of Americans, which entails using energy at a rate comparable with that of Americans. There is no way that they can be successful. The U.S. has about 5% of the world's population and uses about 20% of the world's energy. If the people of the world emulated Americans, world energy usage would increase by a factor of 4 even without an increase in world population. Increasing population would drive energy usage still higher. World primary power consumption in 2006 was 12 TW, of which ~85% was fossil-fueled. By 2050, world power demand may grow to as much as 30 TW. By 2100, it will likely be considerably higher. The main problem facing humankind in the 21st century is providing itself with these enormous energy requirements. If one were to assume that oil, gas, and coal resources are unlimited, and fossil fuels were produced in sufficient quantity to supply the impending demand during the 21st century with a mix typical of today, carbon emissions during the 21st century would be very high and the threat of global warming would likely be significant. If the role of coal increases in the 21st century, it will exacerbate the problem because coal has higher carbon content than oil or gas per unit energy produced. It seems likely however, that a significant fraction of projected future use of coal might be replaced by natural gas. However, the estimated fossil fuel resources remaining are limited. Hydrocarbon production will likely top out before 2025 and coal production will likely top out around 2030. Assuming that greenhouse gas emissions are limited in the 21st century by the available fossil fuel resources, emissions will peak by 2025–2030. The ultimate run-out to depletion of fossil resources would limit increases in the CO_2 concentration, because energy production in the second half of the 21st century will have to undergo significant changes as fossil fuel production (and consumption) inevitably pursues an increasingly downward trend.

The world will face a crisis sometime around or after 2030. But that crisis will not be calamitous global warming. The crisis will be that with oil, gas, and coal production going at full bore, the world will not able to supply the energy that is demanded. This could lead to significantly higher energy costs, resulting in worldwide economic recession or depression. However, on the positive side, it provides the incentive to develop renewable energy to become more competitive. Whether renewable energy can be developed and expanded rapidly enough to stave off economic collapse remains to be seen. Alternatively, Idso and Idso (2007) proposed a massive increase in the use of nuclear power, glossing over the problems inherent in such a strategy.

Brown *et al.* (2011) analyzed the relationship between energy use and economic growth. They concluded:

> "Empirically, the central role of energy in modern human economies is demonstrated by the positive relationship between energy use and economic growth.... To support a projected global population of 9.5 billion in 2050 with an average standard of living equivalent to the current US lifestyle would require about 268 terawatts, 16 times the current global energy use. Even maintaining this increased population at the more modest Chinese standard of living would require 2.5 times more energy than is used today.... The bottom line is that an enormous increase in energy supply will be required to meet the demands of projected

population growth and lift the developing world out of poverty without jeopardizing current standards of living in the most developed countries".

The problem with the over-emphasis on global warming by the alarmists is that they have lost focus on the real problem facing humanity in the 21st century: providing the people of the world with energy as the developing countries gradually industrialize. While some naïve futurists think we can solve our problems with wind and solar energy and electric cars, it is difficult to see how we can make it through the current century without continued heavy use of fossil fuels. Meanwhile, the future looks pessimistic as the world population grows. Maybe we can store CO_2 in water and bottle soda water in huge quantities?

Hoffert *et al.* (2002) emphasized that the problem of stabilizing the CO_2 levels in the atmosphere relates to energy:

"In the 20th century, the human population [of the Earth] quadrupled and primary power consumption increased 16-fold."

Creating a transition toward stabilization of CO_2 levels will require reductions in energy consumption as well as development of primary energy sources that do not emit carbon dioxide to the atmosphere. Mid-century primary power requirements that are free of carbon dioxide emissions could be several times what we now derive from fossil fuels ($\sim 10^{13}$ watts), even with projected improvements in energy efficiency. Hoffert *et al.* (2002) surveyed potential future energy sources with emphasis on their capability to supply massive amounts of carbon emission–free energy. These included terrestrial solar and wind energy, solar power satellites, biomass, nuclear fission, nuclear fusion, fission–fusion hybrids, and fossil fuels from which carbon has been sequestered. They also studied non-primary power technologies that could contribute to climate stabilization including efficiency improvements, hydrogen production, storage and transport, superconducting global electric grids, and geo-engineering. They concluded that all of these approaches currently have severe deficiencies that limit their ability to stabilize the production of CO_2. Furthermore, they suggested that the IPCC (and many others) are overly optimistic regarding the potential for advanced energy technologies to quickly reduce CO_2 emissions significantly. Their conclusion was that we need a drastic expansion of research on renewable energy.

Anon. (J) provides a number of commentaries on Hoffert *et al.* (2002). Some of these came to the defense of their energy technologies (particularly solar-thermal and nuclear) but these appear to be self-serving to some degree. Several comments had to do with the role of energy conservation. Professor Albert Bartlett said:

"Even without the greenhouse problems, the obvious impossibility of continuing these [past energy] growth rates would lead rational people to say that ... the world's first order of business should be to stop the growth of populations and the growth of per capita primary power consumption. Instead of advocating the obvious, the authors paint a picture of all manner of technological fixes that, at enormous expense, may provide some answers to the need to stop the growth in emissions of greenhouse gases that are associated with energy production. As is so often the case, technological fixes are offered without being reviewed in the light of Eric Sevareid's Law: 'The chief cause of problems is solutions.' One can be sure that each technological solution will create new problems that are not indicated by calculations, equations, and technical speculations."

Further analyses of the effects of various future levels of greenhouse gas emissions according to various scenarios continue to be made. Schewe *et al.* (2011) provide references for much of this work. Generally, these studies begin with various models for future emissions of greenhouse gases. From this, estimates are made of the future rise in CO_2 concentration. They can then identify scenarios for future emissions that lead to various levels of CO_2 concentration in 2100 and beyond. From this, they can derive estimates of future temperature rise from global climate models, assuming that these models are representative of reality (for which there is no proof). Until recently, the U. N. had concentrated on future emission scenarios designed to keep "global warming below 2°C" although it is not clear what baseline is taken at the end of the LIA to define "global warming". However, Schewe *et al.* (2011) indicated that the U. N. is now preoccupied with limiting "global warming" to 1.5°C "because climate change impacts associated with 2°C are considered to exceed tolerable limits for some regions". The scenario labeled "RPC3-PD" was selected to limit "global warming" to 1.5°C according to climate models. In this scenario, carbon emissions peak at around 9 Gt/year in 2010, decrease essentially linearly to zero by around 2070, and continue to decrease so they become negative at –2 Gt/year after 2100. This is something like the "down ramp" in Figure 5.3, except that in RPC3-PD, the emission rate continues to drop beyond 2060. According to Schewe *et al.* (2011), "global warming" is presently slightly greater than 1°C, and it would rise to 1.5°C by 2050, and fall off slightly after 2050 and finally reaching 1°C again in 2500. Schewe *et al.* (2011) then utilized various climate models to infer the impact on the Earth of this future scenario. They concluded that "Steric sea level rise under the RCP3-PD scenario continues for 200 years after the peak in surface air temperatures, stabilizing around 2250 at 30 cm". Since according to Schewe *et al.* (2011), we have already undergone "global warming" over 1°C, it is difficult to imagine that a further increase to 1.5°C would result in a rise in sea level of 30 cm. No credence can be put in such alarmist publications.

7.2 Constraints on CO_2 Production Imposed by the Limits of Fossil Fuels

The increase in CO_2 concentration in the atmosphere from the pre-industrial estimate of 280 ppm to the present value of 395 ppm shows that there has been an increase of about 100 ppm. In this period, the world burned a great deal of fossil fuel. Now we must ask how much fossil fuel remains and what is the probable limit to future CO_2 concentration in the atmosphere from burning the available fossil fuels in the 20th century? Unfortunately, it is difficult ot make even rough estimates in this regard.

Deffeyes (2001) provided an analysis of probable U.S. and world oil reserves. The best estimate is that cumulative world oil production will eventually reach about 2.1×10^{12} barrels, and the world has already produced roughly half that amount. Rogner (1997) provides a slightly more optimistic estimate. U.S. oil production was estimated at around 0.22×10^{12} barrels, and about 85% of that has already been consumed. Oil isn't the only fossil fuel; there are also natural gas, natural gas liquids, and coal. It was estimated that world production of total hydrocarbons (oil, gas, and gas liquids) may approach 4.5×10^{12} barrels of oil (equivalent), and the world has already produced roughly 38% of that amount. These resources are finite and significant fractions of their initial endowments in the Earth have already been exploited (Rutledge, 2007).

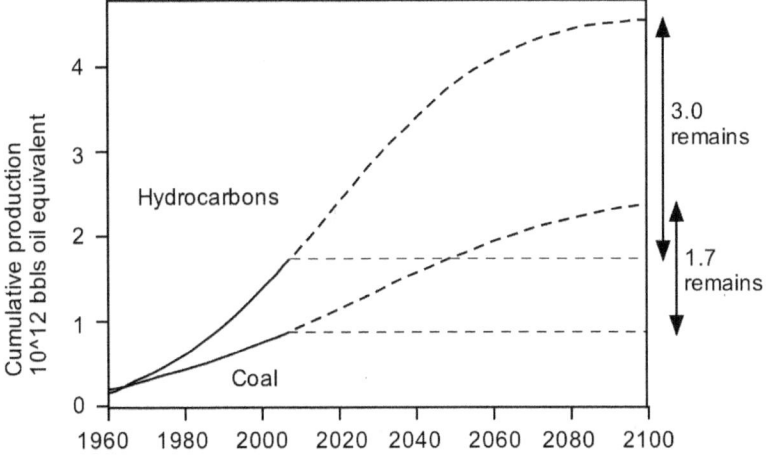

Figure 7.5. Estimated cumulative production of hydrocarbons (oil, gas, gas liquids) and coal to date and projected ultimate production by 2100. Adapted from Rutledge (2007).

A simplistic analysis would suggest that since burning ~38% of the world's fossil fuels has increased the CO_2 concentration in the atmosphere by about 100 ppm, burning the remainder is likely to add another ~150 ppm, bringing the ultimate level to perhaps 545 ppm by the end of the 21st century, at which time the world will have limited amounts of economically recoverable fossil fuels remaining. This suggests that there may be barely enough fossil fuel to "double the CO_2 concentration in the atmosphere" (relative to pre-industrial levels), and all the climatologists who continue to deal with a quadrupling of CO_2 appear to be working in fantasy. While new discoveries, new technologies and new sources will undoubtedly increase the emissions beyond this estimate, it seems clear that total emissions will be limited by resource availability.

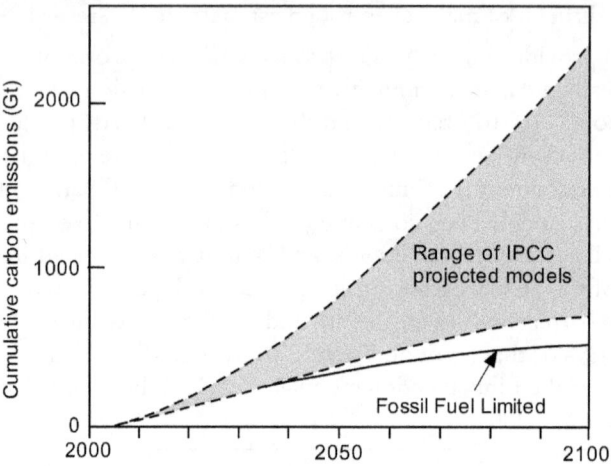

Figure 7.6. Comparison of cumulative CO_2 emissions based on fossil fuel constraints with the range of IPCC projections for future CO_2. Adapted from Rutledge (2007).

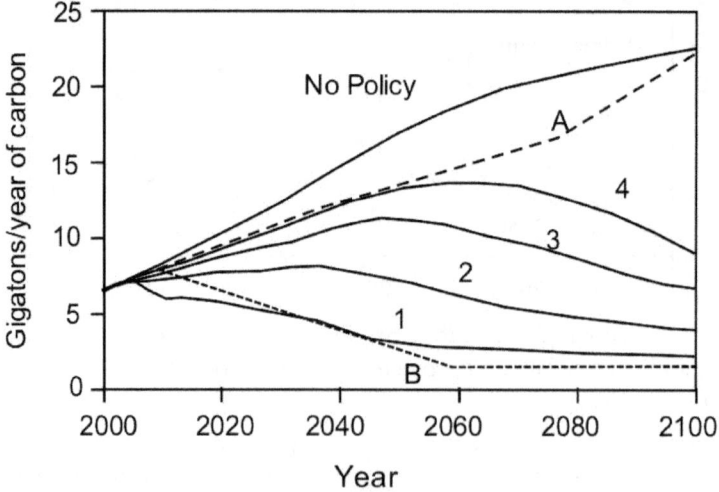

Figure 7.7. MIT estimates of carbon emissions for various levels of control. See text for details. (Webster, *et al.*, 2008)

Estimation of future production of fossil fuels is a tricky business that requires extrapolation from past data. Rutledge (2007) has done a good job of analyzing the (admittedly sketchy) available information and projecting forward to the future. Figure 7.5 shows the expectation for future cumulative production of fossil fuels extrapolated from current levels. Rutledge (2007) estimated CO_2 emissions based on a production rate limited by the finite resources as indicated in Figure 7.5. This is compared with the range of projections made by the IPCC in Figure 7.6. It can be seen that most projections of future CO_2 emissions are overly generous in their implied estimate of remaining fossil fuel resources.

Webster *et al.* (2008) provided estimates of future CO_2 emissions correlated to ultimate concentration stabilization targets as shown in Figure 7.7. Their curves correspond to: no policy for control of CO_2 emissions, 1 = stabilization at 450 ppm, 2 = stabilization at 550 ppm, 3 = stabilization at 650 ppm, and 4 = stabilization at 750 ppm. Curves A and B refer to IS92A, and the "down ramp" in Figure 5.3, respectively.

7.3 The Climate Debate Revisited

Table 7.1. Dimensions of the climate debate.

Aspect	Results
Only remedy is immediate draconian reduction in fossil fuel usage	Immediate draconian reduction in carbon emissions will be difficult to achieve because of political, technical and economic impediments. It would require close cooperation and agreement between all governments. Even so, it is unlikely that the world would have sufficient energy with which to operate, plunging the world into a deep economic depression.
Draconian reduction in fossil fuel usage can be accommodated by renewable energy supply	It does not appear to be possible to provide the world with needed energy while implementing draconian reduction in fossil fuel usage through expansion of renewable energy and sequestration of emissions. Intermittent energy sources would be problematic and economic consequences would be disastrous.

It may be impossible to predict the specific impacts of rising CO_2 concentration in the 21st century. It has been argued that nevertheless, we must prepare for the worst-case scenario. This will require immediate draconian reductions in CO_2 emissions during a period when the Earth's population and energy usage is rapidly expanding. Two critical issues crop up. One is whether such draconian reductions in emissions can be achieved in the face of political, technical and economic impediments. The second is whether renewable energy has thye potential to supply the energy lost by reducing fossil fuel consumption. Table 7.1 summarizes our findings in Chapter 7.

Donald Rapp

8 IMPACTS OF GLOBAL WARMING

8.1 Sea Level Rise

Radic (2008) pointed out:

"In 1990, the near-coastal population ... was 1.2 billion people .... Human settlements are also preferentially located close to the world's shoreline, including most of the largest cities, which means that the world's economy is also concentrated in the coastal zone. Thus, sea level rise has a major impact on coastal cities, deltaic lowlands, small islands, and coastal ecosystems. The potential threat has triggered studies on impacts and responses to sea-level rise which are focused on a range of direct and indirect socio-economic impacts such as loss of land and buildings, loss of tourist amenity, increasing flood risk, impact on variety of commercial infrastructure, coastal process plants and offshore oil and gas production"

Hansen (2004) said:

"The dominant issue in global warming, in my opinion, is sea-level change and the question of how fast ice sheets can disintegrate."

A number of climatologists have predicted the sea level rise for the remainder of the 21st century. This partly depends on projected scenarios for future CO_2 emissions, but the general benchmark question is what is the expected sea level rise if the CO_2 concentration rises from 280 ppm to 560 ppm?

8.1.1 Measurement of Sea Level

Historic sea level has been estimated by several methods. The most extensive source of data is tide gauges, some of which date back more than

100 years. Over the past three decades or so, measurements were made from orbit using altimeters. The contribution of shrinking ice sheets in Greenland to rising sea level has been estimated by modeling, and in the past decade, by satellite-based gravity detectors.

Most of our knowledge of historical variations in sea level is based on tidal gauges. Yet, interpretation of tide gauge data is problematic.

Douglas and Peltier (2002) emphasized that the measurements of relative sea level (RSL) at any location have large annual and decadal fluctuations that tend to obfuscate the long-term trends with noise. Consequently, only very long–term records have the potential to accurately provide the underlying trend. A CNES analysis showed that use of only four decades of data at the tide gauge sites led to an overestimate of the global sea level (GSL) rise. Other studies showed that the extreme dependence of trend on record length is real, and not an artifact of the tide gauge. To remain accurate over periods of a century or more, tide gauges, which over time may be repaired, moved, upgraded, and so on, must be kept consistent. But tide gauges, no matter how accurate and consistent, make local measurements. And they measure only relative sea level with respect to the surface of the solid Earth. Without independent estimates of vertical land movement, tide gauges cannot determine whether the water level is rising, the land is sinking, or both. Even though most estimates of the GSL rise in the past decade or so have used records that are as long as possible, the various estimates still differ significantly. The origin of the differences probably lies in the methods used to correct for vertical land movements at tide gauge sites. Hence, all data on sea level rise must be examined critically.

Houston and Dean (2011) pointed out that sea level measurements via tide gauges record the sum of worldwide eustatic sea level + glacial isostatic adjustment + local effects + noise. The glacial isostatic adjustment is the "rebound" from massive ice sheets at the LGM. It is found to be an uplift of up to several meters per year at high latitudes, and a downtrend of up to about a meter per year in temperate zones. Local effects include compaction of sediments, earthquakes, withdrawal of ground fluids, and building heavy structures on weak sediments. The ideal location for tide gauge data provides very long continuous records in geologically stable regions.

As Fjeldskaar (2008) pointed out, sea level changes are commonly termed eustasy, meaning globally uniform sea level changes. However, sea level changes are caused by:

(1) Changes in the ocean water volume caused by glaciations and deglaciations of ice sheets.

(2) Variations in the ocean basin volume caused by sedimentation, changes in the volume of ocean ridge systems, and hydro-isostasy.

(3) Variations in the Earth's gravity field as, for example, in mountain formation.

(4) Removal of ground water that eventually makes its way into the oceans.

The geoid is defined an equipotential surface of the Earth's gravity field that would establish local sea level at equilibrium in the absence of atmospheric forces. Fjeldskaar (2008) also provided the following insights:

"A mathematical figure representing the sea level surface with all irregularities removed is named the spheroid. The spheroid would be the sea level surface of an Earth with no lateral variations in density. The difference in elevation between the measured geoid and the spheroid is called the geoid anomaly. Some of the major geoid anomalies, like the geoid high over New Guinea (+70m) or the geoid low over India (–100 m) are probably related to mantle convection.

Any sea level change causes deflection of the ocean floor, hydro-isostasy, to attain isostatic equilibrium. The hydro-isostasy is approximately 1/3 of the sea level change. Simultaneously the continents are deflected, with a mean magnitude over the continents twice the deflection of the ocean floor. This is due to the fact that the oceanic area is double the land area. An interesting implication of hydro-isostasy is the fact that the sea level history will differ between oceanic islands and continental margins. An island moving with the sea floor will record the full sea level change, while points near the continents record quite different sea level changes. Thus hydro-isostasy is an important factor in determining relative sea level fluctuations."

Evidently, eustasy is complex, and is certainly not globally uniform.

Relative sea level is the change in sea level relative to the surrounding land, taking into account changes in the elevation of the land due to factors such as glacial loading and rebound and ocean floor subsidence. But Fjeldskaar (2008) emphasized that "relative sea level changes are not the same as eustatic changes" and "it is difficult to imagine where a eustatic change can be measured realistically."

The issues involved in contemplating the effect of global warming on sea level rise illustrate the complexity and uncertainty involved in predicting the future. Kolker and Hameed (2007) said:

"Determining the rate of global sea level rise (GSLR) during the past century is critical to understanding recent changes to the global climate system. However, this is complicated by non-tidal, short-term, local sea-level variability that is orders of magnitude greater than the trend."

8.1.2 Historic Sea Level Change

According to Singer and Avery (2007), when the great ice sheets began to melt at the end of the last period of glaciation, the initial rapid rise of sea level was about 200 cm per century. This gradually changed to a slower rate of rise (15 cm–20 cm per century) about 7,500 years ago, once the large ice masses covering North America and North Europe had melted away. But the slow melting of the West Antarctic Ice Sheet continued and will continue, barring another ice age, until it has melted away perhaps 6,000 years from now. This means that the world will continue to endure a sea level rise of about 18 cm per century (1.8 mm per year), just as it has been in previous centuries. And it is likely that there is nothing we can do about it. Thus, Singer and Avery (2007) attribute the continuing slow sea level rise to the natural consequences of the post-glacial period but make no allowance for anthropogenic global heating adding to this rather bland picture. Their summary of sea level rise since the Last Glacial Maximum is shown in Figure 8.1.

Of all the potential impacts of global warming, the potential rise of sea level would appear to be the most serious because of the possibility of much greater effects if the Greenland Ice Sheet is seriously impacted by modest temperature increases. However, it is not clear that much can be done about this.

Jevrejeva *et al.* (2008) carried out a reconstruction of global sea level since 1700 from tide gauge records (see Figure 8.2). They performed a quadratic fit to the data, and concluded that sea level rise began accelerating at the end of the 18th century at the rate about 0.01 mm/yr². According to them, sea level rose by 6 cm during the 19th century and 19 cm in the 20th century. Superimposed on the long-term acceleration were quasi-periodic fluctuations with a period of about 60 years. It was concluded that if the conditions that established the acceleration continue into the future, then sea level would rise 34 cm over the 21st century. However, acceleration, like beauty, lies in the eye of the beholder. One could argue alternatively from Figure 8.1 that sea level oscillated about a linear trend from 1920 to 2000 with a slope of 2.0 mm/yr with no acceleration at all, and that the acceleration that did occur took place in the 19th century as the Earth came out of the LIA.

Wopplemann *et al.* (2008) claimed that tide gauge records at Brest, France were stable over the period 1889–2007. They found that the rate of sea level rise at Brest was constant over that period. This would suggest that rising CO_2 levels did not play a role in this case. They also found a "close matching of the Brest and Liverpool [UK] time series over more than 200 years." They found that both instrumental records "showed a roughly coincident increase in the rate of relative sea-level rise around the end of the 19th century" – well before the great increase in CO_2 concentration. This increase in the rate of sea level rise in the 19th century is also evident in the results of Jevrejeva *et al.* (2008) as shown in Figure 8.2.

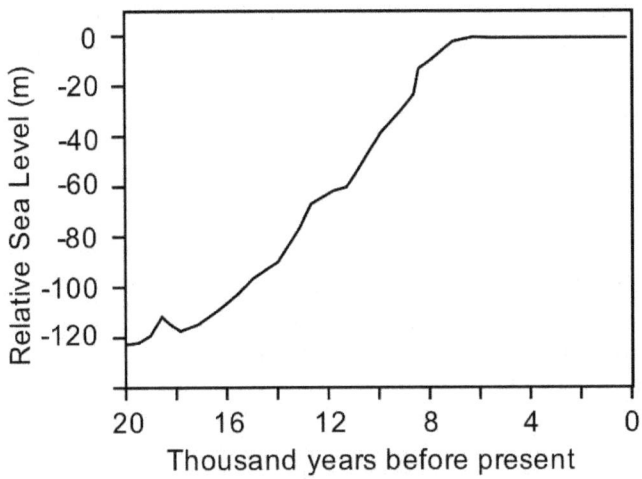

Figure 8.1. Sea-level rise since the Last Glacial Maximum, as deduced from coral and peat data. Adapted from NIPCC.

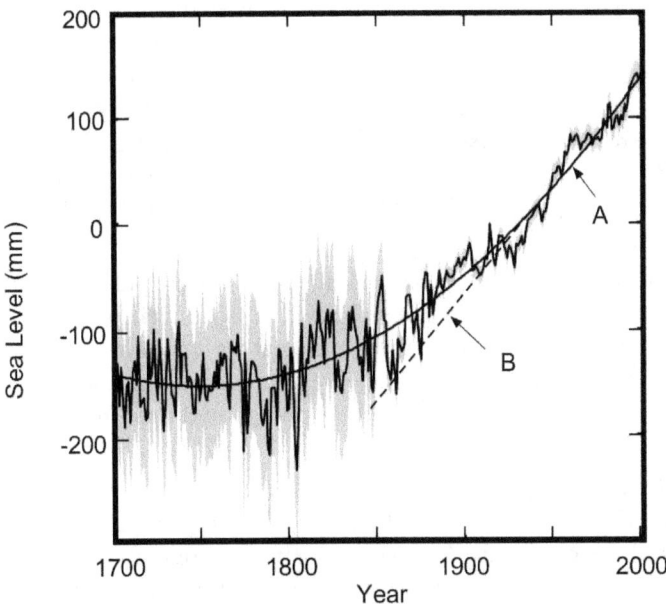

Figure 8.2. Reconstruction of relative sea level since 1700 according to Jevrejeva *et al.* (2008). Absolute numbers are meaningless. Curve "A" is a quadratic fit to the variable sea level. Line "B" is a linear fit to the period from 1920 to 2000. The gray area is an estimate of uncertainty.

Douglas and Peltier (2002) discussed many aspects of measurements of sea level rise and fall. At the height of the last glacial maximum, 21,000 years ago, so much of the Earth's water was tied up in great high-latitude ice sheets that the oceans were about 120 meters lower than they are today. Compared with those of previous millennia, the changes in global sea level (GSL) occurring today are tiny. The GSL increased by about 120 m as a result of the deglaciation that followed the last glacial maximum. By about 5,000–6,000 YBP (years before present), the melting of the great high-latitude ice masses was essentially completed. Thereafter, the GSL rise was small, and appears to have almost ceased by 3,000-4,000 YBP (Douglas and Peltier, 2002).

Douglas and Peltier (2002) claimed that although the long-term average global sea level (GSL) rise for the past few millennia has been stable at a level near zero, the GSL abruptly began to rise around the mid-19th century. No studies, however, have detected any significant acceleration of GSL rise during the 20th century. In the last dozen years, published values of 20th-century GSL rise ranged from 1.0 mm/yr to 2.4 mm/yr, even though all investigators used essentially the same database of tide gauge measurements. Douglas and Peltier (2002) made a significant distinction between estimates of a GSL rise of 2 mm/yr and 1 mm/yr. If the correct value of GSL rise is near 1 mm/yr, then it is argued that global warming provides an explanation: 1 mm/yr of the GSL rise corresponds to that expected from the thermal expansion of the oceans and the melting of small ice sheets and mountain glaciers caused by the 0.6°C increase in global surface temperature during the last 100 years. This explanation would further imply that melting of the great Greenland and Antarctic ice sheets is not currently contributing significantly to the GSL rise. But if, as Douglas and Peltier (2002) argued, the true rate of contemporary GSL rise is probably closer to 2 mm/yr, it is then likely that these ice sheets are contributing.

A detailed analysis of sea level rise was provided by Church and White (2006). They reconstructed global sea level from a variety of data sources back to 1870. From 1870 to 2004 (135 years), the total GSL rise was 195 mm, an average of 1.44 mm per year. For the 20th century, the rise was about 160 mm (1.7 ± 0.3 mm per year) indicating a slight acceleration during the 20th century.

Holgate and Woodworth (2004) estimated the sea level rise from 1952 to 1997 (45 years) based on 177 tide gauges divided into 13 regions with near global coverage, and using a Glacial Isostatic Adjustment model to correct for land movements. Sea level rise over these 45 years was estimated to have averaged 1.7 ± 0.2 mm/year, although the curve showed periodic oscillations. Furthermore, the curve of sea level vs. time seemed to be accelerating upward in the 1990s. In a more recent study, Holgate (2007) chose nine long and nearly continuous sea level records from around the world to explore rates of change in sea level for 1904–2003. The lack of high-quality, long-life gauge records was circumvented by finding representative gauges that matched the data for 1952 to 1997 when more data were available. These records were found to capture the variability found in a larger number of stations over the last half-century studied in their 2004 paper. The addition of new data not only extended the time period back to 1904, but it also extended the time period forward to 2003. The new results indicated that the apparent acceleration noted in the 1990s tailed off and now appears to have just been another oscillation, while the extended curve indicated that the rate of rise of sea level was slightly higher early in the century than it was later in the century.

According to Holgate (2007):

"The rate of sea level change was found to be larger in the early part of last century (2.03 ± 0.35 mm/yr 1904–1953), in comparison with the latter part (1.45 ± 0.34 mm/yr 1954–2003). The highest decadal rate of rise occurred in the decade centered on 1980 (5.31 mm/yr) with the lowest rate of rise occurring in the decade centered on 1964 (1.49 mm/yr). Over the entire century the mean rate of change was 1.74 ± 0.16 mm/yr."

According to this result, the rate of increase of sea level slowed in the latter part of the century.

Woppelmann et al. (2007) provided what appears to be a credible estimate of 20th-century sea level rise. As they discussed:

"Two important problems arise when using tide gauges to estimate the rate of global sea-level rise. The first is the fact that tide gauges measure sea level relative to a point attached to the land that can move vertically at rates comparable to the long-term sea-level signal. The second problem is the spatial distribution of the tide gauges, in particular those with long records, which are restricted to the coastlines."

Corrections for land movement so far have included corrections for one of the many processes that can affect land stability, namely glacial–isostatic adjustment (GIA). However, different GIA models provide very different values in magnitude and sign. Moreover, GIA models do not account for other sources of vertical land motion. Woppelmann et al. (2007) utilized a dedicated GPS measurement system to estimate vertical land movement at 224 stations over a 7.7-year period.

Two important hypotheses were adopted for combining tide gauge and GPS results to derive "absolute" trends in sea level: (a) land motions are extremely low-frequency in character so that the current GPS vertical velocities can be applied for the last century, (b) the vertical velocity observed at the GPS station applies to the tide gauge site. These assumptions were supported by other evidence.

Donald Rapp

The poor spatial distribution of historical gauges is problematic because of the evidence of regional variability of sea level trends, this being confirmed by satellite altimetry results. By selecting only tide gauges with long records (e.g., >60 years) it is hoped that some of the errors might cancel out. The final result of Woppelmann *et al.* (2007) was that the global average sea level rise was estimated to be 1.3 mm/yr for the 20th century.

Domingues *et al.* (2008) pointed out "Climate models ... do not reproduce the large decadal variability in globally averaged ocean heat content inferred from the sparse observational database, even when volcanic and other variable climate forcings are included." They claimed to provide "improved estimates of near-global ocean heat content and thermal expansion for the upper 300 m and 700 m of the ocean for 1950–2003. They added their observational estimate of upper-ocean thermal expansion to other contributions to sea level rise and found that the sum of contributions from 1961 to 2003 was about 1.5 mm/yr. For the period from 1993 to 2003, they estimated sea level rise to be about 2.4 mm/yr.

Houston and Dean (2011) (H&D) provided a review of previous estimates of 20th century sea level rise (see Table 8.1). The IPCC has projected a sea level rise of 180-590 mm from 1990 to 2100 based on rising temperatures. Melting ice sheets might contribute another 200 mm. This wide range of predictions underlines the uncertainty in future sea-level estimates. As H&D pointed out, continuation of the current rate of sea level rise of 1.7 mm/yr would lead to a total sea level rise by 2100 of only 19 mm. To achieve a total rise of 790 mm acceleration of about 0.10 mm/yr^2 is required in the rate of sea level rise over the 110-year period. However, tide-gauge records do not reveal such large accelerations. H&D pointed out the "lack of long-term tide gauge records and their concentration in the northern hemisphere, strong worldwide spatial variations of sea-level rise, vertical land movements, and seasonal-to-decadal temporal variations that can be large compared to sea-level trends and accelerations".

A comparison of estimates of sea level rise for the 20th century is given in Figure 8.3.

H&D analyzed tide gauge data at 44 U. S. sites and 7 long-term Florida sites that met their criteria for acceptability. They fitted a function to the data:

Sea Level = $a_0 + a_1t + \frac{1}{2} a_2 t^2$

Here, t = time, a_0 is a constant that varies with location, a_1t represents a constant rate of rise of sea level, and $\frac{1}{2} a_2 t^2$ represents a constant acceleration in the rate of rise of sea level. They found that many sites yielded $a_1 = 1.7$ mm/yr and $a_2 \sim 0$. The range of a_1 was from 1.25 mm/yr to 1.90 mm/yr. They concluded:

"The results of all of our analyses are consistent - There is no indication of an overall world-wide sea level acceleration in the 20th Century data. Rather, it appears that a weak deceleration is present".

If sea level continues to rise at the rate of 1.7 mm/yr through the 21st century, the total rise from 2000 to 2100 would be 17 cm.

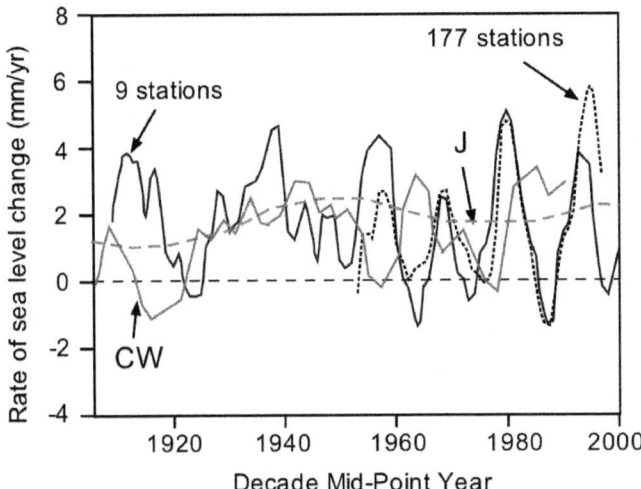

Figure 8.3. Comparison of estimates of sea level rise for the 20th century. The curves marked "9 stations" and "177 stations" are from Holgate (2007). "CW" is from Church and White (2006). "J" is from Jevrejeva *et al.* (2008).

Watson (2011) concluded

"The Australasian region has four very long, continuous tide gauge records which are invaluable for considering whether there is evidence that the rise in mean sea level is accelerating over the longer term at these locations in line with various global average sea level time-series reconstructions. These long records have been converted to relative 20-year moving average water level time series and fitted to second-order polynomial functions to consider trends of acceleration in mean sea level over time. The analysis reveals a consistent trend of weak deceleration at each of these gauge sites throughout Australasia over the period from 1940 to 2000. Short period trends of acceleration in mean sea level after 1990 are evident at each site, although these are not abnormal or higher than other short-term rates measured throughout the historical record".

Table 8.1. Estimates of acceleration of the of sea level rise (Houston and Dean, 2011)

Reference	Time Period	Content	Acceleration (mm/yr^2)
Woodworth (1990)	1870-1990	Oldest European tide gauges - a few gauges back to 1700s	+0.004
Jevrejeva *et al.* (2008)	1800-2000	Old European tide gauges	+0.01; greatest increase 1920-1950
Douglas (1992)	1905-1985 1850-1991	Worldwide tide gauges	-0.01 +0.001
Church *et al.* (2004)	1950-2000	9 yrs of TOPEX data + historical tide gauge data	No detectable acceleration #
Church and White (2006)	1870-2004	12 yrs of altimetry data + historical tide gauge data	+0.013
Woodworth *et al.* (2009)	1870-2004	Review paper	Slight acceleration; mostly before 1930
University of Colorado	1993-2010	Altimeter data *	Rate ~ 3 mm/yr; No acceleration given
Holgate (2007)	1910-2000	10-yr mean sea level trends	Rates varied widely from period to period #
• Vermeer & Rahmsdorf (2009) • Jevrejeva, *et al.* (2010) • Grinsted, *et al.* (2010)	1990-2100	Semi-empirical models based on assumed future scenarios	+0.07 to +0.28 resulting in sea level rise in 2100 of 600-1900 mm
Houston and Dean (2011)	20th century	Claimed consensus prior to their work	Rate = 1.7 to 1.8 mm/yr Acceleration uncertain but apparently small
Houston and Dean (2011)	1930-2010	Long-term U.S. tide gauges	Small deceleration (-0.001 to -0.01)

* H&D emphasized the "many uncertainties and sources of error in satellite-altimeter measurements.

H&D emphasized that Holgate (2007) and Church *et al.* (2004) showed that there have been several periods with rate > 3 mm/yr even though there was no net acceleration due to the cyclic pattern of rate vs. year

Kemp *et al.* (2011) estimated sea level variations over the past 2,000 years based on salt-marsh sedimentary sequences from the North Carolina coast. It was claimed that "salt-marsh sediments and assemblages of foraminifera record former sea level because they are intrinsically linked to the frequency and duration of tidal inundation and keep pace with moderate rates of sea-level rise". They used a dataset of foraminifera (193 samples) from 10 salt marshes in North Carolina as proxies for sea level. microfossils from sediment cores taken in the coastal salt water marshes of mainland North Carolina. These were then compared with North Carolina tidal gage records from 1920 to 2000. The paper is very terse and does not provide details on how they made the connection between sea level and foraminifera in salt marshes. The comparison between modeled sea level and measured sea level is shown in Figure 8.4.

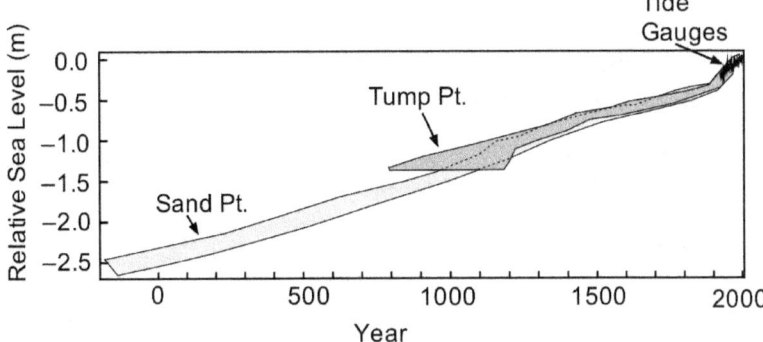

Figure 8.4. Calibration curve for sea level model based on salt-marsh sediments compared to measurements with tidal gauges. The overlap between model and measurements is 80 years. (Adapted from Kemp *et al.*, 2011)

The calibration period is very short and the extrapolation of the model is very long. This calibration is highly suspect for several reasons. One obviously is the short duration of overlap between model and measurements. A second is that, as we have amply demonstrated, tidal gauges are notoriously inaccurate. A third factor was provided by Ken Haapala:

> "Environmentalists generally refer to these coastal saltwater marshes as 'fragile wetlands' and these wetlands have a number of interesting characteristics. They are broad, flat, generally marshy lands made of plants, silt, and sand that were formed by sediments from the long term erosion of the Appalachian Mountains and other uplands.... These wetlands may stretch as far as 50 miles deep into main part of the state. As with most coastal areas built up by sediments, they are probably subject to subsidence.... During the last Ice Age, streams and rivers cut channels through these sediments, but as the sea levels rose by about 120 meters after the last Ice Age, the channels became tidal estuaries resulting in wide rivers and bays. The areas are subject to erosion and accretion caused by the tides and storms such as hurricanes and northeasters. The areas are partially protected from ocean waves by a series of barrier islands made of sand that shift over the years. As the islands shift, they change the influence that tidal currents and storms have on these wetlands. To suggest that a model of global sea levels can be based on studies of such unstable lands is highly questionable." (Haapala, Ken (2011) The Week That Was: June 25, 2011, *http://www.SEPP.org*).

In comparing their results to previous results based on tide gauges, Kemp *et al.* (2011) failed to refer to Alley *et al.* (2005), Huybrechts (2002), Wopplemann *et al.* (2007, 2008), Douglas and Peltier (2002), Holgate (2007),

Domingues *et al.* (2008), Houston and Dean (2011) or Watson (2011). They did refer to Jevrejeva *et al.* (2008), Church and White (2006).

It is worth noting that one of the authors of this paper (Mann) is the progenitor and advocate of the "hockey stick" and another (Rahmstorf) is a well-known alarmist.

8.1.3 Recent Sea Level Change

Lombard *et al.* (2005) found large oscillations in decadal changes in sea level due to the El Niño–Southern Oscillation and the Pacific Decadal Oscillation, and cautioned against extrapolating short-term sea level data from satellite measurements.

Jevrejeva *et al.* (2006) also found significant oscillations in sea level with periods ranging up to 30 years. For the period 1993–2000, they found a sea level rise of 2.4 ± 1.0 mm/year, comparable with the value they estimated for 1920–1945. Since the period 1993–2000 was during an uptrend in the oscillatory pattern, the longer term value will presumably be lower.

Wunsch *et al.* (2007) estimated regional patterns of global sea level change from a 1° horizontal resolution general circulation model based on about 100 million ocean observations and many more meteorological estimates during the period 1993–2004. Regional variability was found to be significant. They estimated a global mean of about 1.6 mm/yr, of which about 70% is from the addition of fresh water. They concluded however: "Useful estimation of the global averages is extremely difficult given the realities of space–time sampling and model approximations. Systematic errors are likely to dominate most estimates of global average change: published values and error bars should be used very cautiously".

Fjeldskaar (2008) presented a summary of a number of studies of sea level rise. Among the references that he cited are the following.

Monaghan *et al.* (2008) found statistically insignificant positive trends in sea level rise over most regions and months during 1960–2005. By contrast, 1970–2005 trends were weakly negative overall.

Harrison and Carson (2007) reported on subsurface temperature trends in the better-sampled parts of the oceans from 1950 to 2000. They found a large spatial variability with some regions showing cooling in excess of 3°C, and others warming of similar magnitude. They concluded: "The ocean neither cooled nor warmed systematically over the large parts of the ocean for the entire analysis period [1950– 2000]."

Tisdale (2009) provided an extensive review of sea level measurements. Global sea level rose at an average rate of about 3 mm/year from 1993 to 2005 but flattened out after 2005. He also presents regional sea level data. Annual data show sharp up and down variations from year to year. The Indian Ocean data show a sharp spike in 1998 due to the strong El Niño.

It should be noted that a number of papers have bandied about the TOPEX-POSEIDON result that the rise in sea level from 1993 to 2003 was 3.0 mm/year (e.g., Shepherd and Wingham, 2007), and Hansen claims it is 3.5 mm/year, but it now seems clear that even if these measurements were accurate (which is still uncertain) this was just part of an upward cycle and cannot and must not be extrapolated. According to Radic (2008), "the error in the instrumental calibration dominates the error budget".

Since measurements of sea level appear to be problematic, estimates have been made of the amounts of ice contained in the great ice sheets on Greenland and Antarctica. If the volumes of ice at these sites diminish, and it is assumed that the lost ice appears in the oceans as liquid water, the sea rise resulting from any volume change in ice can be estimated (360 Gt of ice is equivalent to ~1 mm of sea level).

The contribution of melting ice sheets to sea level rise was estimated by Shepherd and Wingham (2007) based on 14 different satellite-based estimates of the imbalances of the polar ice sheets since 1998. These studies included standard mass budget analyses, altimetry measurements of ice sheet volume changes, and measurements of ice sheets' changing gravitational attraction. As might be expected, they have yielded a diversity of values, ranging from an implied sea level rise of 1.0 mm/year to a sea level fall of 0.15 mm/year. Based on their evaluation of these diverse findings, they estimated that the East Antarctica Ice Sheet (EAIS) is gaining some 25 Gt/year, the West Antarctica Ice Sheet (WAIS) is losing about 50 Gt/year, and the Greenland Ice Sheet (GIS) is losing about 100 Gt/year. These trends provide a modest contribution to sea level rise of about 0.35 mm/year. However, these short-term results since 1998 should not be extrapolated because of the oscillatory behavior of ice sheet loss. In 2012, a new NASA report[44] showed:

> "During 2003 to 2008, the mass gain of the Antarctic ice sheet from snow accumulation exceeded the mass loss from ice discharge by 49 Gt/yr (2.5% of input), as derived from ICESat laser measurements of elevation change. The net gain (86 Gt/yr) over the West Antarctic (WA) and East Antarctic ice sheets (WA and EA) is essentially unchanged from revised results for 1992 to 2001 from ERS radar altimetry.... A slow increase in snowfall with climate warming, consistent with model predictions, may be offsetting increased dynamic losses".

---

[44] Zwally, H. Jay; Li, Jun; Robbins, John; Saba, Jack L.; Yi, Donghui; Brenner, Anita; Bromwich, David (2012) "Mass Gains of the Antarctic Ice Sheet Exceed Losses" NASA Report GSFC.ABS.6573.2012, *http://ntrs.nasa.gov/search.jsp?R=20120013495&utm_source=twitterfeed&utm_medium=twitter*

Donald Rapp

Alley *et al.* (2005) estimated that for Greenland, between 1993/1994 and 1998/1999, the ice sheet lost 54 Gt per year of ice, equivalent to a sea level rise of 0.15 mm/ year (the excess of melt water runoff over surface accumulation was about 32 Gt/year, leaving ice flow acceleration responsible for a loss of 22 Gt/year).

Rignot and Kanagaratnam (2006) used satellite radar interferometry observations of Greenland to detect widespread glacier acceleration below 66°N between 1996 and 2000, which rapidly expanded to 70°N in 2005. Accelerated ice discharge in the west and particularly in the east doubled the ice sheet mass deficit in the last decade from 90 km^3 to 220 km^3 per year. This provides a less optimistic outlook for Greenland.

Divine and Dick (2006) found evidence of persistent ice retreat in the Arctic since the second half of the 19th century. However, it was not clear whether this was a trend that will continue, or whether it was part of a cycle ("a similar shrinkage of ice cover was observed in the 1920s–1930s, during the previous warm phase of the LFO, when any anthropogenic influence is believed to have still been negligible").

Wingham *et al.* (2006), using radar altimetry, found that mass gains for East Antarctica slightly outweighed mass losses for West Antarctica, "exacerbating the difficulty of explaining twentieth century sea-level rise." Chen *et al.* (2006) using the Gravity Recovery and Climate Experiment (GRACE) satellite mission during its first 3.5 years (April 2002–November 2005), found that mass gains for East Antarctica roughly balanced mass losses for West Antarctica. Velicogna and Wahr (2006) found similar results. Davis *et al.* (2005) also found growth in East Antarctica, but of lesser magnitude.

Conway *et al.* (1999) found:

"The history of deglaciation of the West Antarctic Ice Sheet (WAIS) gives clues about its future. Southward grounding-line migration was dated past three locations in the Ross Sea Embayment. Results indicate that most recession occurred during the middle to late Holocene in the absence of substantial sea level or climate forcing. Current grounding-line retreat may reflect ongoing ice recession that has been under way since the early Holocene. If so, the WAIS could continue to retreat even in the absence of further external forcing."

This would seem to suggest that the process of disintegration of the WAIS has been underway for some time, independent of anthropogenic influences. According to Dasgupta *et al.* (2007):

"Until recently, studies of sea level rise (SLR) typically predicted a 0–1 meter rise during the 21st century. The three primary contributing factors have been cited as: (i) ocean thermal expansion; (ii) glacial melt from Greenland and Antarctica (plus a smaller contribution from other ice

312

sheets); and (iii) change in terrestrial storage. Among these, ocean thermal expansion was expected to be the dominating factor behind the rise in sea level. However, new data on rates of deglaciation in Greenland and Antarctica suggest greater significance for glacial melt, and a possible revision of the upper-bound estimate for SLR in this century. Since the Greenland and Antarctic ice sheets contain enough water to raise the sea level by [7.2 m and 61.1 m, respectively] small changes in their volume would have a significant effect. Since the IPCC Report in 2001, there has been an increased effort to improve measures of mass loss for the Greenland ice sheet and its contribution to SLR."

Dasgupta *et al.* (2007) claimed that satellite interferometry observations led to an estimate that the contribution of the Greenland ice sheet to SLR is double the rate assumed in the IPCC Report. Dasgupta *et al.* (2007) also indicated that the rate of loss of the West Antarctic Ice Sheet (WAIS) is several times greater than that assumed in the IPCC Report.

Morner (1973, 2004) provided a contrary view:

"Sea level rose for glacial eustatic reasons up to about 5,000 years before present (YBP). After that, global sea level has been dominated by the redistribution of ocean water masses (and by ocean-stored heat). This redistribution of water masses is driven by the interchange of angular momentum between the solid Earth and the hydrosphere (in feedback coupling) primarily expressed as changes in the oceanic surface current systems. In view of this, it has been very hard to define any global eustatic signal. This is where and why a dialectic between models and observations enter the sea level debate. According to the glacial loading models, global sea level is now rising by 1.8 to 2.4 mm/year. The IPCC models have hypothesized a very rapid rise in the near future ...Both the glacial loading models and the IPCC scenarios are strongly contradicted by observational data for the last 100–150 years that cannot have exceeded a mean rate of 1.0–1.1 mm/ year. In the last 300 years, sea level has oscillated close to the present with peak rates in the period 1890–1930. Sea level fell between 1930 and 1950. The late 20th century lacks any sign of acceleration. Satellite altimetry indicates virtually no changes in the last decade. Therefore, observationally based predictions of future sea level in the year 2100 give a value of ±10 cm [in contradiction to] model outputs by IPCC as well as global loading models. In conclusion, there are firm observationally based reasons to free the world from the condemnation to become extensively flooded in the 21st century AD."

Morner (2004) also concluded that sea level at the Maldive Islands had actually fallen between 1950 and 2001. However, Church, White, and Hunter (2006) contradicted this finding.

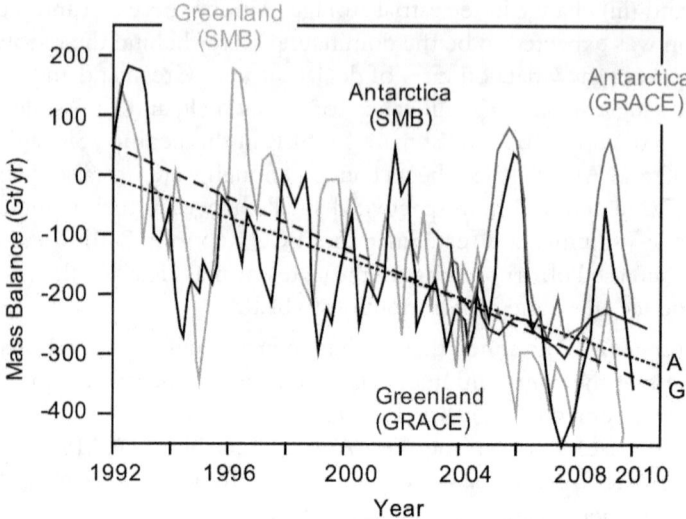

Figure 8.5. Ice sheet mass balance by surface mass balance and GRACE methods. The lines "A" and "G" represent the downward acceleration trends reported by Rignot *et al.* (2011).

Rignot *et al.* (2011) reported that they resolved the disparities between two approaches to estimate ice sheet mass balance at Greenland and Antarctica over the past eight years. The surface mass balance (SMB) method utilizes "the sum of snowfall minus surface ablation reconstructed from regional atmospheric models with perimeter loss calculated from a time series of glacier velocity and ice thickness to deduce the rate of mass change". "The gravity method employs a monthly time series of time‑variable gravity data from the Gravity Recovery and Climate Experiment (GRACE) to estimate the relative mass as a function of time. However, the details in their procedures are very complex and appear to require many assumptions of uncertain veracity. From this, they estimated mass loss rates from Greenland and Antarctica ice sheets over the past 18 years based on the SMB method. Over this period, the annual rate of mass loss has oscillated about a trend that accelerated at both sites. The annual loss in 2010 was about 250 Gt/yr of ice at each site, totaling about 500 Gt/yr for both sites. This corresponds to a rise in sea level of about 1.5 mm/yr that seems a bit high when compared to independent estimates of the rise of sea level by direct measurement. Rignot *et al.* (2011) also noted that the annual rate of heat loss had accelerated over the 18-year period. They estimated that the acceleration for the sum of both sites was about 36 Gt/yr^2. If this acceleration persists into the future, it would imply that total sea level rise would be 15 cm by 2050 and 56 cm by 2100.

However, there appears to be some alarmist chartsmanship in this conclusion. As Figure 8.5 shows, there are large oscillations in the data, and the mass gain in 1992-3 skews the slope of the acceleration lines. Furthermore, the large positive loops in Antarctica since 2005 do not seem to be adequately considered. A strong case can be made for much flatter acceleration lines. In addition, measurements of sea level suggest that there are long-term oscillations, and extrapolations of short-term data are not justified. One of the problems in estimating the rate and acceleration of sea level rise is that some satellite data and climate models indicate larger increases than we observe. If these estimates are correct, why haven't the oceans risen faster?

Boening *et al.* (2012) reported that "Global mean sea level from altimetry from 1992 to 2012 with annual and semi-annual variations removed and smoothed with a 60-day running mean filter" showed an upward trend of 3.2 mm/year. They also emphasized that "a strong 2010/2011 La Niña, which affected precipitation patterns world-wide dropped lobal mean sea level (GMSL) by 5 mm between the beginning of 2010 and mid 2011."

To add to our consternation, Wada *et al.* (2010) provided a global overview of groundwater depletion in sub-humid and arid areas. When more groundwater is removed than is replenished, most of that water ends up in the oceans. Between 1960 and 2000, depletion of groundwater increased by more than a factor of two. It was estimated that in 2000, groundwater depletion added 0.8 mm/yr to the oceans. Yet, the measurements from tide gauges do not seem compatible with this result.

According to Scott K. Johnson (2012):[45]

"In many places, the water table is dropping as groundwater is depleted. When groundwater is pumped up for use, whether for drinking water or irrigation, some portion of it fails to infiltrate back down into the ground. (In drier regions, the portion that infiltrates approaches nil). Instead, the water evaporates into the atmosphere or ends up in surface streams. In either case, most of it eventually makes its way to the ocean. In many places, the amount of precipitation that infiltrates into the ground is too small to make up for that loss. And as the volume of groundwater decreases, sea level must rise in turn."

Although construction of dams on rivers creates large reservoirs (or lakes) behind them, increasing the storage of water on land, this has been estimated to be much smaller than the loss of groundwater to the oceans.

The 2007 IPCC Report assumed that dams and groundwater depletion roughly cancelled each other out.

---

[45] http://arstechnica.com/science/2012/06/groundwater-responsible-for-nearly-half-of-sea-level-rise/

Church *et al.* (2011) found the following results for the time period 1972 to 2008:

Source	Rise in sea level (mm/yr)
Ocean thermal expansion	0.8
Melting of glaciers and ice caps on Greenland and Antarctica	0.4
Melting of other glaciers and ice caps	0.7
Loss of ground water to oceans	0.3
Storage of water behind dams	− 0.4
Total	1.8

These authors found that storage of water behind dams more than offset the loss of ground water to the oceans.

There are two basic methods to estimate loss of ground water contribution to sea level rise. One, used by Wada *et al.* (2010), used an indirect, flux-based water budget approach that assumed that groundwater depletion is equal to the difference between natural recharge and withdrawals— rather than an approach based on actual observations of groundwater conditions. According to Konikow (2011), while the flux-based method is global in nature, it does not take into account the fact that as aquifers become depleted, they undergo reductions in natural discharge from the system (such as to springs and oases). Konikow also argued that the global modeling approach does not account for "non-natural" non-diffuse recharge, such as leakage from canals, sewers, or pipelines, or from artificial recharge. Hence, he argued that the flux-based water budget approach of Wada *et al.* (2010) "can substantially overestimate groundwater depletion". Konikow used a volumetric approach rather than a flux-based approach, in which he analyzed sequential changes in volume stored in various aquifers. Unfortunately, worldwide data are not widely available and Konikow had to extrapolate US data to much of the rest of the world. He estimated that ground water depletion contributed about 6 to 7 percent of sea level rise from 1900 to 2008. However, he noted that there was a rapid acceleration in the contribution of ground water depletion to sea level rise in the latter half of the 20th century. Over the period 2000 to 2008, the contribution of ground water depletion to sea level was estimated to be 0.4 mm/yr.

Wada *et al.* (2012) defended the flux based approach and pointed out that "volume-based assessments are only available for a limited number of aquifers and regions in the world, such that global estimates can be obtained only through extrapolation under assumptions, such as fixed depletion to abstraction ratios, that are difficult to verify".

Not only did Wada *et al.* (2012) estimate the contribution of ground water depletion to sea level rise in the 20th century, but they also estimated this quantity for the remainder of the 21st century using three possible future scenarios and three global climate models to project future climate (see Figure 8.5a).

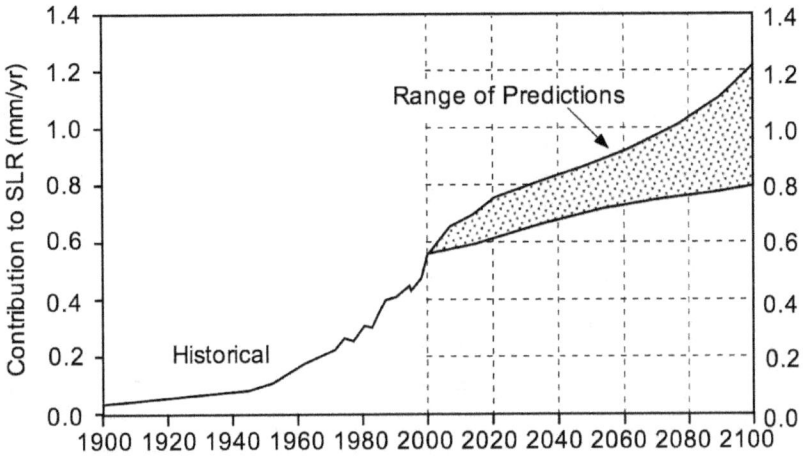

Figure 8.5a. Projection of future contribution to sea level rise from ground water (Wada *et al.*, 2012).

Pokhrel et al. (2012) "estimated sea-level change in response to human impacts on terrestrial water storage by using an integrated model that simulates global terrestrial water stocks and flows (exclusive of Greenland and Antarctica) and especially accounts for human activities such as reservoir operation and irrigation". They found that, "unsustainable groundwater use, artificial reservoir water impoundment, climate-driven changes in terrestrial water storage and the loss of water from closed basins have contributed a sea-level rise of about 0.77 mm/yr to sea level rise between 1961 and 2003, about 42% of the observed sea level rise". They noted that "of these components, the unsustainable use of groundwater represents the largest contribution". Rahmstorf, who predicted a sharp future rise in sea level (see Figure 8.6) was "shocked" to find these results for groundwater.

8.1.4 Global Warming and Future Sea Level Change

According to Hoffman (1984):

"Future global sea level will depend primarily on three factors: (1) the total quantity of water filling the oceans' basins; (2) the temperature of the oceans' layers, which determines the density and volume of their waters; and (3) the bathymetry (shape) of the ocean floor, which determines the water-holding capacity of the basins. A rise in global temperature can, by a variety of physical mechanisms, transfer snow and ice from land to the sea, increasing the quantity of water in the ocean

317

basins, and can raise the oceans' temperatures, causing the thermal expansion of their volumes. Changes in the bathymetry of the oceans' floors occur independently of climate change. Because geological changes in the ocean floor could not raise or lower global sea level by more than a centimeter or two by 2100, this factor is not considered in constructing global scenarios. An evaluation of the impacts of sea level rise at specific coastal sites, however, will require consideration of local uplift or subsidence, which by 2100 could cause changes in land elevation that are large enough to be of significance to local planning. Projecting sea level rise requires the means to estimate future changes in atmospheric composition, to relate these changes to global warming, and then to determine how the warming can cause land-based snow and ice to enter the sea and the oceans to expand thermally."

Hoffman (1984) estimated the sea level rise expected from various degrees of global warming. Low, medium, and high scenarios were based on ultimate temperature rises of 1.5°C, 3°C, and 4.5°C by 2100, respectively, for a doubling of CO_2 concentration. The projected rise in sea level is summarized in Table 8.2.

Table 8.2. Projected rise in sea level (cm). From Hoffman (1984).

Year	Low $\Delta T \sim 1.5°C$	Medium $\Delta T \sim 3°C$	High $\Delta T \sim 4.5°C$
1986	0		0
2000	5	11	17
2025	13	32	55
2050	24	66	117
2075	38	114	213
2100	56	180	345

According to Titus (1990):

"Since 1979, there has been a general consensus that a doubling of carbon dioxide would raise global temperatures 1.5 to 4.5°C, and that such a doubling is likely to occur over the next century. More recent assessments have pointed out that emissions of methane, nitrous oxide, and numerous other gases that absorb infrared radiation could further increase this warming, and that warmer temperatures may increase the rate of natural emissions of these gases. Although national policy makers are beginning to formulate strategies to slow global warming, there is an emerging consensus that at least a one or two degree warming is inevitable, due to past emissions and the time it will take to change production practices and retire existing machinery. In the late 1970s, some scientists suggested that the projected global warming might cause a 5 to 7 meter rise in sea level over the next few decades, due to a disintegration of the West Antarctic Ice Sheet. However ...such a deglaciation would take at least 200–500 years. As a result, most recent assessments have focused on other contributors to future sea level rise: expansion of ocean water and the melting of mountain glaciers and parts of the ice sheet in Greenland."

Titus (1990) provided the estimates of future sea level rise shown in Table 8.3. IPCC (2001) said:

"Disintegration of the West Antarctic Ice Sheet or melting of the Greenland Ice Sheet could raise global sea level up to 3 m each, over the next 1,000 years, submerge many islands, and inundate extensive coastal areas ... The projected sea-level rise of 5 mm/yr for the next 100 years would cause enhanced coastal erosion, loss of land and property, dislocation of people, increased risk from storm surges, reduced resilience of coastal ecosystems, saltwater intrusion into freshwater resources, and high resource costs to respond to and adapt to these changes (high confidence) ...Many coastal areas will experience increased levels of flooding, accelerated erosion, loss of wetlands and mangroves, and seawater intrusion into freshwater sources as a result of climate change. The extent and severity of storm impacts, including storm-surge floods and shore erosion, will increase."

Table 8.3. Predicted rise (cm) in sea level by 2100 (Titus, 1990).

Source	Year	Low estimate	Mid–Low	Mid–High	High
World Meteorological Organization	2050	20			170
Environmental Protection Agency	2100	70	160	210	340
National Research Council	2100	50			190

Alley *et al.* (2005) suggested that the Greenland Ice Sheet may melt entirely from future global warming, whereas the East Antarctic Ice Sheet (EAIS) is likely to grow through increased accumulation (for warming less than 5°C). The future of the West Antarctic Ice Sheet (WAIS) remains uncertain, with its marine-based configuration raising the possibility of important losses in the coming centuries. Alley *et al.* (2005) said:

"Despite these uncertainties, the geologic record clearly indicates that past changes in atmospheric CO_2 were correlated with substantial changes in ice volume and global sea level. Recent observations of startling changes at the margins of the Greenland and Antarctic ice sheets indicate that dynamical responses to warming may play a much greater role in the future mass balance of ice sheets than previously considered. Models are just beginning to include these responses, but if they prove to be important, sea-level projections may need to be revised upward. Also, because sites of global deepwater formation occur immediately adjacent to the Greenland and Antarctic ice sheets, any notable increase in freshwater fluxes from these ice sheets may induce changes in ocean heat transport and thus climate."

The projection of 5 mm/yr is far higher than most other projections, and reflects the extreme alarmist views of the IPCC Report. Figure 8.6 shows the range of IPCC projections for the 21st century. Also shown is Rahmstorf's 2007 projection based on an assumed proportionality of 3.4 mm/year per °C, and a hypothesized range of temperature increases of 1.5°C to 3.5°C. However, note that for the 20th century the proportionality was about 120 mm per 0.8°C temperature rise, or 1.5 mm/year per °C increase. Fortunately for Rahmstorf, he won't live long enough to see his projections tested.

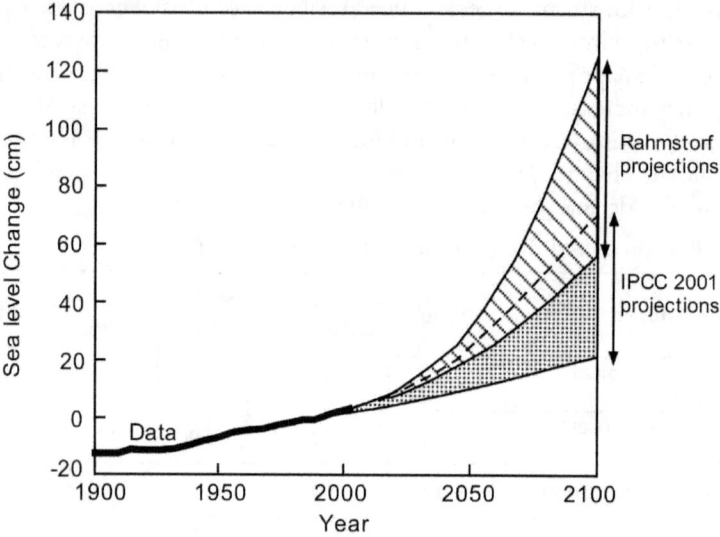

Figure 8.6. Projections of future sea level rise (IPCC, 2001; Rahmstorf, 2007).

According to Anon. (D):

"Over the past century, sea level has slowly been rising. This is in part due to the addition of water to the oceans through either the melting of or the 'calving' off of icebergs from the world's land ice. Many individual mountain glaciers and ice caps are known to have been retreating, contributing to the rising sea levels. It is uncertain, however, whether the world's two major ice sheets—Greenland and Antarctica—have been growing or diminishing. This is of particular importance because of the huge size of these ice sheets, with their great potential for changing sea level. Together, Greenland and Antarctica contain about 75% of the world's fresh water, enough to raise sea level by 70 m–75 m, if all the ice were returned to the oceans. Measurements of ice elevations are now being made by satellite radar altimeters for portions of the polar ice sheets, and in the future measurements will be made by a laser altimeter as part of NASA's Earth Observing System (EOS). The laser altimeter will provide more accurate measurements over a wider area. The Greenland ice sheet is warmer than the Antarctic ice sheet and as a result, global warming could produce serious melting on Greenland while having less effect in the Antarctic. In the Antarctic, temperatures are far enough below freezing that even with some global warming, temperatures could remain sufficiently cold to prevent extensive surface melting. Where ice sheets extend outward to the ocean, the ice tends to move out over the surrounding water, forming ice shelves. There is concern that, with global warming, the water under the ice shelves would be warmer and cause them to break up more readily, forming very large icebergs. If the ice shelves of West Antarctica were to break up, this would release more inland ice in an irreversible process, possibly leading to sea level rises of several meters."

Alley *et al.* (2005) emphasized:

"Future sea-level rise is an important issue related to the continuing buildup of atmospheric greenhouse gas concentrations. The Greenland and Antarctic ice sheets, with the potential to raise sea level ~70–75 meters if completely melted, dominate uncertainties in projected sea-level change. Freshwater fluxes from these ice sheets also may affect oceanic circulation, contributing to climate change. Observational and modeling advances have reduced many uncertainties related to ice-sheet behavior, but recently detected, rapid ice-marginal changes contributing to sea-level rise may indicate greater ice-sheet sensitivity to warming than previously considered. Over the last century, sea level rose ~1.0 to 2.0 mm year, with water expansion from warming contributing 0.5 ± 0.2 mm (steric change) and the rest from the addition of water to the oceans (eustatic change) due mostly to melting of land ice. By the end of the 21st century, sea level is projected to rise by 0.5 ± 0.4 m in response to additional global warming, with potential contributions from the Greenland and Antarctic ice sheets dominating the uncertainty of that estimate."

The projections by Hoffman (1984), Titus (1990), the IPCC (2001), Alley (2005) and Rahmstorf (2007) represent the alarmist viewpoint. In each case, extreme projections are made reprsenting worst-case scenarios. More sober projections have also been made. As Fjeldskaar (2008) summarized, "There is no doubt that sea level is currently rising on a global scale ...but a key question is not whether sea level is actually rising, but rather, has there been any acceleration in its rise during recent decades?" Examining all the conflicting data, he reached "the first conclusion" that "we have not reached a consensus on the rate at which sea level rises."

It has been widely hypothesized that a warmer climate in Greenland would increase the volume of lubricating surface meltwater reaching the ice–bedrock interface, accelerating ice flow and increasing mass loss (Joughlin *et al.*, 2008). Alarmists have postulated that there may be a tipping point in which the temperature rises to a point of no return, and an irreversible melting of the Greenland ice sheet would ensue, raising the oceans by some 7 meters. Several alarmists have projected the tipping point to be a 3°C global temperature rise (of which about 0.6 to 0.8°C has already occurred), although recent prognostications by James Hansen would have you believe that the tipping point is lower than that. However, Carlin (2009) commented on several recent papers that suggest that the mass loss has been lower than predicted. (Joughlin *et al.*, 2008; van de Val *et al.*, 2008). The paper by van de Val *et al.* is particularly intriguing because the title "Large and rapid melt-induced velocity changes in the ablation zone of the Greenland ice sheet" belies the finding: "The overall picture obtained by averaging all stake measurements at all sites for individual years indicates a small but significant (r = 0.79, P < 0.05) decrease of 10% in the annual average velocity [of ice movement in western Greenland] over 17 years." Most recently, Jonathan Bamber, an ice sheet expert at the University of Bristol, told the Copenhagen Climate Congress in 2009 that previous studies had misjudged the so-called Greenland tipping point, at which the ice sheet is certain to melt completely. He said the tipping point is closer to 6°C than 3°C. Murray *et al.* (2010) found that the "the early 2000s speedup of SE Greenland tidewater outlet glaciers was followed by a widespread and synchronous slowdown event.... Runoff lubrication of the glaciers does not provide the explanation for this speedup/slowdown event.... They indicated "that the speedup was the result of warm ocean waters coming into contact with the glaciers". They suggested "a negative feedback that currently mitigates against continued very fast loss of ice from the ice sheet in a warming climate ... namely the cold melt waters of the coastal Eastern Greenland Coastal Current act to stem further melting. "Since these SE Greenland outlet glaciers have dominated recent changes in ice sheet mass loss, the negative feedback we identify will also help regulate Greenland's contribution to sea level rise against a background of increasing ocean temperatures. We should expect similar speedup and slowdown events of these glaciers in the future, which will make it difficult to elucidate any underlying trend in mass loss resulting from changes in this sector of the ice sheet". Additional discussion and references are given in NIPCC (2011).

Shepherd and Wingham (2007) reviewed what is known about sea-level contributions arising from wastage of the Antarctic and Greenland Ice Sheets, focusing on the results of 14 different satellite-based estimates of the imbalances of the polar ice sheets that have been derived since 1998. The conclusion is that the current 'best estimate' of the contribution of polar ice wastage to global sea level change is a rise of 0.35 mm/yr.

Hansen (2005) claims that even this relatively benign scenario might nevertheless lead to destruction of the Greenland Ice Sheet. His arguments were based on a previous paper (Hansen, 2004) in which he pointed out that ice sheet growth is a slow, dry process, inherently limited by the snowfall rate, but disintegration is a wet process, driven by positive feedbacks, and once well under way it can be explosively rapid. A moulin, a near-vertical shaft worn in the ice by surface water, carries water to the base of the ice sheet. There the water acts as a lubricating fluid that speeds motion and disintegration of the ice sheet. Hansen believes that the Earth is "now out of energy balance by close to $+1$ W/m^2, i.e., with that much more energy absorbed from sunlight than the energy emitted to space as thermal radiation" which is "due mainly to rapid growth of greenhouse gases, especially CO_2 and CH_4, and the thermal inertia of the ocean." The greenhouse gases produce a downward forcing while the thermal inertia of the oceans prevents a rapid temperature rise, thus limiting re-radiation from Earth to space. Most of this putative energy imbalance goes into warming the oceans. However, Hansen believes that mechanisms exist to transfer some of this energy to the ice sheets. He believes that a further 1°C temperature rise might be enough to do significant damage to the Greenland Ice Sheet. Such mechanisms are likely to have occurred during the rapid disintegration of the ice sheets after the last glaciation (14,000–11,000 years ago). According to Hansen (2005):

"The net effect of these processes, which eventually will include a positive feedback from lowering of the ice surface altitude, is the potential for a highly nonlinear response, a process that could run out of control, possibly to the ultimate demise of the entire south dome (64°N) of the Greenland ice sheet, if the strong planetary forcing is maintained long enough. The question is: how long is long enough?"

Hansen (2005) suggested:

"Three time constants play critical roles in creating a slippery slope for human society: T_1, the time required for climate, specifically ocean surface temperature, to respond to a forced change of planetary energy balance; T_2, the time it would take human society to change its energy systems enough to reverse the growth of greenhouse gases; T_3, the time required for ice sheets to respond substantially to a large relentless positive planetary energy imbalance ... T_1, the climate response time, is 50–100 years, as a result of the large thermal inertia of the ocean. T_2, the energy infrastructure time constant, also is perhaps 50–100 years ... T_3, the ice sheet response time, is the time constant of issue."

"T_3 is of the order of centuries, not millennia, as commonly assumed. Growth of ice sheets requires millennia, as growth is a dry process limited by the snowfall rate. Ice sheet disintegration, on the other hand, is a wet process that can proceed more rapidly, as evidenced by the saw-toothed shape of glacial–interglacial temperature and sea level records."

Hansen (2005) summarized:

"The likelihood that T_3 is comparable to $T_1 + T_2$ has a staggering practical implication. $T_3 >> T_1 + T_2$ would permit a relatively complacent "wait and see" attitude toward ice sheet health. If, in the happy situation $T_3 >> T_1 + T_2$, we should confirm that human forcings were large enough to eventually alter the ice sheets, we would have plenty of time to reverse human forcings before the ice sheets responded. Unfortunately, $T_3 \sim T_1 + T_2$ implies that once ice sheet changes pass a critical point, it will be impossible to avoid substantial ice sheet disintegration. If T_3 indeed is not very much larger than $T_1 + T_2$, it becomes of high priority to detect as early as possible beginnings of ice sheet disintegration. High precision measurements of ice motion and sea level change are needed for early detection of any acceleration in the global rates of ice movement and sea level rise."

Oppenheimer and Alley (2005) commented on Hansen (2005).

"If Hansen is right about ice sheet response to the global energy imbalance and if IPCC's projections of future greenhouse gas concentrations prove correct, [a 1°C rise might disintegrate the Greenland Ice Sheet] it would be too late to stem a catastrophic sea level rise, given the commitment to future warming already in the pipeline."

But Oppenheimer and Alley (2005) indicated that, based on evidence from the last interglacial, the ice sheets may be stable to a 3°C temperature increase.

Graversen et al. (2010) applied a wide range of existing climate models to a detailed dynamical model of mass loss from Greenland expected in the 21st century. Unfortunately, the variations from model to model are so wide that the final result is a prediction that sea level will rise between 0 cm and 17 cm in the 21st century – a result that is so broad that it is quite useless.

The bottom line on sea level rise is that there is no bottom line. The data for the past and present are uncertain, but seem to indicate a steady rise since the 19th century of about 1.5 to 2.0 mm/yr. While some satellite measurements indicate an increased rate toward the end of the 20th century, the evidence in favor of acceleration of sea level rise is fragmentary. Projections for the future vary widely. In the worst case scenario, sea level rise could be a significant problem toward the end of the 21st century.

8.1.5 Evidence from Previous Deglaciations

Interglacial periods are of great interest because we are presently in an interglacial period, and there is widespread concern that rising CO_2, generated by human activity, may amplify and extend this climate period with negative consequences for civilization. It is therefore relevant to review data on past interglacials with particular emphasis on CO_2 levels and prevailing sea levels. It has been estimated that past interglacials were warmer than the present interglacial, and therefore it has been hypothesized that such past interglacials may provide an analog for the future if global warming continues.

Holden *et al.* (2009) reported:

> "Ice core evidence indicates that even though atmospheric CO_2 concentrations did not exceed 300 ppm at any point during the last 800,000 years, East Antarctica was at least 3–4°C warmer than pre-industrial ($CO_2 \sim$ 280 ppm) in each of the last four interglacials. During the previous three interglacials, this anomalous warming was short lived (~3,000 years) and apparently occurred before the completion of Northern Hemisphere deglaciation".

Sime *et al.* (2009) presented oxygen isotope data from three sites in Antarctica showing that the change in isotope content was considerably greater in the previous three deglaciations than in the present deglaciation. They analyzed the relationship between isotope index and temperature at the three sites and concluded "that maximum interglacial temperatures over the past 340 kyr were between 6°C and 10°C above present-day values" and "there are serious deficiencies in our understanding of warmer than present day climates". A simplified version of their data is shown in Figure 8.7.

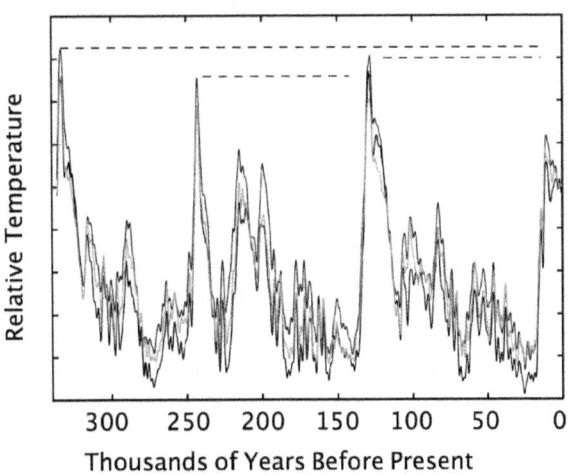

Figure 8.7. Relative temperatures of the last four interglacials. The previous three interglacials were significantly warmer than the present one, even though the CO_2 concentration was comparable (Sime *et al.*, 2009).

Rohling *et al.* (2008) quoted previous studies that indicated that "The last interglacial period ... was characterized by global mean surface temperatures that were at least 2°C warmer than present [and] mean sea level stood 4–6 m higher than modern sea level, with an important contribution from a reduction of the Greenland ice sheet". They used "a combination of a continuous high-resolution sea level record, based on the stable oxygen isotopes of planktonic foraminifera from the central Red Sea and age constraints from coral data to estimate rates of sea-level change during [the previous interglacial]". Rohling *et al.* (2008) estimated average rates of sea-level rise of 1.6 m per century. From this, they inferred that such a rate of sea-level rise might occur in the next century. However, the previous interglacial was very short lived compared to the present interglacial, and is not directly comparable.

Clark and Huybers (2009) asked: "Why was sea level so much higher 125,000 years ago?" They suggested "one possibility is that ice sheets have multiple potential steady states for a given climate". They mention "the global temperature was apparently 1.5–2°C warmer than the pre-anthropogenic global average of the past 10,000 years despite there being essentially no difference in atmospheric greenhouse-gas concentrations". However, this conclusion, which seems amazing to me, did not seem to cause them much consternation. If the Earth can be up to 2°C warmer while the CO_2 concentration is unchanged, then how can CO_2 be the cause of all climate change as proposed by alarmists? Clark and Huybers went on to say "that the climate of the last interglacial might, by coincidence, provide a reasonable analog for establishing ice-sheet sensitivity to global warming" and this implies "that the equilibrium response of sea level to 1.5–2°C of global warming could be an increase of 7–9 meters". They didn't seem to see that there is a logical impasse here. If changes in CO_2 concentration accompany glacial – interglacial transitions, why indeed did the CO_2 concentration not rise sharply above 300 ppm during warmer interglacials? And if rising CO_2 toward 400 ppm has produced the global warming of the past century, how is that connected to the warming of past interglacials when CO_2 remained below 300 ppm? Nor did they consider the possibility that the present interglacial has not yet reached its maximum temperature independent of CO_2, and perhaps (who knows?) is now extending temperatures upward to emulate previous interglacials regardless of CO_2. Ultimately, the relationship between CO_2 concentration and global temperature seems to be poorly understood.

Sime *et al.* (2009) presented oxygen isotope data from three sites in Antarctica showing that the change in isotope content was considerably greater in the previous three deglaciations than in the present deglaciation. They analyzed the relationship between isotope index and temperature at the three sites and concluded "that maximum interglacial temperatures over the past 340 KYBP were between 6°C and 10°C above present-day values" and "there are serious deficiencies in our understanding of warmer than present day climates". Applying the "factor of two" rule for global average temperature change vs. Antarctic temperature change (Hansen and Sato,

2011), this would imply that global average temperatures were 3° to 5°C higher in previous interglacials.

Kopp *et al.* (2009) concluded that sea level during the previous interglacial was 6-9 m higher than during the present interglacial. This was presumably due to the warmer temperatures. (They estimated that polar temperatures were 3-5°C warmer in the previous interglacial). Kopp *et al.* projected that such a warming today could raise the oceans by a similar amount, 6-9 m. Clark and Huybers (2009) echoed this theme. However, it is not clear why previous interglacials had higher temperatures with about the same concentration of CO_2, and therefore it is far from clear that increased CO_2 should produce conditions like that of the past interglacial.

8.2 Sea Ice Extent

Goody (1980) described sea ice as a "bizarre substance". He went on to say:

> "The minimum extent differs greatly in the two hemispheres. In the north more than half of the sea ice survives the summer; in the south less than one-eighth.... The ice margin is also subject to year-to-year changes and to large secular changes, although the geological record indicates that sea ice has never completely disappeared from the Arctic Ocean during the last million years.
>
> The thickness of sea ice does not differ greatly in the two polar regions, and averages ~2 m. Ice can grow almost to this thickness in one year. Thereafter, it stabilizes at ~3 m with a seasonal variation of 1 m. The ice is broken up by leads and open areas or polynyas and also by long pressure ridges up to 25 m high, which, like icebergs, are indicators of much larger changes at the bottom surface. Sea ice first forms from a slush which ultimately aggregates into a matrix of pure ice with brine inclusions. The physical properties of the sea ice are governed by the phase equilibrium between ice and brine giving rise to a substance with no definite melting point and with strongly temperature-dependent physical properties such as specific heat and conductivity. The freezing of the ocean surface dramatically changes both its thermal and mechanical properties.
>
> ... [It was estimated that] the heat flux (radiation, sensible and latent heat together) through 3 m of ice is only ~ 1.5% of that from the open ocean. Leads form 1% or so of even old sea ice; they, therefore, compete in thermal importance with the rest of the ice surface."

Humlum (2011) described sea ice as follows:

> "Sea ice occupies about 7% of the surface area of planet Earth. The sea ice thickness, its spatial extent, and the fraction of open water within the ice pack can vary rapidly and profoundly in response to weather and

climate. Sea ice typically covers about 14 to 16 million square kilometers in late winter in the Arctic and 17 to 20 million square kilometers in the Southern Ocean around the Antarctic. The seasonal decrease is much larger around the Antarctic, with only about three to four million square kilometers remaining at summer's end, compared to approximately seven to nine million square kilometers in the Arctic. The main reason for this difference is that the Arctic Ocean is centered on the Pole, while the Southern Ocean is not.

Sea ice variations have recently attracted much public interest. Part of the reason for this is the high albedo (c. 80%) of sea ice, which reflects much of the incoming solar short-wave radiation during the summer time. If not reflected, this radiation may instead be consumed by warming ocean water, thereby initiating a positive feedback, leading to more warming. This simple analysis however ignores that evaporation will increase from the ocean when the total sea ice cover are reduced in size. Increased evaporation usually results in an increased cloud cover and increased reflectance of incoming solar radiation, which tend to counteract the above process. The decrease or increase of sea ice has no effect on the global sea level."

Two important parameters are the extent of Arctic sea ice coverage and the average thickness of Arctic sea ice. Serreze *et al.* (2007) showed that Arctic sea ice extent has declined in every year from 1979 to 2006. The most pronounced rate of loss occurred in September of each year. However they said "evidence for accompanying reductions in ice thickness is inconclusive". While there are some indications that the sea ice has thinned, "sparse sampling complicates interpretation". According to Serreze *et al.* (2007):

"The observed decline in ice extent reflects a conflation of thermodynamic and dynamic processes. Thermodynamic processes involve changes in surface air temperature, radiative fluxes, and ocean conditions. Dynamic processes involve changes in ice circulation in response to winds and ocean currents".

Divine and Dick (2006) found evidence of persistent ice retreat in the Arctic since the second half of the 19th century. However, it was not clear whether this was a trend that will continue, or whether it was part of a cycle ("a similar shrinkage of ice cover was observed in the 1920s–1930s, during the previous warm phase ... when the anthropogenic influence is believed to have still been negligible").

Koberle and Gerdes (2003) concluded that wind forcing significantly contributes to the decadal variability in the Arctic ice volume. They concluded: "these results make connecting 'global warming' to Arctic ice thinning very difficult for two reasons." The two reasons are (i) large decadal and longer term variability masks any trend, and (ii) there might be long-term trends in wind stress that tend to either increase or decrease ice volume that would mask the effects of temperature change.

Lindsay and Zhang (2005) pointed out that recent observations of summer Arctic sea ice over the satellite era show that record or near-record lows for the ice extent occurred in the years 2002–2005. It was hypothesized that:

"The thinning since 1988 is due to preconditioning, a trigger, and positive feedbacks: 1) the fall, winter, and spring air temperatures over the Arctic Ocean have gradually increased over the last 50 years, leading to reduced thickness of first-year ice at the start of summer; 2) a temporary shift, starting in 1989, of two principal climate indexes (the Arctic Oscillation and Pacific Decadal Oscillation) caused a flushing of some of the older, thicker ice out of the basin and an increase in the summer open water extent; and 3) the increasing amounts of summer open water allow for increasing absorption of solar radiation, which melts the ice, warms the water, and promotes creation of thinner first-year ice, ice that often entirely melts by the end of the subsequent summer. Internal thermodynamic changes related to the positive ice–albedo feedback, not external forcing, dominate the thinning processes over the last 16 years. This feedback continues to drive the thinning after the climate indexes return to near-normal conditions in the late 1990s. The late 1980s and early 1990s could be considered a tipping point during which the ice–ocean system began to enter a new era of thinning ice and increasing summer open water because of positive feedbacks. It remains to be seen if this era will persist or if a sustained cooling period can reverse the processes."

The website *http://arctic.atmos.uiuc.edu/cryosphere/* provides extensive data on Arctic and Antarctic sea ice extent since 1980. Data from this website were utilized in preparing Figure 8.8. Ice coverage has particularly diminished in the summer period.

Perovich *et al.* (2011) said:

"... observed changes [in sea ice extent] are intricately linked to sea-ice dynamics and thermodynamics and are driven by atmosphere and ocean forcing. Several factors have been established as contributors to the decline in the ice cover: a general warming, changes in atmospheric circulation patterns, changes in cloudiness, advected ocean heat from lower latitudes, increased ice export from the Fram Strait, and increased solar heating of the upper ocean."

Serreze *et al.* (2007) indicated that surface air temperatures in the Arctic increased from 1979 to 1997, and they increased as well during the period 2000 to 2006 relative to 1979 to 1999. While many climatologists assume automatically that this was due to the effect of greenhouse gases, Serreze *et al.* (2007) pointed out:

> "... at least part of the recent cold-season warming ... is itself driven by the loss of ice, because this loss allows for stronger heat fluxes from the ocean to the atmosphere. The warmer atmosphere will then promote a stronger long wave flux to the surface".

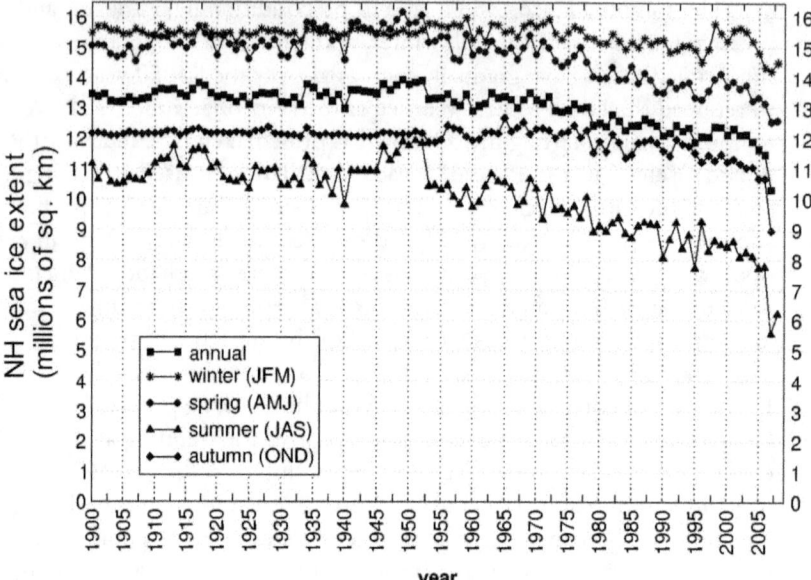

Figure 8.8. Northern Hemisphere seasonal sea ice extent (*http://arctic.atmos.uiuc.edu/ cryosphere/*).

In other words, it is possible that declining sea ice could be caused by factors other than greenhouse gases, and this, in turn, would produce higher surface air temperatures.

There is some evidence that changes in surface air temperatures during the 20th century were cyclic with an upswing in the cycle after 1970 (Frolov *et al.* 2009). There was a significant upswing from 1910 to 1940 while CO_2 levels were much lower. Chylek *et al.* (2009) said:

> "Temperature trend reversals in 1940 and 1970 separate two Arctic warming periods (1910–1940 and 1970–2008) by a significant 1940–1970 cooling period. Analyzing temperature records of the Arctic meteorological stations we find that (a) the Arctic amplification (ratio of the Arctic to global temperature trends) is not a constant but varies in time on a multi-decadal time scale, (b) the Arctic warming from 1910– 1940 proceeded at a significantly faster rate than the current 1970–2008 warming, and (c) the Arctic temperature changes are highly correlated with the Atlantic Multi-decadal Oscillation (AMO) suggesting the Atlantic

Ocean thermohaline circulation is linked to the Arctic temperature variability on a multi-decadal time scale".

Arctic sea ice loss was very high in years 2005 and 2007 (Kay *et al.* (2008)). Their results suggest that variable cloudiness may be a potent force for Arctic climate variability, but it does not reveal what causes the change in cloudiness. However, more recently, Screen and Simmonds (2010) did not think that changes in cloud cover have contributed strongly to recent warming.

Francis and Hunter (2007) asserted "while increases in greenhouse gases are believed to be the underlying cause of the melting, interactions among the Arctic's changing thermodynamic and dynamic processes driving ice loss are poorly understood". They found that in areas dominated by low clouds containing liquid water the emission of infrared radiation from the atmosphere to the surface during spring around the periphery of the Arctic Ocean down-welling long-wave flux is driven primarily by increasing cloud fraction and more abundant water vapor. In ice-cloud dominated regions they found that changing water vapor is more important than changing cloud fraction.

According to Winton (2006), "it is difficult to associate the Arctic-global differences with specific features of the atmosphere's CO_2 response" because multiple processes contribute feedbacks and forcing. He claimed "there are reasons to expect significant contributions to the Arctic-global long-wave feedback difference from cloud, water vapor and temperature feedbacks.

Polyakov *et al.* (2003) provided data on sea ice in various regions of the Arctic dating back to 1900. Figure 8.9 shows the annual change in extent of sea ice summed over four Arctic areas. The long-term trend was downward (shown as straight line in figure) but oscillations about this trend were huge. Polyakov *et al.* (2003) also provided data on fast sea ice thickness at five Arctic locations. Here, the trend was less clear, but after 1990 it was predominantly downward. In addition, Vinje (2001) provided data showing a gradual withdrawal of sea ice extending back to 1860.

Polyak *et al.* (2010) compared estimates of historical sea ice extent in the Nordic Seas based on ice core and tree ring data with more recent measurements of maximum sea ice extent in the Arctic-wide area. They concluded that this showed an anomalous reduction in sea ice extent in the 20th century. However, the estimates of past sea ice extent based on ice cores and tree rings appears somewhat fragile. Furthermore, the analysis only went back as far as year 1200. Had it been carried back through the MWP to say, year 850, it is possible that the recent decrease in sea ice extent might not appear so anomalous. It is noteworthy that the work by Polyak *et al.* (2010) was funded by U. S. Government agencies that support the alarmist agenda.

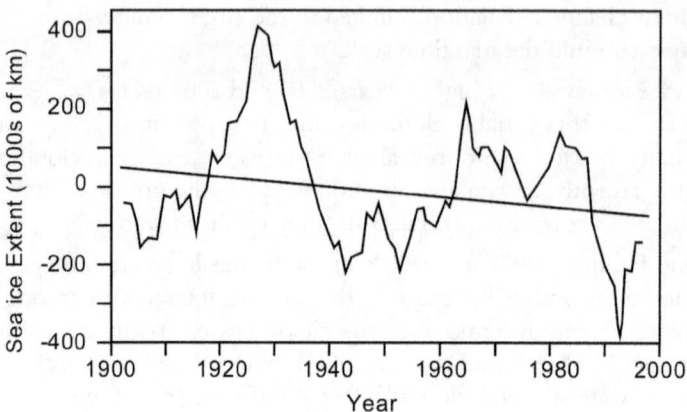

Figure 8.9. Six-year running mean of sum of changes in sea ice extent in four Arctic seas (Kara, Laptey, East Siberian and Chukchi). (Adapted from Polyyakov, 2003).

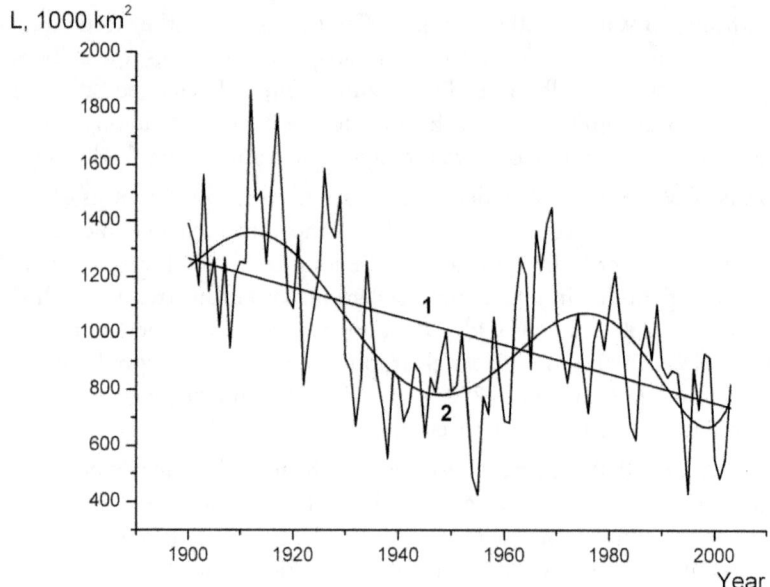

Figure 8.10. Variability of the total ice extent in the Greenland, Barents, and Kara Seas for the period 1900–2003 (August): 1 = linear trend and 2 = polynomial fit. (Frolov *et al.*, 2009).

Frolov *et al.* (2009) wrote a detailed monograph on Eurasian Arctic seas. They utilized direct measurements where available, as well as anecdotal data on ice navigation conditions early in the 20th century. Charts showing ice conditions with the routes of ships and ice edge locations, with coordinates or orientation markers were used to fill in missing quantitative data. They were thus able to reconstruct Arctic climate and sea ice extent over the past ~100 years for a vast area of about 5×10^6 km². They found that sea ice variability was quite different for two Arctic regions: (1) Greenland, Barents, and Kara Seas, and (2) Laptev, East Siberian, and Chukchi Seas. Their results for these two regions are shown in Figures 8.10 and 8.11. It can be seen that in the Greenland, Barents, and Kara Sea region, there has been an overall descending extent of sea ice throughout the 20th century, although there have been significant cycles about the main trend line. The trend for the Laptev, East Siberian, and Chukchi Sea region has been predominantly flat with significant variations about this trend line.

Over the past decade or so, it has become fashionable to make exaggerated claims about the dire state of declining Arctic sea ice. For example, according to Wang and Overland (2009):

"Summers in the Arctic may be nearly ice-free in as few as 30 years, not at the end of the century as previously expected. The updated forecast is the result of a new analysis of computer models coupled with the most recent summer ice measurements."

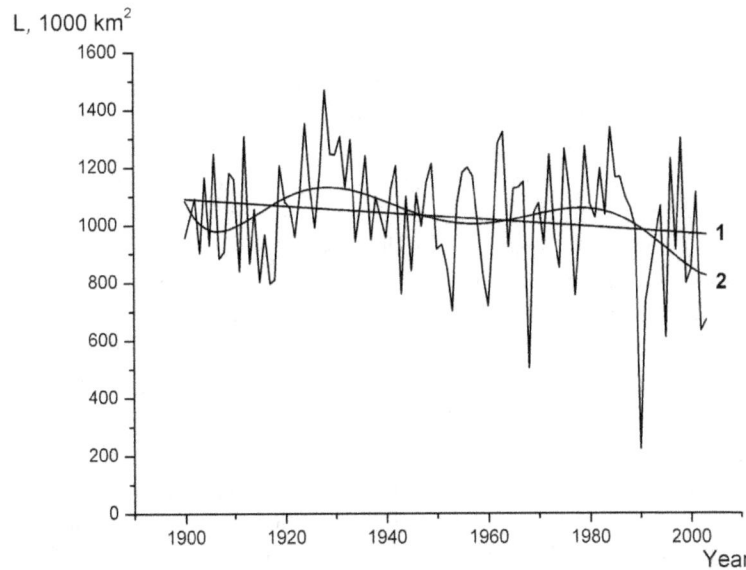

Figure 8.11. Variability of total ice extent in the Laptev, East Siberian, and Chukchi Seas for the period 1900–2003 (August): 1 = linear fit and 2 = polynomial fit. (Frolov *et al.*, 2009).

Polyak *et al.* (2010) in their "History of sea ice in the Arctic" said: "Arctic sea-ice extent and volume are declining rapidly. Several studies project that the Arctic Ocean may become seasonally ice-free by the year 2040 or even earlier".

The Internet is full of similar dire predictions. On the other hand, Frolov *et al.* (2009) argue that we are in the midst of the down leg of a natural cycle and all we need to do is wait for the next expansion of sea ice. According to Holloway and Sou (2000):

> "Previous reports that Arctic sea ice volume decreased nearly by half in recent decades have been widely cited in popular media and scientific considerations, e.g., IPCC 2001. We find instead ...that Arctic sea ice volume has decreased more slowly. Misleading inferences of rapid ice loss were a result of variable wind stress forcing a natural component of sea ice variability. In particular a dominant mode of variability moves ice between the central Arctic and the Canadian sector, suggestive of the modes of Arctic ice motion ... We conclude that observations and model results together suggest only modest reduction of ice volume, like the modest decline in ice areal extent. Previously inferred rapid loss of ice volume is unlikely."

A more recent report (Lindsay *et al.*, 2009) concluded:

> "[although] the minimum of Arctic sea ice extent in the summer of 2007 was unprecedented in the historical record ... the loss in total ice mass was not. Rather, the 2007 ice mass loss is largely consistent with a steady decrease in ice thickness that began in 1987. Since then, the simulated mean September ice thickness within the Arctic Ocean has declined from 3.7 to 2.6 m at a rate of 0.57 m per decade. Both the area coverage of thin ice at the beginning of the melt season and the total volume of ice lost in the summer have been steadily increasing. The combined impact of these two trends caused a large reduction in the September mean ice concentration in the Arctic Ocean. This created conditions during the summer of 2007 that allowed persistent winds to push the remaining ice from the Pacific side to the Atlantic side of the basin and more than usual into the Greenland Sea. This exposed large areas of open water, resulting in the record ice extent anomaly."

Arctic sea ice extent passes through an annual curve with maximum extent around mid-March and minimum extent around mid-September. Arctic sea ice extent bottomed out from 2004 to 2007, but more recent data than that reported by Lindsay *et al.* (2009) show that sea ice extent expanded in 2008 and 2009 to levels about 5% greater than in 2007. While the prospects for Arctic sea ice remain gloomy, alarmists have exaggerated the threat. More to the point, they uniformly blame Arctic sea ice diminution on greenhouse gases and rarely, if ever, mention black carbon. On the other hand, those who argue in favor of natural cycles have no credible explanation for the source of these putative cycles. Similarly, those who argue for the rise in CO_2 as the sole cause cannot explain the changes that occurred earlier in the 20th century. Furthermore, Sedlaek and Mysak (2008) pointed out that "The wind-driven changes in sea-ice area are about twice as large as those due to thermodynamic (i.e., radiative) forcing".

The National Snow and Ice Data Center[46] provides data on Arctic sea ice. These data indicate a significant diminution in sea ice extent over the past decade and predictions are now abundant that the Arctic sea ice might disappear in 20 to 30 years. Two of their graphs are presented in Figures 8.11a and 8.11b.

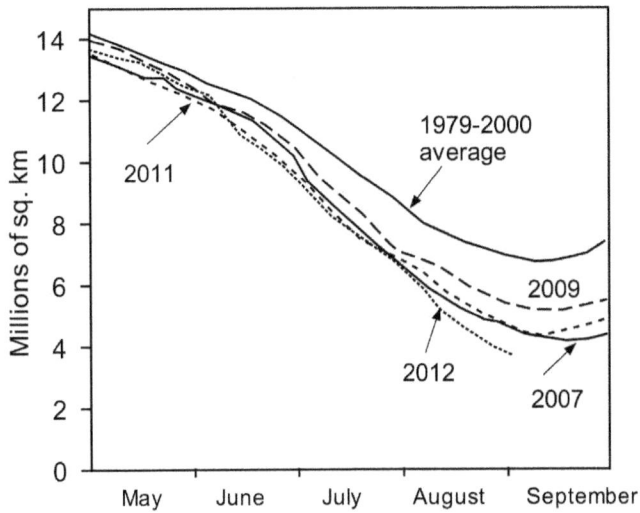

Figure 8.11a. Summer Arctic sea ice extent in recent years. *(http://nsidc.org/arcticseaicenews/2012/09/)*

As Figure 8.11a shows, the summer sea ice extent in the past few years has been considerably lower than the 1979-2000 average. The sea ice extent dropped to a new low in 2007, rose somewhat in 2009, and then decreased again after 2009. Figure 8.11b shows August sea ice extent from 1980 to 2012. The NSDIC used a vertical scale starting at 4 to exaggerate the apparent falloff. With a vertical scale starting at 0, the curve is not as impressive but remains neverhtless as source of concern.

By contrast, Antarctic sea ice has been expanding. Ozsoy-Cicek *et al.* (2009) report that "Antarctic sea ice cover has shown a slight increase (< 1%/decade) in overall observed ice extent as derived from satellite mapping from 1979 to 2008, contrary to the decline observed in the Arctic regions." It has been claimed that this increase in Antarctic sea ice extent resulted from the ozone hole that strengthened surface winds around Antarctica and

---

[46] *http://nsidc.org/arcticseaicenews/2012/09/*

deepened the storms in the South Pacific area of the Southern Ocean that surrounds the continent. This resulted in greater flow of cold air over the Ross Sea (West Antarctica) leading to more ice production in this region (Turner *et al.*, 2009). Turner *et al.* said:

"The only thing we can conclude at this point in time, therefore, is that for some still-unproven reason, and in spite of the supposedly unprecedented increases in mean global air temperature and CO_2 concentration that the planet has experienced since the late 1970s, Antarctica sea ice extent has stubbornly continued to just keep on growing."

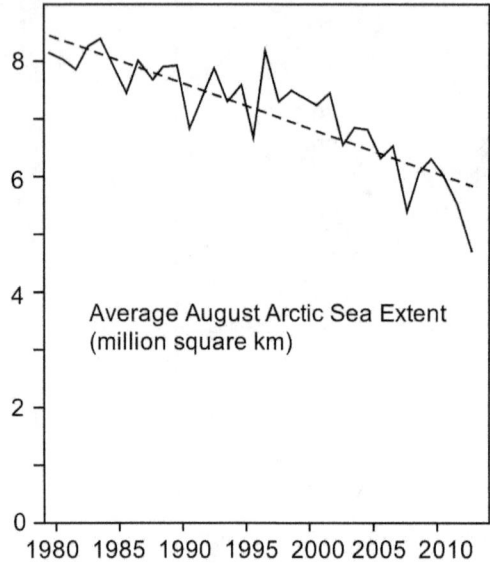

Figure 8.11b. Average August sea ice extent (1980-2012).
(http://nsidc.org/arcticseaicenews/2012/09/)

8.3 Future Increases in Global Temperature

There have been many estimates of the putative rise in global average temperature that will result from increased CO_2 concentration above the pre-industrial level of roughly 280 ppm. Three key steps are involved in this process: (1) estimating the future increase in CO_2 concentration, (2) estimating the putative temperature increase due to the increase in CO_2, and (3) estimating the additional heating due to feedback, primarily from water vapor, clouds, aerosols, and other secondary processes. Most such models limited their study to a future doubling of CO_2 to approximately 560 ppm.

Nozawa *et al.* (2007) utilized three projections for future greenhouse gas concentrations in a coupled ocean–atmosphere general circulation model. One projection assumed 1% annual increases in CO_2 for the entire 21st century, leading to a concentration of over 800 ppm by 2100. The most moderate model had an asymptote of 540 ppm in 2100. The estimated temperature increases in the 21st century ranged from 2.4°C to 4°C.

Anon. (F) utilized several projections of future CO_2 concentrations that increase by 2100 over a range varying from 470 ppm to well over 1,000 ppm, leading to temperature increases by 2100 ranging from 1.4°C to 5.8°C. They presented an ultimate *hockey stick* temperature profile with almost no temperature variation over the past millennium.

Anon. (H) projected temperature increases by 2100 ranging from 2°C to 6°C.

At the heart of almost all alarmist beliefs are claims that we are already in an unprecedented, incredibly high–temperature cycle, and projections of significant future increases in global temperature will lead us still higher. For example, Mann, Bradley, and Hughes (1999) said:

"...our results suggest that the latter 20th century is anomalous in the context of the last century. The 1990s was the warmest decade and 1998 the warmest year at moderately high levels of confidence."

Of course, this was the first MBH paper where the *hockey stick* was revealed, and these claims should be treated with caution.

Singer and Avery (2007) quote a number of alarmist claims from various sources:

"Nineteen ninety-nine was the most violent year in the modern history of weather. So was 1998. So was 1997. And 1996 ...A nine-hundred-year-long cooling trend has been suddenly and decisively reversed in the past fifty years ... Scientists predicted that the Earth will shortly be warmer than it has been in millions of years. A climatological nightmare is upon us. It is almost certainly the most dangerous thing that has ever happened in our history."

"Climate extremes would trigger meteorological chaos-raging hurricanes such as we have never seen, capable of killing millions of people; uncommonly long, record-breaking heat waves; and profound drought that could drive Africa and the entire Indian subcontinent over the edge into mass starvation."

"From sweltering heat to rising sea levels, global warming's effects have already begun ... We know where most heat-trapping gases come from: power plants and vehicles. And we know how to limit their emissions."

"No matter if the science of global warming is all phony ... climate change [provides] the greatest opportunity to bring about justice and equality in the world."

In technical press releases, there is a strong bias to portray data in the worst possible light. Temperatures have meandered since the hot year of 1998 induced by a huge El Niño. As it turns out, 2008 was an unusually cold year in which much of the warming of the previous 20 years (or more) was mitigated. But in early 2009, the U.S. National Weather Service (NWS) published a news release that said: "2008 was the 39th warmest year since 1895." Of course, 1895 was a very cold baseline year. Since 1930, 39 out of 79 years were warmer than 2008—hardly a basis for alarm. Yet the NWS made it seem as if 2008 was an exceptionally hot year! Similarly, the NOAA website reports 2009 temperature as "the nth warmest year on record out of 130 years" where n is typically in the range 6 to 9. But the 6 to 9 years that were warmer than 2009 all occurred since the hot year of 1998 induced by a huge El Niño; hence 2009 is stacking up as a relatively cold year compared with the previous decade!

Fortunately, Hansen (2004), written by a leading alarmist, provided a more balanced and credible picture. He pointed out:

"The IPCC scenarios may be unduly pessimistic, however. First, they ignore changes in emissions, some already under way, because of concerns about global warming. Second, they assume that true air pollution will continue to get worse, with ozone, methane and black carbon all greater in 2050 than in 2000. Third, they give short shrift to technology advances that can reduce emissions in the next 50 years."

Furthermore, limitations on the availability of fossil fuels will have a dramatic effect as the 21st century wears onward. Hansen (2004) went on to say:

"Observed global carbon dioxide and methane trends for the past several years show that the real world is falling below all IPCC scenarios. It remains to be proved whether the smaller observed growth rates are a fluke, soon to return to IPCC rates, or are a meaningful difference."

Note that Khalil, Butenhoff, and Rasmussen (2007) found that increases in methane concentration have stagnated. Quirk provided Figure 8.12.

The projection made by Hansen (2004) was:

"...at the low end of the IPCC range of two to four watts per square meter. The IPCC four watts per square meter scenario requires 4 percent a year exponential growth of carbon dioxide emissions maintained for 50 years and large growth of air pollution; it is implausible."

The bottom line (as usual) is that there is not a bottom line. As we discussed in Section 5.5.2, the predictions of climate models hinge upon feedback effects for water vapor, clouds and aerosols, and our present understanding of these factors is minimal.

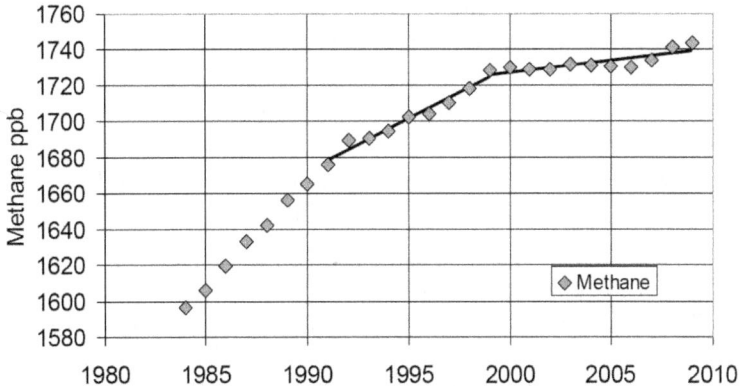

Figure 8.12. Atmospheric methane concentrations at Cape Grim (latitude 40°S). Values before 1991 are affected by fugitive natural gas from pipeline leakage (Quirk, 2012).

8.4 Changes in Precipitation: Floods, Drought and Storms

8.4.1 Drought

A number of climatologists of the alarmist persuasion have made dire predictions regarding the occurrence of drought, floods, and intense storms. In a sort of "Can you top this?" mode, each new report hypothesizes more and more dire outcomes. For example, recently, the ardent alarmist Andrew Dessler has suggested that the current drought in Texas is related to global warming.

In contrast, Idso (2008) wrote a 133-page rebuttal on the putative relationship between global warming and extreme weather. Although his approach is clearly biased, he provides many references to support his claim that (1) many climate models are unable to predict recent African Sahel drought, and (2) severe drought in Africa is not unique to the 20th century; there is evidence for severe drought dating back many hundreds of years. Hence the current drought is not necessarily a byproduct of global warming, and predictions of future drought are based on models that don't work in the recent past. Idso (2008) also provides evidence from Asia, North America, and Europe that global warming does not necessarily lead to the occurrence of more frequent or more severe drought.

The particular case of the continental U. S. is interesting. As Idso (2008) pointed out,

> "Andreadis and Lettenmaier (2006) examined 20th-century trends in soil moisture, runoff and drought over the conterminous United States with a hydro-climatological model forced by real-world measurements of precipitation, air temperature and wind speed over the period 1915–2003. This work revealed, in their words, that 'droughts have, for the most part, become shorter, less frequent, less severe, and cover a smaller portion of

the country over the last century' ... Van der Schrier et al. (2006) constructed maps of summer moisture availability across a large portion of North America and concluded that over the area as a whole, 'the 1930s and 1950s stand out as times of persistent and exceptionally dry conditions, whereas the 1970s and the 1990s were generally wet'."

They also said: "no statistically significant trend was found in the mean summer drought severity index over the 1901–2002 period."

Fye et al. (2003) used tree ring proxies to infer that periods of sustained U.S. drought have been recurrent phenomena over hundreds of years. Both Fye et al. (2003) and Stahle et al. (2000) concluded that drought in the 16th century was far worse than the severe drought of the "dust bowl" in the 1930s. Other papers are cited by Idso (2008) that indicate that occurrences of periodic U.S. drought date back 1,000 years or more.

More recently NIPCC (2011) updated Idso's analysis with additional references that indicate that periodic droughts have occurred in the past and there is nothing unique about the 20th century. The reader is referred to NIPCC (2011) for details. Sheffield *et al.* (2012) "show that the previously reported increase in global drought is overestimated because [previous models used] a simplified model of potential evaporation that responds only to changes in temperature and thus responds incorrectly to global warming in recent decades. More realistic calculations, ... that take into account changes in available energy, humidity and wind speed, suggest that there has been little change in drought over the past 60 years".

8.4.2 Floods

If the alarmists are correct that global warming is producing increased flooding in the world, the data should reflect this. Idso (2008) examined data for flooding during the 20th century when global warming took place. He examined data from the literature for Asia, Europe, and North America. In general, the data show no enhanced flooding in the 20th century compared with the past, and periodic flooding has occurred for centuries. As Idso pointed out, every time there is a flood somewhere, the alarmists immediately blame it on global warming and a hue and cry is raised to politically constrain emissions. But Idso quotes from many articles (e.g., Mudelsee et al., 2003) that conclude there was no increase in the occurrence of floods during the 20th century.

More recently, NIPCC (2011) provided an extensive discussion of evidence regarding flooding based on references as recent as 2010. They began by pointing out that it is difficult to derive accurate trends for the effect of climate on flooding because there has been a multitude of modifications to navigable rivers such as dams, dikes, weirs, bridges, etc. Nevertheless, examining the data such as it is, there is no evidence that flooding was unusually frequent or severe as the Earth warmed in the 20th century.

The reader is referred to Idso (2008) and NIPCC (2011) for details.

8.4.3 Storms

8.4.3.1 Tropical Hurricanes

Idso (2008) reviewed hurricane activity in the Atlantic, Pacific, and Indian Ocean basins, as well as for the globe as a whole, beginning with Atlantic Basin hurricanes. He provided a very large number of references.

Climate alarmists contend that global warming increases the frequency and intensity of hurricanes. Al Gore's film An Inconvenient Truth makes the contention that global warming produces more fierce storms. His film shows whistling winds, flooded cities, and people in despair from storm damage. In fact, Al Gore specifically claims that globally warmed gulf waters were responsible for the increase in Katrina's strength after it left Florida. The implication of the film is that the Earth is besieged by waves of storms of unprecedented intensity.

But Idso (2008) and Pielke et al. (2005) have made the point that even if storm damage does increase due to global warming, the gross damage to the biosphere in the 21st century will be dictated far more by population growth and industrialization than it will by intense storms. Yet these topics are often ignored while global warming commands attention from the media.

There is a huge literature on the question of the degree to which increased sea surface temperatures (SSTs) produce more frequent and more intense hurricanes. Closely related to this are the questions whether such increases in SST are due generally to anthropogenic activity, and specifically to the greenhouse effect from CO_2 emissions.

As is the case for many areas of climatology related to global warming, there is a diversity of opinion and one must be careful in reading papers to discern biased views, whether of the alarmist persuasion or the skeptical persuasion. Knutson (2008) provided a review that appears to be comprehensive and evenhanded. The intensity of tropical cyclones is measured by the power dissipation index (PDI) that includes contributions by frequency and intensity of storms. Emanuel (2005, 2007) studied the occurrence of tropical cyclones and compared the historical record of sea surface temperature (SST) with the PDI from 1870 to 2005. He found a significant correlation between SST and PDI over this time span but craftily only quoted the correlation coefficient after 1970. Emanuel (2005) claimed a 90% increase in the PDI, due to a 67% increase in the frequency and a 19% increase in intensity from 1870 to 2004. Actually, however, this is a biased view. The PDI varied around 6 prior to 1930, and jumped to about 8 from 1930 to 2000. The spike in the curves after 2000 may represent a fluctuation (as in 1950), swayed by Katrina and a few other hurricanes. And, indeed, more recent data show that the PDI dropped sharply after 2005.

Nevertheless, Knutson (2008) concluded that observed records of Atlantic hurricane activity show a strong correlation, on multi-year time scales, between local tropical SSTs and the PDI. He said: "Both Atlantic SSTs and PDI have risen sharply since the 1970s, and there is some evidence that PDI levels in recent years are higher than in the previous active Atlantic hurricane era in the 1950s and 60s." He went on to say that if the projected temperature increases in the 21st century (made by alarmists) are used to infer corresponding increases in SSTs, such a correlation between SST and PDI would be of grave concern.

Michaels *et al.* (2006) compared SST with wind speed of storms for Atlantic Basin tropical systems from 1982 to 2005. They found that a SST higher than a threshold value of 28.25°C is a virtual necessity for attaining category-3 or higher winds. However, they also found that an SST greater than 28.25°C does not act to further increase the intensity of tropical cyclones. In other words, they found a bimodal distribution: (a) below 28.25°C very few strong hurricanes occur; (b) above 28.25°C strong hurricanes do occur but the frequency is roughly independent of temperature above 28.25°C (at least up to 30.5°C where measurements ended). This is in contrast to conclusions reached by some alarmists that the frequency of strong hurricanes increases continuously with SST. They also concluded that the actual rise in SST was far too small to explain the post-1994 changes in tropical cyclone characteristics relative to prior decades. Other factors must cause the increase in hurricane intensity.

Knutson (2008) attempted to analyze much longer (>100 yr) records of Atlantic hurricane activity to determine at the century scale any increase in global and tropical Atlantic SSTs accompanied by a long-term rising trend in PDI. Although existing records of past Atlantic tropical storm numbers since 1878 show a pronounced upward trend, correlated with rising SSTs, it is believed that the reporting of storms was incomplete in the historical era. After adjusting for an estimated number of missing storms, there is a small nominally positive upward trend in tropical storm occurrence from 1878 to 2006 that it is not significantly distinguishable from zero. Thus the historical tropical storm count record does not provide compelling evidence for a greenhouse warming–induced long-term increase in PDI.

When Knutson (2008) analyzed Atlantic Basin hurricanes, rather than all Atlantic tropical storms, the result was similar. The evidence for an upward trend was even weaker for U.S. land-falling hurricanes.

Knutson's conclusions were:

"In summary, neither our model projections for the 21st century nor our analyses of trends in Atlantic hurricane and tropical storm counts over the past 120+ yr support the notion that greenhouse gas-induced warming leads to large increases in either tropical storm or hurricane numbers in the Atlantic.

Therefore, we conclude that despite statistical correlations between SST and Atlantic hurricane activity in recent decades, it is premature to conclude that human activity—and particularly greenhouse warming—has already had a discernible impact on Atlantic hurricane activity.

Similarly, efforts to project future levels of Atlantic hurricane activity using observed SST–PDI statistical relations derived from recent decades should be treated as highly speculative at this stage."

Some other relevant papers are mentioned below.

Webster *et al.* (2005) found a doubling of the number of category-4 and category-5 hurricanes in the 15-year period 1990–2004 as compared with 1975 – 1989. However, this article pointed out that they

"...deliberately limited this study to the satellite era because of the known biases before this period, which means that a comprehensive analysis of longer-period oscillations and trends has not been attempted. There is evidence of a minimum of intense cyclones occurring in the 1970s, which could indicate that our observed trend toward more intense cyclones is a reflection of a long-period oscillation."

Klotzbach (2006) found only a 10% growth in global category-4 and category-5 hurricanes from 1986–1995 to 1996–2005, of which most was in the Southern Hemisphere. In another publication, Klotzbach said:

"These findings indicate that there has been very little trend in global tropical cyclone activity over the past twenty years, and therefore, that a large portion of the dramatic increasing trend found by Webster and Emanuel is likely due to the diminished quality of the datasets before the middle 1980s. One would expect that if the results of Webster et al. and Emanuel were accurate reflections of what is going on in the climate system, then a similar trend would be found over the past twenty years, especially since SSTs have warmed considerably (about 0.2°C– 0.4°C) during this time period."

Klotzbach (2011) showed that over the period 1900 to 2009, tropical cyclone activity in the Atlantic was reduced by roughly a factor of two in El Niño years.

Some alarmists have pointed out that dollar losses in the U.S. from hurricanes rose sharply in the 20th century and particularly since 1990. However, Lonborg (2007) pointed out: "just comparing costs over long periods of time does not make sense without taking into account the change in population patterns and demography as well as economic prosperity. [Today] there are many more people, residing in much more vulnerable areas, with many more assets to lose." He showed that if one evaluates past hurricanes assuming the population distribution of today, one finds that hurricane damage would not have changed in the 20th century. Furthermore, the great 1926 Miami hurricane would have created far more devastation than Katrina. Ackerman (2008) presented a lengthy criticism of Lonborg's work, and indeed he found a number of errors and imperfections in Lonborg's analysis, as well as a bias toward skeptics' publications. However, Lonborg's conclusions remain valid: (1) the connection between human activity and global warming is uncertain, (2) the effects of global warming have been grossly exaggerated, (3) the economic problems faced by the world are severe, and a severe worldwide reduction in carbon emissions would add measurably to these problems, (4) there are bigger problems than global warming and we are not dealing with them effectively.

Pielke, Jr. *et al.* (2008) studied the total economic damage related to hurricane landfalls along the U.S. Gulf and Atlantic coasts from 1900 to 2005. They normalized the data by estimating damage that historical storms would have caused had they made landfall under contemporary levels of societal development by adjusting historical damages by three factors: inflation, wealth, and population. They said: "As people continue to flock to the nation's coasts and bring with them ever more personal wealth, losses will continue to increase". They concluded: "... that while 2004 and 2005 were exceptional from the standpoint of the number of very damaging storms, there is no long-term trend of increasing damage over the time period covered by this analysis. Even Hurricane Katrina is not outside the range of normalized estimates for past storms." Neumayer and Barthel (2011) pointed out that in assessing the change of hurricane damage over many years, "affected areas become wealthier over time and rational individuals and governments undertake defensive mitigation measures, which requires normalizing economic losses if one wishes to analyze trends in economic loss from natural disasters for detecting a potential climate change signal". They argued "that the conventional methodology for normalizing economic loss is problematic since it normalizes for changes in wealth over time, but fails to normalize for differences in wealth across space at any given point of time". They introduced "an alternative methodology that overcomes this problem in theory, but faces many more problems in its empirical application". In general, they found "no significant upward trends in normalized disaster damage over the period 1980–2009 globally, regionally, for specific disasters or for specific disasters in specific regions".

Bender *et al.* (2010) produced a model that predicts "nearly a doubling of the frequency of category 4 and 5 storms by the end of the 21st century". They also estimated that it would take 60 years to discern the putative anthropogenic influence above and beyond normal fluctuations. However, Knutson *et al.* (2010) reviewed the data and models regarding the possible anthropogenic influence on tropical cyclones. They said:

"Whether the characteristics of tropical cyclones have changed or will change in a warming climate — and if so, how — has been the subject of considerable investigation, often with conflicting results. Large amplitude fluctuations in the frequency and intensity of tropical cyclones greatly complicate both the detection of long-term trends and their attribution to rising levels of atmospheric greenhouse gases. Trend detection is further impeded by substantial limitations in the availability and quality of global historical records of tropical cyclones. Therefore, it remains uncertain whether past changes in tropical cyclone activity have exceeded the variability expected from natural causes."

They pointed out that models predict that greenhouse warming will produce fewer storms (6-34% less) but stronger storms (2-11% greater). However, the wide range of modeling results suggests that the issue is not well understood. Crompton *et al.* (2011) revised the analysis of Bender *et al.* (2010) and concluded that it would take between 120 and 550 years to discern the putative anthropogenic influence.

Landsea *et al.* (2010) said:

"Records of Atlantic basin tropical cyclones (TCs) since the late nineteenth century indicate a very large upward trend in storm frequency. This increase in documented TCs has been previously interpreted as resulting from anthropogenic climate change. However, improvements in observing and recording practices provide an alternative interpretation for these changes: recent studies suggest that the number of potentially missed TCs [in the past] is sufficient to explain a large part of the [apparent] recorded increase in TC counts".

Their study examined TC duration, using a widely used Atlantic hurricane database (HURDAT). It was found that "the occurrence of short-lived storms (duration of 2 days or less) in the database has increased dramatically, from less than one per year in the late nineteenth–early twentieth century to about five per year since about 2000, while medium- to long-lived storms have increased little, if at all". The authors went on to say:

"While it is possible that the recorded increase in short-duration TCs represents a real climate signal, ... it is more plausible that the increase arises primarily from improvements in the quantity and quality of observations, along with enhanced interpretation techniques. These have allowed National Hurricane Center forecasters to better monitor and detect initial TC formation, and thus incorporate increasing numbers of very short-lived systems into the TC database".

The authors made "quantitative estimates of the frequency of missed TCs, focusing just on the moderate to long-lived systems with durations exceeding 2 days in the raw HURDAT. Upon adding the estimated numbers of missed TCs, the time series of moderate to long-lived Atlantic TCs showed substantial multidecadal variability, but neither time series exhibited a significant trend since the late nineteenth century".

It is instructive that Kerry Emanuel, a leading exponent of the interpretation that increased storm activity derived from global warming, backed off this position quite sharply in Emanuel (2010). He said:

"When applied to the 1908-1958 reanalysis, the derived global frequency of tropical cyclones shows no significant trend over the period, while the frequency of events in the southern hemisphere shows a statistically significant decline and that of the northern hemisphere shows a marginally significant increase. There are statistically significant increases in frequency over the period in the North Atlantic, eastern North Pacific, and northern Indian Oceans, while frequency declines in the western North Pacific.... Finally, while it is tempting to believe that specification of sea surface temperature is sufficient for capturing most aspects of the general state of the atmosphere relevant to tropical cyclones, we show, using simple arguments, that failure to account for changing radiative properties of the atmosphere can distort the response of tropical cyclone activity to changing distributions of sea surface temperature; moreover, models appear to systematically underestimate the response of near-tropopause temperatures to changing surface temperature, and this too can affect the response of potential intensity".

Smith *et al.* (2010) claimed some progress in predicting hurricane activity in the tropical Atlantic. Klotzbach and Gray (2011) developed a methodology for predicting hurricane activity in the next season. In this report, they have a section entitled: "Why CO_2 Increases are not Responsible for Atlantic SST and Hurricane Activity Increases". Amongst other things, they said:

"Theoretical considerations do not support a close relationship between SSTs and hurricane intensity.... We have no plausible physical reasons for believing that Atlantic hurricane frequency or intensity will significantly change if global or Atlantic Ocean temperatures were to rise by 1-2°C. Without corresponding changes in many other basic features, such as vertical wind shear or mid-level moisture, little or no additional tropical cyclone activity should occur with SST increases".

They went on to say

"Any potential CO_2 influence on tropical cyclone activity is deeply buried as turbulence within the tropical atmospheres' many other energy components. It is possible that future higher atmospheric CO_2 levels may cause a small influence on global tropical cyclone activity. But any such potential influence would likely never be able to be detected, given that our current measurement capabilities only allow us to assess tropical cyclone intensity to within about 5 mph."

Villarini *et al.* (2011) reexamined the historical record of tropical storm observations from the Atlantic Ocean since 1880, dividing storms into those that lasted less than two days and those that lasted longer than two days. They found no increase in the longer storms but there was a significant increase in the number of shorter storms. They attributed most of this increase to better observing techniques and questionable data, rather than an actual increase in short storm frequency.

Villarini *et al.* (2011a) began with the introductory statement:

"The impact of future anthropogenic forcing on the frequency of tropical storms in the North Atlantic basin has been the subject of intensive investigation. However, whether the number of North Atlantic tropical storms will increase or decrease in a warmer climate is still heavily debated and a consensus has yet to be reached".

To shed light on this issue, Villarini *et al.* (2011a) "used a recently developed statistical model, in which the frequency of North Atlantic tropical storms was modeled by a ... rate of occurrence parameter that is a function of tropical Atlantic and mean tropical sea surface temperatures (SSTs)". Their "results do not support the notion of large (~200%) increases in tropical storm frequency in the North Atlantic basin over the twenty-first century in response to increasing greenhouse gases (GHGs)". The authors also modeled "projected changes in U.S. land-falling tropical storm activity under a variety of different climate change scenarios and climate models. These results were similar to those for the overall number of North Atlantic tropical storms, and do not point to a large increase in U.S. land-falling tropical storms over the twenty-first century in response to increasing GHGs".

The blog: *http://devoidofnulls.wordpress.com/* presented an analysis of category 4 and 5 tropical cyclones from 1987 to 2011 and found no trend in the frequency of occurrence.

NIPCC (2011) provided a more general discussion of storms, going beyond tropical hurricanes to include storminess in the Arctic, Europe, Australia, New Zealand and China. As in other work, there is no evidence of more intense storminess in the late 20th century than other times.

8.4.3.2 Tornados

Donald Prothero provided a short article on tornados. Tornados form when

"... warm moist air mass rising from the Gulf of Mexico that moves north and meets cooler, drier air from the northern Plains and the Rockies. When these collide, a strong front develops which causes a big horizontal cylindrical vortex to form. The warm air slides beneath the cold air and thunderheads grow. If there is also strong shear from the jet stream, the horizontal cylindrical spiral of air will tilt into a vertical funnel. If it continues to grow, it will touch the ground and become a tornado".

"The U.S. has by far most of the world's tornadoes due to its favorable geography.... The only other country in the world with significant tornadoes is Bangladesh."

The U.S. tornado season got off to "a rip-roaring start in 2011 ... with more than 322 deaths in the April 27 outbreak and the more than 144 deaths in the May 22 outbreak. However, ...these numbers don't yet approach the deadliest tornadoes in U.S. history, including over 700 killed and 2,027 injured by the Tri-State Tornado of March 18, 1925.

So, the question arises: Has the incidence and severity of tornados increased as the U. S. warmed in the 20th century? Unfortunately, this is very difficult, and may even be impossible to answer. According to H. E. Brooks:[47]

" The historical records of the occurrence of and losses from severe thunderstorms and tornadoes present significant challenges in attempting to establish trends, if they exist.... It is likely that the highest-quality dataset of significant length is the tornado dataset of the United States, which began in the early 1950s. Even these data have serious problems with consistency."

The data, such as they are, indicate that "the number of tornadoes reported in the United States per year has been increasing steadily over the past half century". However, they state that determining whether this is a robust trend in tornado occurrence is difficult because the historical record is both relatively short and non-uniform in space and time. In addition, the increase in yearly tornado numbers runs parallel with the concurrent increase in the country's population, which makes for that much better geographical coverage and more complete (i.e., numerous) observations". On the other hand, the data indicate the number of severe tornados per year has decreased over the past few decades. Attempts have been made to estimate changes in tornado occurrence from data on damages, but these are obfuscated by changes in population, increases in wealth, and changes in reporting.

In his "Concluding Remarks", H. E. Brooks said:

"Problems in the reporting databases mean that it is extremely unlikely that climate change will be detectable in severe thunderstorm and tornado reports, even if there is a physical effect.... There is no evidence to date to suggest that changes in damage are related to anything other than changes in wealth in the U.S.... In the absence of databases of at

---

[47] Tornado and Severe Thunderstorm Damage,
cstpr.colorado.edu/sparc/research/projects/...events/.../brooks.pdf

least reasonable quality, it is extremely difficult to say much of substance".

8.4.3.3 Extreme Weather

The UKCIP Climate Digest recently raised the question: "Has human-induced global warming increased the risk of heavy rainfall and flooding in the UK?" They answered this question by claiming "two new papers published in *Nature* (Min *et al.* (2011) and Pall *et al.* (2011)) add to the body of evidence that suggests that human-induced global warming is leading to an increased risk of heavy rainfall and flooding. They showed that human activity has contributed to an increased intensity of heavy rainfall events in the Northern Hemisphere over the second half of the 20th century and substantially increased the risk of flood occurrence in England and Wales in autumn 2000". This report concludes: "These results show that human activity has contributed to an increased intensity of heavy rainfall events in the northern hemisphere over the second half of the 20th century.... This implies that extreme precipitation events may strengthen more quickly in the future than projected and therefore have more severe impacts than previously estimated." However, the papers by Min *et al.* (2011) and Pall *et al.* (2011) are utterly unconvincing to this writer.

John Christy discussed extreme events. He reviewed a number of recent extreme climate events and compared the events with the historical record. He showed that recent flooding in Australia is not out of character with events in the 19th century and to a lesser extent, even earlier in the 20th century. In regard to flooding in England, he said:

> "For the Thames River, there has been no trend in floods since records began in 1880, though [there was] a lull in flooding events from 1965 to 1990.... Flooding events on the Thames since 1990 are similar to, but generally slightly less than those experienced prior to 1940. One wonders that if there are no long-term increases in flood events in England, how could a single event (Fall 2000) be pinned on human causation as in Pall et al. (2011), while previous, similar events obviously could not? Indeed, on a remarkable point of fact, Pall *et al.* 2011 did not even examine the actual history of flood data in England to understand where the 2000 event might have fit".

He also showed that the 2010 heat wave in Russia, and high snowfall in the U. S. in 2009-2010 and 2010-2011 can be explained by natural variability. Then he went on to show why occurrence of extreme events is not a good measure of climate change, using an example of extreme high and low temperatures as recorded by the states of the U. S.

> "For each of the 50 states, there are records kept for the extreme high and low temperatures back to the late 19th century. In examining the years in which these extremes occurred (and depending on how one deals

with "repeats" of events) we find about 80 percent of the states recorded their hottest temperature prior to 1955. And, about 60 percent of the states experienced their record cold temperatures prior to that date too. One could conclude, if they were so inclined, that the climate of the U. S. is becoming less extreme because the occurrence of state extremes of hot and cold has diminished dramatically since 1955.... Then, one might look at the more recent record of extremes and learn that no state has achieved a record high temperature in the last 15 years (though one state has tied theirs.) However, five states have observed their all-time record low temperature in these past 15 years (plus one tie.) This includes last month's record low of 31°F below zero in Oklahoma, breaking their previous record by a rather remarkable 4°F. If one were so inclined, one could conclude that the weather that people worry about (extreme cold) is getting worse in the U. S. (Note: this lowering of absolute cold temperature records is nowhere forecast in climate model projections, nor is a significant drop in the occurrence of extreme high temperature records.)"

8.5 Species Extinction

The media are replete with alarmists' claims that global warming has produced or will produce extensive extinction of species. In recent years, the number of voices raised in this regard has increased significantly.

The CO_2-induced global-warming extinction hypothesis claims that as the world warms in response to the ongoing rise in the air's CO_2 content, many species of plants and animals will not be able to migrate either poleward in latitude or upward in elevation fast enough to avoid extinction as they try to escape the stress imposed by the rising temperature.

Anon. (R) claims that "climate change has led to some 25 per cent of the world's mammals and 12 per cent of birds being at significant risk of extinction" and "that climate warming scenarios for 2050 could lead to extinction of approximately 18% to 35% of species" for low, mid, and high scenarios. Anon. (R) provides references to a number of recent papers that make such dire predictions. However, these all seem to utilize terms such as "at risk of," "could lead to," and "could possibly lead to."

Idso, Idso, and Idso (2003) argued however that

"... as long as the atmosphere's CO_2 concentration rises in tandem with its temperature, most species will not 'feel the heat,' as their physiology will change in ways that make them better adapted to warmer conditions. Hence, although Earth's plants will likely spread poleward and upward at the cold-limited boundaries of their ranges in response to a warming-induced opportunity to do so, their heat-limited boundaries will probably remain pretty much as they are now or shift only slightly. Consequently, in a world of rising atmospheric CO_2 concentration, the ranges of most of Earth's plants will likely expand if the planet continues to warm, making plant extinctions even less likely than they are currently. Animals should react much the same way. In response to concurrent increases in atmospheric temperature and CO_2 concentration, they will likely migrate poleward and upward, where cold temperatures prevented them from going in the past, as they follow Earth's plants ... A goodly portion of earth's plants and animals should actually expand their ranges and gain a stronger foothold on the planet as the atmosphere's temperature and CO_2 concentration continue to rise."

NIPCC (2011) provided an extensive discussion of the impact of global warming on species but their detailed analysis is beyond the scope of this book. Only a short discussion based on NIPCC (2011) is given here. One approach for estimating the impact of gobal warming on species is to determine the "climate envelopes" of over 1,000 species. "Each of these envelopes represented the current climatic conditions under which a given species was found in nature. Then, after seeing how the habitat area of each of the studied species would be expected to change in response to an increase in temperature, they used an empirical power-law relationship that relates species number to habitat area size to determine extinction probability calculations". However, in a warmer climate, the potential habitat of many species would be expanded (e.g. the climate would be milder at higher altitudes) which would open up new living space for species presently threatened with extinction by non-climatic factors, "due to increasing habitat loss attributable to expanding urbanization and agricultural activities; while it may help other species that are threatened with extinction by habitat fragmentation to cross geographical barriers that were previously insurmountable obstacles to them". In general, as NIPCC (2011) pointed out in considerable detail, analysts seem to too ready to pounce on any and all extinctions of amphibians, birds, butterflies, insects, lizards and animals, with the claim that global warming was the cause. NIPCC (2011) refutes these claims with references to studies that show that these were either caused by non-climatological human impact on the habitats or changes in other climatological parameters than temperature (winds, precipitation, ...). In some cases, such extinctions were counterbalanced by evolution of new species.

8.6 Vegetation

NIPCC (2011) discusses the impacts of climate change on terrestrial plants and soils in great detail in about 120 pages with many references. This included: plant growth responses to atmospheric CO_2 enrichment, below-ground biotic responses to atmospheric CO_2 enrichment, transpiration and water use efficiency, ecosystem responses to elevated temperature, responses of plants under stress to atmospheric CO_2 enrichment, ecosystem biodiversity, soil carbon sequestration, extinction, evolution, food production, greening of the Earth, nitrogen, dating of arrival of spring, and range expansion. In general, the effect of increased CO_2 was beneficial.

8.7 Coral Reefs

Silverman *et al.* (2009) pointed out that the recent anthropogenic increase in atmospheric CO_2 is increasing the acidity of the surface oceans. Models predict that doubling CO_2 from the pre-industrial value of ~ 280 ppm will reduce the ocean pH by about 0.5 units. They claim that this may produce "a severe global decline of coral reef abundance but not a complete extinction".

The NIPCC (2011) said:

"According to the IPCC, CO_2-induced global warming is increasing the temperatures of Earth's oceans and seas and lowering their pH values, a process called acidification. Both processes, according to the IPCC, are likely to harm aquatic life. The IPCC said: 'Many studies incontrovertibly link coral bleaching to warmer sea surface temperature ... and mass bleaching and coral mortality often results beyond key temperature thresholds' Modeling, the IPCC goes on to say, 'predicts a phase switch to algal dominance on the Great Barrier Reef and Caribbean reefs in 2030 to 2050'. The IPCC further claims that "coral reefs will also be affected by rising atmospheric CO_2 concentrations ... resulting in declining calcification".

NIPCC (2011) took issue with these claims. According to the NIPCC:

"While some corals exhibit a propensity to bleach and die when sea temperatures rise, others exhibit a positive relationship between calcification, or growth, and temperature. Such variable bleaching susceptibility implies that there is a considerable variation in the extent to which coral species are adapted to local environmental conditions.... The latest research suggests corals have effective adaptive responses to climate change, ... that allow reefs in some areas to flourish despite or even because of rising temperatures. Coral reefs have been able to recover quickly from bleaching events as well as damage from cyclones.... Bleaching and other signs of coral distress attributed to global warming are often due to other things, including rising levels of nutrients and toxins in coastal waters caused by runoff from agricultural activities on land and associated increases in sediment delivery.... The IPCC expresses concern that rising atmospheric CO_2 concentrations are lowering the pH values of oceans and seas, a process called acidification, and that this could harm aquatic life. But the drop in pH values that could be attributed to CO_2 is tiny compared to natural variations occurring in some ocean basins as a result of seasonal variability, and even day-to-day variations in many areas. Recent estimates also cut in half the projected pH reduction of ocean waters by the year 2100".

8.8 Food Production

In the spring of 2011, several reports claimed that consequences of human-caused global warming are more significant than previously projected in the 2007 IPCC Report. These new claims were quickly repeated by many news organizations. In one of these papers, Lobell *et al.* (2011) claimed

"... global average temperatures have risen by roughly 0.13°C/decade since 1950, yet the impact this has had on agriculture is not well understood. An even faster pace of roughly 0.2°C decade of global warming is expected over the next 2-3 decades, with substantially larger trends likely for cultivated land areas. Understanding the impacts of past trends can help to gauge the importance of near-term climate change for supply of key food commodities. In addition, identifying which particular crops and regions have been most impacted by recent trends would assist efforts to measure and analyze ongoing efforts to adapt".

Ken Haapala[48] discussed the paper by Lobell *et al.* As Haapala pointed out:

"Lobell *et al.* (2011) claim that global warming is restricting world food production. They analyze four major crops: maize (corn), wheat, rice and soybeans, which they state account for 75% of the world's human caloric consumption, either directly or indirectly. To estimate reduction in food production, the authors created a model estimating ... production ... without global warming (climate change) and compared it with actual production. They concluded that global wheat production is 5.5% below and maize production is 3.8% below what they would be without global warming. The main question is: does the model actually measure what the authors claim or are there significant confounding variables that are not identified in the study?"

Haapala made two key points:

[1] The claim that "global average temperatures have risen by roughly 0.13°C decade since 1950" is mistaken. This would imply that world temperatures have increased by 0.91°C from 1950 to 2010 which is considerably greater than the widely accepted range of 0.5 to 0.6°C.

[2] The lower 48 states of the U.S. have experienced a cooling from 1980 to 2008 and since the U.S. experienced a cooling, it is exempt from the conclusions of the study even though the U.S. is the world's major producer of two of the four food commodities studied, maize and soybeans, accounting for some 40% of world production.

Haapala also discussed "the production of wheat and rice in China and India, the two largest producers and consumers of these staples, accounting for

---

[48] Ken Haapala, Science and Environmental Policy Project (SEPP) The Week That Was: 2011-05-14 (May 14, 2011) *http://www.sepp.org/the-week-that-was.cfm*

approximately 29% of world wheat production and 48% of world rice production. Historically, these countries were noted for widespread famines, often due to changing weather patterns (climate change)". Haapala pointed out:

"From 1960 to 2008, in China, wheat production went up by 437% and milled rice production went up by 221%, or an average annual increase through the period of 9% and 5%, respectively. In India, wheat production went up by 661% and rice production by 186%, or an average annual increase through the period of 14% and 4%, respectively.

These remarkable increases in production occurred during a period of global warming, including the great climate shift of the late 1970s. Those inclined to hasty generalizations, without consideration of confounding variables, may conclude that global warming has been a boon to agriculture production in China and India. However, much of the increase is due to the green revolution, carbon dioxide enhancement, and changes in government policies.

Also during this period both China and India rapidly increased maize production becoming the world's second and fourth largest producers, respectively. Chinese maize production increased more than 10-fold....

Are these reductions in the spectacular increases in production due to global warming? No! An analysis of the complete data suggest otherwise, showing that, generally, production increases began to taper off in the 1990s. Basically, China and India became self-sufficient in grain production. Famines are no longer an issue and grain imports in 2008 were less than 0.5% of domestic production. [Lobell *et al.*, 2011] mistakenly attribute to warming the reductions in production increases due to market stabilization from the green revolution, carbon dioxide enhancement, and changes in government policies.

The authors claim that any excess would be available for export. However, export of grains requires an integrated system for such purposes that China and India do not have. Low-cost producers, such as the US and Canada, have such systems, and dominate the world markets. There is no incentive for farmers in China and India to produce more than what they can sell in domestic markets.

Hunger remains a major problem in much of the world, especially in sub-Sahara Africa, parts of which are subject to incessant warfare and political turmoil, resulting in low production of foodstuffs. To falsely attribute this hunger to global warming ignores the real causes and is a disservice to science and humanity."

NIPCC (2011) presented considerable evidence that food production has been higher historically during warm periods.

8.9 The Climate Debate Revisited

Alarmists have published a number of papers, and made various press releases and other media communications claiming that the impacts of future global warming will be disastrous. The U.N. has supported studies of global climate change for a number of years through its Inter-government Panel on Climate Change (IPCC). IPCC (2001) provides not 1—but 19—separate chapters on the impacts of global warming, totaling over 1,000 pages. It is noteworthy that many of these reports include phrases indicating a lack of certainty (could, might, it is possible that, etc.) All of these claims are debatable and none are certain. The claims that global warming will induce high levels of extreme weather are refutable. The only impact of global warming that should concern us is the possibility of a significant rise in sea level. Even here, the projections of the future are highly speculative.

9 GLOBAL CLIMATE CHANGE AND PUBLIC POLICY

9.1 The Kyoto Protocol

The Kyoto Protocol was the first of several agreements made under auspices of the U.N. whereby countries that ratify this protocol commit to reduce their emissions of carbon dioxide and five other greenhouse gases, or engage in emissions trading if they maintain or increase emissions of these gases. Governments are separated into two general categories.

Developed countries, referred to as "Annex I countries" (who have accepted GHG emission reduction obligations and must submit an annual greenhouse gas inventory); and

Developing countries, referred to as "Non-Annex I countries" (who have no GHG emission reduction obligations but may sell emission rights if they do reduce emissions).

According to this agreement, between 2008 and 2012, Annex I countries have to reduce their GHG emissions by an average of ~5% below their 1990 levels (for many countries, such as the EU member states, this corresponds to some 15% below their expected GHG emissions in 2008). While the average emissions reduction is 5%, specific national limitations vary widely.

Kyoto includes "flexible mechanisms" which allow Annex I economies to meet their GHG emission limitation by purchasing GHG emission reductions from elsewhere. What this means in practice is that Non–Annex I nations have no GHG emission restrictions so they can produce as much GHG emissions as they prefer, but if they do go about implementing GHG reductions, they can sell the rights to these reductions to Annex I nations, thus enabling them to generate the CO_2 instead.

The treaty was negotiated in Kyoto, Japan in December 1997, and signed by most countries by March 15, 1999. The treaty could not take effect until it included enough Annex I countries to account for at least 55% of emissions from Annex I countries. The agreement came into force on February 16, 2005 following ratification by Russia on November 18, 2004 that drove the total over 55%. As of December 2006, a total of 169 countries and other governmental entities had ratified the agreement (representing over 61.6% of emissions from Annex I countries). Emissions from non–Annex I countries were not stated. Notable exceptions include the U.S. and Australia. Other countries, like India and China, being Annex II countries, have ratified the protocol but are not required to reduce carbon emissions under the present agreement. In late 2011, Canada withdrew from the agreement, and several other countries were rumored to be falling short of their obligations.

Some public policy experts who are skeptical of global warming alarmism see Kyoto as a scheme to either slow the growth of the world's industrial democracies or to transfer wealth to the Third World in what they claim is a global socialism initiative. Others claim that the costs of the Kyoto Protocol outweigh the benefits, and the standards that Kyoto sets are too optimistic. Others see a highly inequitable and inefficient agreement that would do little to curb greenhouse gas emissions, and they are right. While the conspiracy theories of Jaworowski (2007) seem to be rather paranoid, his quotations by a high-level U.N. diplomat:

> "We may get to the point where the only way of saving the world will be for industrial civilization to collapse";

and by U.S. State Department officials:

> "We have got to ride the global warming issue. Even if the theory of global warming is wrong, we will be doing right thing in terms of economic policy and environmental policy."

> "A global warming treaty must be implemented even if there is no scientific evidence to back the [enhanced] greenhouse effect"

cannot but help his case. There are idealists in the U.N. who want to save the world from calamity, real or imagined. And there are politicians everywhere, willing to "go with the flow" if it is politically expedient.

Nevertheless, if action were taken to reduce greenhouse gas emissions in the 21st century, the number one culprit is clearly China. With its rapid growth in greenhouse gas emissions, China is now the world's greatest generator of greenhouse gases. It defies logic to exempt China from any controls at all under the Kyoto Protocol.

But there are other forms of pollution besides greenhouse gases, and China is clearly the greatest source of pollution in the world. The Kyoto Protocol does nothing about that. It is noteworthy that China already is the

world's #1 polluter. The satellite picture of Chinese pollution at
http://visibleearth.nasa.gov/view_rec.php?id=1036 demonstrates the extent of
pollution from Chinese coal burning. But with all the hoopla about global
warming, hardly anyone worries about conventional pollution anymore.

It is also noteworthy that even the *Los Angeles Times* (editorial June 11,
2007) a loyal believer in the theory that greenhouse gases cause excessive
global warming, believes that the Kyoto Protocol is fatally flawed. As a *Los
Angeles Times* editorial pointed out:

"The choice of 1990 as a base year simply rewards countries whose
economies have shrunk since then and punishes growth. Russia, Eastern
Europe, Germany and Britain are strong backers of Kyoto, and if one
looks at the costs and benefits of the pact, it's no wonder. Today, these
countries emit either less than they did in 1990 or just a little bit more. In
Britain, that's because the privatization of the coal industry led to a
decline in coal-fired power plants in favor of natural gas; elsewhere, it's
because the collapse of the Soviet Union was followed by the closing of
filthy Soviet-era industrial plants, while economies in Russia and much of
Eastern Europe stagnated. The U.S. economy, meanwhile, has grown
significantly since 1990, with a corresponding rise in power demand that
... has caused carbon dioxide emissions to jump 20.4%. What a global
carbon-trading scheme boils down to, then, is a massive wealth transfer
from the U.S. to Russia.

U.S. polluters would pay billions of dollars to buy carbon credits from
other countries, mostly Russia, because it would have the most to sell.
Why should we inject huge sums into a country with a rotten human
rights record, rampant corruption and opposing geopolitical views? And
what did Russia do to earn the cash, other than shrink?

Further, because there is no world body that polices greenhouse gas
emissions, countries and polluters are on the honor system—we have to
trust them to be honest about how much they're polluting. Governments
in Russia or Ukraine aren't capable of monitoring emissions from every
pollution source even if they wanted to, and under Kyoto, there's no
reason for them to want to. After all, if Ukraine claims to be cleaner than
it really is, rich countries such as the U.S. and Japan will shower it with
money for carbon credits. And corrupt governments will tend to
distribute credits unfairly, using them to reward political supporters and
reducing the market's effectiveness.

India and China are left out ...If China keeps growing at its current
rate, its per capita income is expected to reach U.S. levels within 25 years.
Once that happens ...there might be 1.1 billion vehicles in China.
Currently, there are only 800 million vehicles in the entire world.
Meanwhile, China builds a new coal-fired power plant every week to

stoke its growth, and ...in a quarter of a century, its CO_2 emissions will be double those of the other industrialized nations combined. India, the world's fourth-biggest polluter, is also growing at a blistering pace.

Unless these two countries can be persuaded to embrace green power, they will render Kyoto and every other attempt to reduce greenhouse gases moot. Unfortunately, they're not eager to change their ways. Last Monday, China unveiled its climate-change plan in response to pressure from the G-8; it made no commitments to any quantifiable carbon reductions and rejected international efforts to impose them. India also refuses to consider anything that might slow its economic development."

The *Los Angeles Times* then went on to propose

"...a new, improved version of Kyoto that brings India and China onboard and commits them to 'grow green,' but still leaves the tougher cuts up to those nations better able to make them, such as the U.S., Canada, Japan and Europe."

It is not clear to this writer why China, the world's worst polluter (other than CO_2) should get any favoritism. The *Los Angeles Times* also said:

"A better treaty would scrap the unworkable carbon-trading scheme and instead impose new taxes on carbon-based fuels."

This sounds like a good idea but can it be made to work? Or, is it like the mice deciding that they need to put a bell around the cat's neck?

In a recent review of the Kyoto Protocols, Adam (2008) reported on figures released by the U.N. suggesting that the world is on track to meet its Kyoto targets for greenhouse gases. It was claimed that emissions by the 40 industrialized nations that agreed to binding cuts in pollution are down 5% compared with 1990 levels. But it turns out that the drop had little to do with climate policies: the bulk of the decline was due to the collapse of the Soviet Union and the subsequent economic decline in eastern Europe in the 1990s. Without these so-called "economies in transition", greenhouse gas emissions have grown by almost 10% since 1990. Furthermore, emissions are once again rising in Eastern Europe.

Adam (2008) reports that, among industrialized nations, 16 are on target to meet their Kyoto obligations, including France, the U.K., Greece, and Hungary, while 20 countries are off-course, including Canada, Germany, Ireland, Italy, Japan, New Zealand, and Spain. Nations that miss their Kyoto target in 2012 will incur a penalty of an additional third added to whatever cut they agree under a new treaty being drafted in Copenhagen.

According to Yvo de Boer, executive secretary of the U.N. Climate Secretariat, Kyoto has been successful at establishing an architecture for future reductions, but has not been successful at actually reducing emissions significantly.

9.2 Economics: Will it cost more to do nothing?

9.2.1 The Stern Report

Over the past few years, a number of economists have attempted to analyze the costs associated with global warming – comparing business-as-usual (BAU) with draconian cuts in emissions.

The Stern Report (CCSP, 2007) was one of the early studies. It produced a lengthy report (about 600 pages) on the economics of global climate change. The Report presupposed that (1) drastic climate impacts will occur if no action is taken, (2) we have the means to take action to prevent climate change, and (3) global warming is the biggest threat faced by humanity in the 21st and 22nd centuries. Based on this, the economic study reached the conclusion that it will cost less to invest in climate control now, than to pay later for the problems that will be created by future climate change. The study was conducted as a cost–benefit analysis for the 200-year period ending around 2200 that assessed the likely cost of climate impacts in a BAU scenario without active control of greenhouse gas emissions, and compared this with the cost and benefits of various levels of control of greenhouse gas emissions.

However, the Stern Report was based on a number of scientific and technical assertions that at best are quite uncertain, and at worst, may simply be wrong. The Stern Report estimated that the current CO_2 equivalent level ("CO_{2e}") is about 430 ppm (375 ppm of CO_2 + about 55 ppm equivalent from other greenhouse gases). A fundamental basis of the Stern Report is stated as:

> "The level of 550 ppm CO_{2e} could be reached as early as 2035. At this level there is at least a 77% chance—and perhaps up to a 99% chance, depending on the climate model used—of a global average temperature rise exceeding 2°C ... On current trends, average global temperatures will rise by 2–3°C within the next fifty years or so."

A glance at Figure 5.4 shows that the level of 550 ppm CO_{2e} will certainly not be reached by 2035, and probably won't be reached until about 2060. The temperature rise associated with this increase in greenhouse gas concentrations is difficult to nail down. Assigning numerical probabilities to various temperature anomalies is purely subjective.

The Stern Report delineates the putative impacts of global warming in great detail: floods, reduced water supplies for one-sixth of the world's population, declining crop yields, leaving hundreds of millions without the ability to produce or purchase sufficient food, spread of disease, displacement of 200 million people by rising sea levels, extinction of 15% to 40% of species, destruction of the Amazon rainforest, etc. Various large-scale discontinuities are hypothesized. The Stern Report predicts a temperature rise of 3.9°C to 4.3°C by 2100 and 7.4°C to 8.6°C by 2200 in the BAU scenario, assuming that the world supply of fossil fuels is limitless, and that climate models are essentially correct. This is, of course, an absurd scenario because the world will take action long before such temperature increases are approached. There are fundamental flaws in the underlying basis for the Stern Report. A fundamental problem facing humanity in the 21st century is how to provide the burgeoning population with energy? Oil supplies, then gas, will gradually run down by mid-century. Coal is problematic for many reasons. There is no presently known technical fix for the energy problem, but we can be fairly sure of one thing: if there is a fix, it will have to involve ramping down use of fossil fuels as the century wears on. Not because of fear of global warming, but because fossil fuel supplies are finite. Hence the challenge we face in the coming century is not global warming, but rather, providing the people of the world with energy. In the short run, plans to urgently reduce carbon emissions will exacerbate the bigger problem of providing energy. Fossil fuels (and carbon emissions) must be viewed as a bridge to a new energy paradigm, but premature curtailing of fossil fuel use can only bring on economic ruin. The Stern Report promulgates the false belief widely held by economists (but not by energy specialists) that technology exists for replacing fossil fuels by non–carbon emitting technologies, and all we need to do is implement such technologies at an affordable cost.

In the past, an increase in CO_2 (not equivalent) of 100 ppm from the pre-industrial era to the early 21st century used up about one-third of original world fossil fuel resources. The temperature rise during that period was about 0.6°C, but only part (possibly a small part) of that rise can be attributed to greenhouse gases. Why should we believe that further increases in CO_2 will produce a 2°C to 3°C temperature rise in 50 years, and an 8°C rise in 200 years?

According to the Stern Report,

"... the poorest countries and people will suffer earliest and most. And if and when the damages appear, it will be too late to reverse the process. Thus we are forced to look a long way ahead ... Climate change may initially have small positive effects for a few developed countries, but is likely to be very damaging for the much higher temperature increases expected by mid-to late-century."

However, the Stern Report admits that "there is much to learn about these risks" but insists that "the temperatures that may result from unabated climate change will take the world outside the range of human experience." Since the human experience includes ice ages, this is certainly an exaggeration.

There are three problems with the impacts listed by the Stern Report: (1) we don't know what the future CO_{2e} will rise to, (2) we don't know what temperature changes will result from a putative increase in CO_{2e} and how they will vary regionally, and (3) the speculations on how any given temperature rise will affect the world economies are quite subjective.

The Stern Report carried out detailed academic exercises predicting future reductions in gross domestic products due to global warming. However, the Stern Report cautions:

> "Economic forecasting over just a few years is a difficult and imprecise task. The analysis of climate change requires, by its nature, that we look out over 50, 100, 200 years and more. Any such modeling requires caution and humility, and the results are specific to the model and its assumptions. They should not be endowed with a precision and certainty that is simply impossible to achieve. Further, some of the big uncertainties in the science and the economics concern the areas we know least about (for example, the impacts of very high temperatures), and for good reason—this is unknown territory."

But the Stern Report maintains that despite these uncertainties,

> "The main message from these models is that when we try to take due account of the upside risks and uncertainties, the probability-weighted costs look very large. Much (but not all) of the risk can be reduced through a strong mitigation policy, and we argue that this can be achieved at a far lower cost than those calculated for the impacts. In this sense, mitigation is a highly productive investment."

Until now, North America and Europe have produced around 70% of all the CO_2 emissions due to energy production, while developing countries have accounted for less than one-quarter. However, most future emissions growth will come from today's developing countries.

The Stern Report insists: "the world does not need to choose between averting climate change and promoting growth and development." It argues for "decarbonization" of energy technologies to achieve climate stabilization. This requires that annual emissions be brought down to the level that balances the Earth's natural capacity to remove greenhouse gases from the atmosphere. The Stern Report indicates that the rate of CO_{2e} emission in 2000 was about 41 Gt/yr (equivalent to $41/3.67 = 11.2$ Gt/yr of carbon). However, this figure appears to be too high. The Stern Report postulates a number of future scenarios in which annual CO_{2e} emissions rise in the near term (to ~48 Gt/yr to ~63 Gt/yr) and then fall back after 2040, dropping to about 20 Gt/yr by 2100. It indicates that if the peak CO_{2e} emission rate were less than 50 Gt/yr around 2030 to 2040, and it dropped to 30 Gt/yr in 2050 and 20 Gt/yr by 2100, stabilization of CO_{2e} at around 550 ppm might be achievable. However, these cuts will have to be made in the context of a world economy in 2050 that may be three to four times larger than today, so the emissions per unit of GDP would need to be reduced to about 20% to 25% of current levels by 2050.

Note that the Stern Report deals with CO_2 emissions. One mass unit of CO_2 is equivalent to $(44/12) = 3.67$ mass units of carbon. Thus, when the Stern Report postulates a peak CO_2 emission rate of 50 Gt/yr that is equivalent to 13.6 Gt/yr of carbon. Thus, the emission scenarios postulated by the Stern Report would more or less follow IS92a out to 2030 to 2040 and then ramp down to about 5.4 Gt/yr carbon by 2100. This can be seen more clearly in Figure 5.4.

Achieving deep cuts in emissions will have a cost. Greenhouse gas emissions can be cut in four ways.

1. Reducing demand for emissions-intensive goods and services (economic depression?).

2. Increasing efficiency.

3. Action on non-energy emissions, such as avoiding deforestation.

4. Switching to lower carbon technologies for power, heat, and transport.

The Stern Report estimates the annual costs of stabilization at 550 ppm CO_{2e} to be around 1% of GDP by 2050 (3%–4% of present GDP), a level that is claimed to be "significant but manageable".

The Stern Report claims that by 2050, energy efficiency has the potential to be the biggest single source of emissions savings in the energy sector. Non-energy emissions make up one-third of total greenhouse gas emission. A substantial body of evidence suggests that action to prevent further deforestation could be relatively cheap compared with other types of mitigation. Large-scale uptake of a range of clean power, heat, and transport technologies is required for radical emission cuts in the medium to long term. The power sector around the world will have to be decarbonized by at least 60%, and perhaps as much as 75%, by 2050 to stabilize at or below 550 ppm CO_{2e}. Deep cuts in the transport sector are likely to be more difficult in the shorter term, but will ultimately be needed.

The Stern Report says that the shift to a low-carbon global economy will take place despite "an abundant supply of fossil fuels." It says that stocks of hydrocarbons that are profitable to extract (under current policies) are more than enough to take the world to levels of greenhouse gas concentrations well beyond 750 ppm CO_{2e}, which conflicts with the estimates given in Section 7.2. The Stern Report postulated that:

"Even with very strong expansion of the use of renewable energy and other low-carbon energy sources, hydrocarbons may still make over half of global energy supply in 2050. Extensive carbon capture and storage would allow this continued use of fossil fuels without damage to the atmosphere."

It is not clear in the Stern Report how GDP impacts vary with country. Discussions of GDP seem to be limited to a world average. The argument seems to be that the consequences of business as usual in the 21st century are disastrous for the whole world, and the cost of stabilization at, say, 550 ppm are less than the cost of global-warming impacts. To achieve such stabilization, the Stern Report provides hundreds of pages on policy responses for mitigation and adaptation, and international collective action (a la Kyoto). That would certainly be a boon for bureaucracy. The Stern Report appears to be based on unfounded expectations of disaster and the economic analyses of consequences are of dubious credibility. When impending shortages of fossil fuel are taken into account, it is likely that many of the proposed mitigations will take place through market forces without micromanagement by the U.N.

9.2.2 Critiques of the Stern Report

Nordhaus (2006) claimed that the Stern Report's use of a near zero (0.1%) social discount rate is unfounded and leads to very unrealistic conclusions. The social discount rate has to do with social time preference; it measures the importance of the welfare of future generations relative to the present.

> "It is calculated in percent per year, like an interest rate, but refers to the discount in future utility or welfare, not future goods or dollars. A zero social discount rate means that future generations into the indefinite future are treated equally with present generations; a positive social discount rate means that the welfare of future generations is reduced or discounted compared to nearer generations. Philosophers and economists have conducted vigorous debates about how to apply social discount rates in areas as diverse as economic growth, climate change, energy policy, nuclear waste, major infrastructure programs such as levees, and reparations for slavery" (Anon. (K)).

On a far smaller scale, each individual has a similar problem in deciding how much of their current income to sequester into 401K and other tax shelter plans for future retirement vs. taking current income for current expenses in living. Achieving a balance between current needs and preparing for the future is a problem that confronts individuals, institutions, and governments. It is desirable to avoid the extremes of dying as the "richest man in the cemetery" vs. living in poverty in one's old age. As Nordhaus (and others) have discussed, there are many possible philosophies on how to value future gains in the present. Economists have their Greek letters (deltas and etas) that they put into formulas, but it all comes down to how to value savings in the future against costs in the present.

As we stand in 2011, the world economy is in desperate straits due mainly to the fact that in the previous seven years, we borrowed and spent the funds (in advance) that should have been available in in the future. The approach used by world governments to deal with the credit crisis produced by excessive debt is to borrow more. Against this backdrop, the proposal to expend several percent of the world GDP per year to head off a hypothetical problem that might occur a hundred years hence would exacerbate the world economic problem to the point where it might cause a world economic collapse.

On the other hand, if one really believes the threat of global warming is a blight on the lives of our grandchildren, the moral imperative would be to spend the money now to prevent future disaster, at whatever the cost to us in the present. In times of war, when we are fighting for survival, we spend what is necessary, regardless of the cost. But, can economists really be certain of the future? Should they take the views of alarmists verbatim? The answer seems to be that we must face the real problem of the 21st century: providing the people with energy, not reducing carbon emissions. In the process, carbon emissions will gradually diminish, but not as quickly as the alarmists would prefer.

Nordhaus showed that the basis for the extreme economic impacts in the Stern Report rests on "selectively chosen studies that emphasized high damage estimates, some of which are highly speculative." More importantly, Nordhaus demonstrated that the Stern Report used considerable legerdemain in reporting losses from global warming.

Nordhaus posed this illustration:

> "Suppose that scientists discover that a wrinkle in the climatic system will cause damages equal to 0.01 percent of output starting in 2200 and continuing at that rate thereafter. How large a one-time investment would be justified today to remove the wrinkle starting after two centuries? The answer is that a payment of 15 percent of world consumption today (approximately $7 trillion) would pass the Stern Review's cost–benefit test. This seems completely absurd. The bizarre result arises because the value of the future consumption stream is so high with near-zero discounting that we would trade off a large fraction of today's income to increase a far-future income stream by a very tiny fraction ... Hence, the damage puzzle is resolved. The large [reported] damages from global warming reflect large and speculative damages in the far-distant future; the impacts now, as in today, are small; and ... the [proposed] 20 percent cut in consumption from global-warming might be reduced by an order of magnitude if alternative assumptions about discounting are used."

Nordhaus went on to say:

"A further unattractive feature of the Review's near-zero social discount rate is that it puts present decisions on a hair-trigger in response to far-future contingencies. Under conventional discounting, contingencies many centuries ahead have a tiny weight in today's decisions. Decisions focus on the near future. With the Review's discounting procedure, by contrast, present decisions become extremely sensitive to uncertain events in the distant future."

Nordhaus provided further examples of absurd requirements for present investment to cover unlikely small impacts in the distant future if a zero social discount rate is used.

A number of other critiques of the Stern Report have been published, most notably that of Weitzman (2007). In response, there have been quite a number of defenses of the Stern Report by economists, most notably one by Cole. Quiggan (2006) also wrote a defense of the Stern Report. Quiggan (an economist) provides us assurance that the credibility of the doubters regarding the connection between global warming and rising CO_2 "was on the verge of collapse" by 2006, due to the success of the documentary *An Inconvenient Truth*, and the fact that the "scientific controversies have now been resolved." In his view, the Stern Report "outflanked the remaining skeptics. They could either continue denying the results of scientific analysis, or try to salvage the fallback position, undermined by the Stern Report, that although global warming is real, the costs of doing anything significant about it exceed the benefits, at least in the short term." He then proceeded to fill about a dozen pages with economic hash, full of sound and fury, signifying very little. It is nice to hear from an economist that the "scientific controversies have now been resolved," but considering the ineptitude of economists at their own trade, it would be better if they did not venture into unknown territories.[49]

The "Dual Critique" (Part I: Science – Carter et al., 2006 and Part II: Economics – Byatt *et al.*, 2006)) provided a serious independent assessment of the Stern Report. Part I began by taking issue with the quotation:

"... what is not in doubt is that the scientific evidence of global warming caused by greenhouse gas emissions is now overwhelming... [and] ... that if the science is right, the consequences for our planet are literally disastrous... what the Stern Review shows is how the economic benefits of strong early action easily outweigh any costs".

As Part I emphasized, the Stern Report

"... presumes without question that moderate further increases in atmospheric CO_2 levels will give rise to major climatic changes and that

---

[49] The first law of economics is that if you have two economists, they have diametrically opposite viewpoints. The second law is that they are both wrong.

these are likely to be seriously damaging; that the climatic changes observed over recent decades can be reliably blamed on emissions of 'greenhouse gases' in general, and CO_2 in particular; and that climate model projections and forecasts present a sufficiently accurate view of the future at relevant geographic and temporal scales to form a basis for major policy decisions".

Part I argues against the conclusion expressed in the Stern Report that warming in the 20th century was unprecedented "for at least the last 1,000 years". This of course implies acceptance by the Stern Report of the hockey stick and all the claims of alarmists regarding the late 1900s and early 2000s being the hottest since the last Ice Age. Part I mentions the Wegman Report as contradicting this finding. But the Stern Report insists that the hockey stick is "only one of a number of lines of evidence" although these other lines of evidence are not evident.

The confidence expressed in the Stern Review appears to derive heavily from a single published paper (Stott, *et al.*, 2000) that utilizes a global climate model in an attempt to separate natural variations from those induced by human generation of CO_2. Unfortunately, the model predicts a much greater temperature rise in the 20th century than was observed and the modelers had to invoke a significant cooling due to aerosols *ad hoc* to reduce the heating produced by the model.

Part I went on to discuss other aspects of the Stern Review's technical basis for alarmism, as well as the putative impacts of predicted global warming, much of which we have already discussed in this book.

Part II dealt with economics issues. The first point that is made is that the lack of clarity in the Stern Report makes it difficult to determine precisely which procedures were used. Part II is divided into six elements: (1) economic impacts of global warming, (2) costs of mitigation, (3) discounting the future, (4) choice of policy instruments, (5) major omissions from the Stern Review, and (6) a summary and conclusions.

In regard to the economic impacts of global warming, Part II pointed out that 80 to 90 percent of the proposed impacts are subjective, being attributed to "non-market impacts" and "catastrophes with little further definition provided". These impacts are further amplified by the fact that business as usual (BAU) as defined in the Stern Report does not take into account economic pressures for conservation, and adoption of new technologies because they are profitable. Because the Stern Review deals with the long run, such factors will change even in a BAU scenario.

Just as the Stern Report is pessimistic in regard to costs of global warming, it is grossly optimistic in regard to the costs of mitigation. One topic discussed in Part II was revenue recycling in which "some emission pricing policies (taxes, auctioned permits) generate revenue for the government, and this added revenue could be used to finance a cut in other tax rates". Only economists could think that this is a benefit. Wealth is created by efficiently producing things. Capturing CO_2 and storing it provides no wealth and only adds to our costs for living. The fact that revenues are raised by taxing emissions merely transfers the tax burden from one group to another; it creates no wealth.

Part II also discusses the discount rate. However, their discussion is not as clear as that of Nordhaus.

A number of comments were published in response to the "Dual Critique". These were rebutted. The reader is directed to Volume 8 of World Economics (2007).

Pirilä provided an independent commentary on the Stern Review. He said:

> "The analysis assumes that we can calculate the consequences of near-term decisions to very distant future. How impossible that is can be envisioned by thinking, how decisions of the 19th or 18th century influence the present. Some decisions of those periods may have a great influence, but how can we decide, what would be the alternative counter-factual history, and how much worse or better its results would have been."

Pirilä argued that the decisions we make, even very important ones such as going to war, typically only impact humanity for years to decades.

> "For many present issues a history of hundreds or thousands or years can be identified, but even in these cases the influence of any single original decision had hardly significant influence for long, when the comparison is done to potential counter-factual histories. It's mostly a fallacy to say that we are here with the present state of the world, because of some specific decision of distant history, and even in those very few cases, we don't know, how the alternative would be different".

Pirilä emphasized the importance of coal in the world's future. As oil, and eventually gas supplies wind down in the 21st century, expanded use of coal seems to be the only way to maintain industrial economies, particularly in view of prospects for expansion in China and India. The main alternatives of continued or expanded use of coal include widespread introduction of alternative energy production technologies (renewable, nuclear), or reduction of consumption via more efficient energy systems, and/or changes in consumption patterns. Alternatively, carbon sequestration may permit greater use of coal without great buildup of CO_2. Nevertheless, we don't know what limit on CO_2 concentration can be achieved in a practical sense while "offering ... social and economic development of emerging economies like China and India. It cannot present politically unrealistic requirements for the industrial economies either. The present optimism on possibilities expressed by many European states and organizations is not based on solid arguments".

In 2012, a 100-page criticism of the Stern Report was distributed.[50] Lilley pointed out that the three principal conclusions were:

- "If we don't act, the overall costs and risks of climate change will be equivalent to losing at least 5% of global GDP each year now and forever. If a wider range of risks and impacts is taken into account, the estimates of damage could rise to 20% of GDP or more".

- "In contrast, the costs of action – reducing greenhouse gas emissions to avoid the worst impacts of climate change – can be limited to around 1% of GDP each year."

- "Our actions now and over the coming decades could create risks of major disruption to economic and social activity, on a scale similar to those associated with the great wars and the economic depression of the first half of the 20th century."

As an economist, Lilley accepts the basis dogma behind these assertions, but argues against improper economic extrapolation. First, he argues that the 5% figure should actually be 3.1%. Second, Stern's use of the phrase "now and forever" is grossly misleading. Lilley pointed out that in reality, what Stern did was to project

"... how much unrestricted climate change would reduce GDP each year from now (when the reduction is negligible) to infinity (when it will [supposedly] be large), discounting it back to the present, and then calculating what constant percentage reduction in GDP, discounted back to the present at Stern's very low rate of discount, has the same present value. To say this averaged value – which reflects high impacts centuries hence – reflects the impact of climate change on GDP 'now' is simply untrue. In fact, far from experiencing a 5% loss of GDP now, the impact of warming could be beneficial now and for several decades since moderately higher temperatures boost crop yields, as do increased concentrations of CO_2."

Of course, the projected costs of future climate change (whether 5%, 3.1% or whatever) are highly subjective and depend not only on the unknown future climate change, but the even less known impact of the unknown cliamte change. Furthermore, in "Stern's gloomiest scenario global warming reduces GDP now and forever by 14.4%, actually involves losses averaging less than 1% of world GDP over this century, reaching 0.4% by 2060, 2.9% in 2100, and rising to 13.8% in 2200."

---

[50] "What Is Wrong With Stern? The Failings of the Stern Review of the Economics of Climate Change, Peter Lilley, Global Warming Policy Foundation GWPF Report 9, *http://www.thegwpf.org/*.

Lilley also took issue with the claim that costs "can be limited to around 1% of GDP each year" saying that this figure "is below any of the 21 studies monitored by Stanford University. Moreover, Stern has subsequently doubled his cost estimate [in a publication in 2009] saying that we need to aim for the bottom end of his target range of emission cuts 'costing about 2 per cent of global GDP each year'." Costs will start to be incurred "now" and for the foreseeable future. It is not clear to this writer that the steps required by Stern are technically feasible, are not far more costly than he estimates, and won't create a permanent worldwide economic depression.

Lilley's discussion entails 100 pages and it is not possible to summarize it in the space available here. We will content ourselves with a brief review of Lilley's discussion of discount rate. According to Lilley:

"In its 700 pages, the Stern Review does not reveal the discount rate used even though this is its most crucial assumption. It was not until some time after publication and as a result of strenuous enquiries, that it emerged that the Stern Review used a discount rate of just 1.4% pa. This is far lower than rates typically used in most previous studies or the discount rates used by businesses, governments, the World Bank etc. It is the main reason the Stern Review's conclusions are out of line with those of most other studies. Since publishing his Review, Stern has indicated that, on reflection, he would now use a discount rate almost double. This would dramatically scale down his headline figures. This about face received predictably little coverage. Yet according to the Review's belatedly published sensitivity analysis, the effect of doubling the discount rate to 2.8% pa is to reduce his base case estimate of the amount by which unrestricted global warming would reduce global GDP from 5% to 1.4%, 'now and forever'. Using a discount rate of 2.8% pa in Stern's most gloomy case reduces his central estimate of loss of GDP from 14.4% to just 4.2% 'now and forever'. The effect of using the Stern Review's low discount rate is to give huge weight to events in the distant future which are assumed to be the ineluctable consequence of actions taken by this generation. Estimates by Tol and Yohe using Stern's model suggest that nearly half of all damage the Review attributes to global warming relates to events more than two centuries ahead. Using different assumptions, Nordhaus has estimated that, under Stern's methodology, half of all benefits of preventing global warming will accrue to generations living after 2800!"

Lilley also pointed out another fallibility of the assumptions in the Stern Report. As Lilley said, "climate change must slow down in the very long run even if no policies are introduced to reduce emissions" because "fossil fuel resources are assumed to be finite, and global warming is proportional to the natural logarithm of carbon dioxide concentrations. Assumptions about what happens in the very long run are highly uncertain, and relevant only with a low discount rate".

9.3 Investment opportunities in climate change

With the passage of time, the literature on impacts of global warming has expanded considerably. Almost any academic can further his or her career by writing on the dangers and impacts of global warming, always adopting the view that such warming is entirely produced by greenhouse gases, CO_2 in particular. The advent of global warming as an integrative theme for academia has provided new opportunities, new platforms for publication, and what is most important of all, new funding. A representative example (of many) is the 39-page document by the European Climate Forum (Anon. (T)).

The Internet is full of reports by organizations such as the Wall Street Journal and (the now defunct) Goldman–Sachs emphasizing the opportunities for investment in climate change.

The Deutsche Bank Group (2008) has been a leader in arguing for investment in mitigation of global warming (presuming that carbon emissions are the sole culprit). They said: "the growing investment opportunities in climate change ...[will] continue into the foreseeable future" and went on to say:

> "In the energy sector alone, the International Energy Agency estimates that about $45 trillion will be needed to develop and deploy new, clean technologies between now and 2050. This represents nothing less than a low-carbon Industrial Revolution. Writers and policymakers from across the political and intellectual spectrum have recognized the potential this holds for long-term job growth and industry creation. The debate around climate change is shifting away from cost and risk towards the question of how to capitalize on exciting opportunities."

In their view, in seeking new investment capital to renew the world economy, they "believe that for investors, climate change has a built-in advantage over most other sectors. Its regulated markets hold the promise of enormous secular growth. In the long-term, *the earnings of companies and projects that are supported by governments for policy reasons are more trustworthy*." (Emphasis added.)

More recently, the Deutsche Bank distributed a 142-page report entitled "Investing in Climate Change 2011" dealing with how to profit from investments based on climate change policies.

9.4 U.S. Congress: meeting the climate change challenge

> "The whole aim of practical politics is to keep the populous alarmed, and hence clamorous to be lead to safety, by menacing it with an endless series of hobgoblins, all of them imaginary." (H. L. Mencken)

The U.S., although a signatory to the Kyoto Protocol, has neither ratified nor withdrawn from the Protocol. The signature alone is symbolic, as the Kyoto Protocol is non-binding on the U.S. unless ratified. The U.S. was, as of 2005, the largest single emitter of carbon dioxide from the burning of fossil fuels. China hs recently take overtaken this honor in the 2008–2009 time frame.

On July 25, 1997, before the Kyoto Protocol was finalized (although it had been fully negotiated, and a penultimate draft was finished), the U.S. Senate unanimously passed (by a 95–0 vote) the Byrd–Hagel Resolution, which stated that the sense of the Senate was that the U.S. should not be a signatory to any protocol that did not include binding targets and timetables for developing as well as industrialized nations or "would result in serious harm to the economy of the U.S." On November 12, 1998, Vice President Al Gore symbolically signed the protocol. Both Gore and Senator Joseph Lieberman indicated that the protocol would not be acted upon in the Senate until there was participation by the developing nations. The Clinton Administration never submitted the protocol to the Senate for ratification.

President George W. Bush did not submit the treaty for ratification, partly because of the exemption granted to China. Bush also opposed the treaty because of the strain he believed the treaty would put on the economy; he emphasized the uncertainties that are present in the climate change issue.

In October 2007, Chairman Dingell's *Energy and Commerce Committee* of the *U.S. House of Representatives* released its first white paper in a series on "Meeting the Climate Change Challenge"

(*http://energycommerce.house.gov/Climate_Change/White_Paper.100307.pdf*). This was claimed to be "the next step in the legislative process leading to enactment of a mandatory, economy-wide climate change program." The essential basis for this program was the belief that:

> "The United States should reduce its greenhouse gas emissions by between 60 and 80 percent by 2050 to contribute to global efforts to address climate change. To do so, the United States should adopt an economy-wide, mandatory greenhouse gas reduction program ... The central component of this program should be a cap-and-trade program. The cap-and-trade program will have increasingly stringent caps on greenhouse gas emissions, eventually reaching a level that reduces emissions by 60 to 80 percent in 2050. The Government will distribute allowances equal to the level of allowed greenhouse gas emissions. Allowances can then be bought and sold."

It is not clear from the white paper whether the Committee meant a 60%–80% reduction from projected levels in 2050 based on a business-as-usual scenario, or a 60%–80% reduction from levels in 2007. However, the wording seems to suggest a reduction from 2007 levels. Considering that, under a business-as-usual scenario, energy usage in the U.S. is likely to grow between 2007 and 2050 that would imply an even greater percentage reduction from projected levels in 2050 under business as usual. Assuming that energy usage in the U.S. in 2050 under business as usual would increase by, say, 40% compared with 2007, that would imply that the required reduction would be between 72% and 86% compared with energy usage in the U.S. in 2050 under a business-as-usual scenario.

An epidemic of insanity has apparently invaded the U.S. House of Representatives. Such a program will send the U.S. reeling back toward the lifestyle of the 18th century, bringing on a far worse economic depression than that of the 1930s. This program does not require that the U.S. supply its people and its industries with the energy needed to operate. It merely requires that emissions be reduced. Non-emitting energy presently accounts for perhaps 20% of the total energy mix in the U.S., and much of that is hydropower and nuclear power. The prospects for increasing hydropower are limited, and furthermore many environmentalists have turned away from hydropower because of its negative ecological impacts. The prospects for expanded nuclear power are quite uncertain. That being the case, the likely candidates for achieving the putative 72% to 86% reduction compared with energy usage in the U.S. in 2050 under a business-as-usual scenario are:

- Improved efficiency (drastically downsized vehicles, widespread use of hybrid vehicles, new standards for buildings, ...)
- Vast increase in solar, wind, bio-energy, and other renewable energy sources
- Huge cutback in energy usage, a major change in lifestyle.

Improved efficiency can make a contribution, but there are limits. The potential for expansion of use of renewables is highly debatable and remains to be seen. In the opinion of this writer, there is no way that renewables can come close to filling the gap left by a 72% and 86% reduction in emissions in 2050. Inevitably, it appears that the consequence of this program would be to drive America back to a lifestyle reminiscent of the 18th century. It would seem appropriate that before requiring one to sail across the ocean, it would be useful to make sure that one has a boat that does not leak. Meanwhile, governments continue to set quotas for increased reliance on renewables for electric power production, ignoring the technical difficulties in implementing these goals. For example, on May 3, 2011, Governor Jerry Brown signed Senate Bill X1 2, which requires all California utilities to generate 33% of their electricity from renewable energy by 2020. This would proceed in stages: (i) 20% by December 31, 2013, (ii) 25% by December 31, 2016, and (iii) 33% by December 31, 2020. These levels appear to be unattainable.

Aside from the economic consequences of this policy, it is also myopic because it only deals with one country. As bad as the Kyoto Protocol is, at least it made an attempt to deal with emissions on a global scale. Piecemeal policies by individual nations that ignore China are unlikely to be effective.

Senator John McCain authored S. 2191, a bill that required a reduction of CO_2 emissions 70% below current levels by 2050 (about 80% below levels in 2050 in a business-as-usual scenario). Senator Barbara Boxer advocated a 90% reduction (Michaels and Balling, 2009).

President Obama proposed an 80% reduction in emissions (*Los Angeles Times,* November 19, 2008). As part of his plan, he would increase the use of coal and use a cap-and-trade system for emissions. The cap-and-trade system would allow those who can afford it to continue to emit. Thus the percentage reduction for those not buying the right to emit would be even higher than 80%. Obama plans the use of "clean coal." But, coal pollutes in the mines, in the runoff from the mines, in the desecration left behind, in the railroads that transport the coal, in the power plants that burn the coal, in the emissions from the power plants, and in the ash left over. Coal produces a lot more CO_2 per unit energy produced than petroleum or natural gas. In the process of cleaning up coal for combustion, a considerable amount of CO_2 is emitted. The economic impact of such policies will be measured in many trillions of dollars, and the technical and economic challenges in implementing such policies have generally been underestimated (Pielke, Wigley, and Green, 2008). Thus, if one accepts the alarmist view that continued use of fossil fuels will produce unacceptable global warming, humanity is caught between the proverbial "rock and a hard place." According to this belief, we cannot accept the consequences of continuing business-as-usual; however, we have neither technical nor economic capability to do otherwise without creating great financial and operational dislocations.

In late June 2009, the U.S. House of Representatives passed a bill requiring reduction in carbon emissions of 17% by 2020 and by 83% by 2050, and establishment of a cap-and-trade system for exchanging rights to emit CO_2. Apparently, none of the members of the House read the 1,500-page bill. President Obama's views were expressed during his candidacy (widely quoted on the Internet):

> "So if somebody wants to build a coal-powered plant, they can; it's just that it will bankrupt them because they're going to be charged a huge sum for all that greenhouse gas that's being emitted. You know, when I was asked earlier about the issue of coal, uh, you know, under my plan of a cap-and-trade system, electricity rates will necessarily skyrocket. Even regardless of what I say about whether coal is good or bad. Because I'm capping greenhouse gases, coal power plants, you know, natural gas, you name it—whatever the plants were, whatever the industry was, uh, they would have to retrofit their operations. That will cost money. They will pass that money on to consumers."

Donald Rapp

Starting with the Stern Report, the economists have taken over the business of estimating costs for implementing draconian reductions in carbon emissions. Since the economists almost always get things wrong, that is not very encouraging. The *MIT Joint Program on the Science and Policy of Global Change* has been cashing in on the current fear of global warming with a series of funded studies on the economics of carbon emission reduction. One report of note is Paltzev *et al.* (2009). This report considered three scenarios: (1) constant emissions rate from 2008 to 2050 totaling 287 billion metric tons of CO_2-e (CO_2 equivalent in all greenhouse gases), (2) linear reduction in emissions from 2008 to 2050 down to a 50% reduction in 2050 totaling 203 billion metric tons of CO_2-e, and (3) linear reduction in emissions from 2008 to 2050 down to an 80% reduction in 2050 totaling 167 billion metric tons of CO_2-e. It is noteworthy that they mention that because of the current recession, they downgraded their estimates of future GNP, resulting in a reduction in emissions of 20% compounded over 40 years. This suggests that the best (and perhaps only) way to meet the 80% reduction target is to have a permanent recession. The "reference scenario" used by Paltzev *et al.* (2009) is in some ways a rosy picture. It assumes that with no policy at all, the annual U. S. emissions of greenhouse gases in units of CO_2 equivalent will slowly rise between 2008 and 2050 from about 7 to 11 billion tons per year (contrast this with our Figures 7.4 and 7.7). Over this same period, the U. S. GNP is estimated to rise from $12 trillion to $37 trillion.

The problems for politicians in planning policy center around uncertainty. The MIT Group published a series of reports dealing with climate change policy under uncertainty (e.g. Webster *et al.*, 2008, 2009). They pointed out

"Though the climate policy challenge is essentially one of risk management, requiring an understanding of uncertainty, most analyses of the emissions implications of these various policy targets have been deterministic, applying [specific] scenarios of emissions and reference (or at best median) values of parameters that represent aspects of the climate system response, and the cost of emissions control.... These efforts provide insight to the nature of the human-climate relationship, but necessarily they fail to represent the effects of uncertainty in emissions, or to reflect the interacting uncertainties in the natural cycles of CO_2 and other gases or the response of the climate system to these gases".

While these authors utilized a range of possible future emission scenarios, they used specific model results for the ultimate concentrations of greenhouse gases, forcings, and temperature increases resulting from any emission scenario. In other words, they accounted for uncertainty in the emission scenario, but not for uncertainty in the climate impact of such emissions. Their final results are sets of probability distributions for the global mean temperature rise from the average for 1981-2000 to the decadal average for

376

2091-2100. If there is no policy to restrict carbon emissions (business as usual) they end up with a most probable temperature rise of 5°C and a range of about 3.5-7°C based on a most probable CO_2 concentration of about 900 ppm in the last decade of the 21st century. They postulate four possible levels of carbon emission control. The least stringent leads to a most probable temperature rise of 3.2°C and a range of about 2-5°C based on a most probable CO_2 concentration of about 700 ppm in the last decade of the 21st century, and the most stringent leads to a most probable temperature rise of 1.8°C and a range of about 1-3°C based on a most probable CO_2 concentration of about 480 ppm in the last decade of the 21st century. There is no strong reason to believe these results.

Unfortunately, the definition of GNP used by economists includes almost any kind of activity; yet the thing that we are really interested in is activity that produces wealth. Wealth is produced by activity that efficiently produces goods and services that better the quality of life of the people. Consider the hypothetical case of employing 200,000 bean counters to monitor CO_2 emissions at factories and power plants across the country. The government would take credit for creating 200,000 jobs, and the GNP would value this activity at something roughly like $300,000 per person, adding up to $60 billion. But these people would produce nothing of value for the population and would not add to the wealth of the nation. On the contrary, they would be a drag on the wealth of the nation because taxes would be needed to raise the $60 billion needed to pay for the bean counters. Of course, the proponents of CO_2 reduction would argue that these activities would improve the quality of life by stemming global warming, and if the global climate models are correct, and if the alarmist estimates of impacts of global warming are correct, they would have a point. At the same time, the projections of increasing GNP over the next 40 years assure that as the cost of reducing CO_2 emissions builds up, the cost will appear more moderate when written as a percentage of GNP. According to Paltzev *et al.* (2009) the "welfare change" associated with the 80% reduction scenario reaches about 2.5% for the decade 2040 to 2050. Multiplying by an average GNP of 40 trillion dollars during that period implies a cost of a trillion dollars a year. Paltzev *et al.* (2009) also estimate that during the 2040 to 2050 decade, the cost of CO_2 removal is about $200/ton, which when multiplied by ~70% of 10 billion tons per year, amounts to $1.4 trillion dollars per year. This assumes that the results of Paltzev *et al.* (2009) are correct. Actual costs always seem to be much higher than those predicted by economists.

Prior to the late 1990s, one might have winced at the thought of taking on a charge of a trillion dollars a year. However, since then, trillions of dollars are acquired by borrowing. There is an economic nirvana. The U. S. Government can spend as much as it pleases, and if it does not have the tax revenues to cover these expenditures, it simply borrows. As long as there is a more or less permanent recession, the demand for money is low, interest rates are equally low, and the Government can spend trillions more than it takes in. With the prospect of permanent recession facing us, this system should work for the foreseeable future.

Roe and Bauman (2011) described the possibility that the climate sensitivity might be at the upper end of predictions based on climate models. They emphasized that recent economic analyses suggested that a "rational policy strategy" should be based on this worst-case scenario "if the damages associated with such high temperatures are large enough". Some have claimed that damages rise non-linearly with temperature and therefore the worst-case scenario should dictate government policy. Roe and Baker (2007) showed that a "fat tail" to the probability function for future temperature rise is a necessary consequence of uncertainty in parameters that enter into climate models. But Roe and Bauman (2011) were "skeptical of this approach" because they claimed that two factors prevent the high temperatures in the fat tail from being reached for many centuries. One factor is

"... the enormous thermal inertia represented by the deep ocean. The whole climate system cannot reach a new equilibrium until the deep ocean has also reached equilibrium. In response to a positive climate forcing (i.e., a warming tendency), the deep ocean draws heat away from the surface ocean, and so buffers the surface temperature changes, making them less than they would otherwise be. The deep ocean is capable of absorbing enormous amounts of heat, and not until this reservoir has been exhausted can the surface temperatures attain their full, equilibrium values".

As Roe and Bauman (2011) pointed out, "a second key [factor] is the inherent relationship between feedbacks and adjustment time scales in physical systems". If the climate sensitivity is high, that is because there are strong positive feedbacks.

"A positive feedback reflects a tendency to retain energy within the system, inhibiting its ultimate emission to space, and therefore requiring a larger temperature response in order to achieve energy equilibrium. Moreover, it is generally true that, all else being equal, an inefficient system takes longer to adjust than an efficient one. A useful rule-of-thumb is that the relevant response time of the climate system is given by the effective thermal inertia of the deep ocean multiplied by the climate sensitivity parameter".... As time progresses, more and more of the ocean abyssal waters become involved in the warming, and so the effective thermal inertia of the climate system increases. Hansen *et al.* (1985) solved a simple representation of this effect and showed that the adjustment time of climate is proportional to the square of climate sensitivity. In other words, if it takes 50 yrs to equilibrate with a climate sensitivity of

1.5°C, it would take 100 times longer, or 5,000 yrs to equilibrate if the climate sensitivity is 15°C."

The term "climate sensitivity" as used here means the global temperature rise due to a doubling of CO_2 from the pre-industrial value.

Roe and Bauman (2011) then launched into a discussion of the economics of future damage from global warming based on Weitzman (2009 and 2011), taking into account the point that the higher the ultimate temperature rise from greenhouse gases, the longer it will take to reach that temperature. The Weitzman papers are framed in mathematical expressions for economic models, but there are three unknowns for the next few hundred years: (i) the future rise in temperatures due to greenhouse gases, (ii) the impacts of this rise on human welfare, and (iii) the economic health of the world. Since none of these can be estimated with any reliability, all the economic mathematics in the world cannot help us make policy decisions. Humans always face the challenge of how much to spend now to provide for the future but the problem here is the future is far out, and the need for various draconian approaches to remediation is highly uncertain.

9.5 Renewable Energy

As fossil fuel resources dwindle down in the 21st century, it is desirable to rely to a greater extent on renewable energy sources (solar, wind, biomass, ...). At the same time, use of such resources will reduce CO_2 emissions, which will likely provide benefits in reducing the rate of climate change. Federal and state governments have provided funding for development of these resources as well as policies to provide financial incentives for the public to use these resources while they are more expensive than conventional energy. Since these energy sources are intermittent, it remains unclear what percentage of our total energy consumption can ultimately be supplied by renewable energy. As in the case of reducing fossil fuel emissions of greenhouse gases, governments act by setting quotas for the future, regardless of the technical and economic feasibility of achieving such quotas.

Krey and Clarke (2011) discussed the potential role of renewable energy in climate change mitigation by reviewing 162 recent medium- to long-term scenarios from 15 large-scale, energy-economic and integrated assessment models. It seems clear that significant increases in the use of renewable energy are required to reduce CO_2 emissions, but as they pointed out:

"One cannot say with certainty today whether a future heavily reliant on renewable energy will be extraordinarily costly or whether the costs will only be modest. The scenarios in this study demonstrate no meaningful correlation between carbon prices and renewable energy production. Indeed, this sort of variability in indicators of mitigation cost is common in multi-model scenario analyses."

They also concluded that the high level of uncertainty that pervades future scenarios is "unsatisfactory" and that unpacking this uncertainty is a "very challenging task".

The IPCC (2011a) published a 1,544 page report on the relationship between expanded use of renewable resources and climate change mitigation that was heavily based on the scenarios reviewed by Krey and Clarke (2011). One important parameter of interest is the percentage of total world energy consumed that is generated by renewable energy. Despite the 1,544 pages in the IPCC Report, very little mention of this parameter is provided. However, it is briefly mentioned in passing in Chapter 10 that:

> "... the global primary energy supply share of renewable energy (RE) differs substantially among the scenarios. More than half of the scenarios show a contribution of RE in excess of a 17% share of primary energy supply in 2030, rising to more than 27% in 2050. The scenarios with the highest RE shares reach approximately 43% in 2030 and 77% in 2050. RE deployment levels in 2100 are substantially larger than these, reflecting continued growth throughout the century."

In order to estimate the percent of total energy supplied by RE, one must first estimate the total energy consumption. According to IPCC (2011), the various scenarios for future world energy use range from 450 to 1,000 EJ in year 2050, and 500 to 1,500 EJ in year 2100.[51] There is also a wide variation of estimates of RE corresponding to each estimate of total energy. Hence predicting future RE percentage amounts to "any number can play". It is interesting that the IPCC issued a press release prior to releasing the IPCC (2011) report. The press release said (amongst other things):

> "Close to 80 percent of the world's energy supply could be met by renewables by mid-century (2050) if backed by the right enabling public policies a new report shows."

McIntyre pointed out that this was highly misleading. As he said,

> "The report does NOT show that 'close to 80 percent of the world's energy supply could be met by renewables by mid-century if backed by the right enabling public policies'. It does list 'a scenario from Greenpeace in which 77% of world energy is supplied by renewables, but the report itself did not conduct any independent assessment of the validity of the Greenpeace scenario and did not 'show' that the claim in the press release was true'." (*http://climateaudit.org/2011/06/16/responses-from-ipcc-srren/#comments*)

McIntyre also pointed out that the scenario that led to 77% renewable energy by 2050 was the most extremely optimistic of 164 different scenarios.

---

[51] 1 exajoule (EJ) = 10^{18} J.

The least optimistic scenario provides for 15% RE by 2050, up from 12.9% in 2008 (see Figure 9.1). A large number of BLOG enthusiasts responded to McIntyre's report, few of which added anything to McIntyre's presentation.

Trainer (2011) wrote a critical review of the 2011 *IPCC Report on Renewable Energy*. The IPCC Report purported to confirm "the widespread belief that renewable energy can replace about 80% of fossil fuels and more or less meet world energy demand by 2050. It is more than 1000 pages long, has 38 lead authors and input …from over 120 leading experts from all over the world…, reports on 164 studies, and digests 24,766 comments …from more than 350 expert reviewers and government and international authorities".

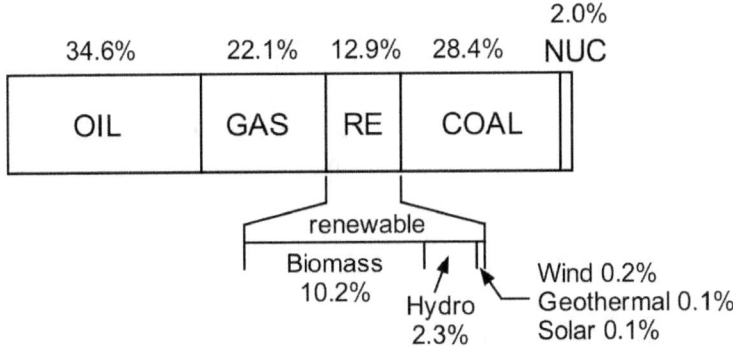

Figure 9.1. Distribution of sources of world energy use in 2008.

Trainer's points included:

[1] *No Case is Made.* "The report does not show that renewable sources can meet future energy demand, or a large fraction of it…. It does not attempt to show what proportion of demand could be met by renewables. It presents much evidence relevant to the issue, but this is not put together into a case that sets out reasoning leading to the conclusion that the necessary quantities could be provided, how they could be provided, and that the difficulties could be overcome. The report merely presents the results of some studies that state conclusions about renewable energy's potential, without attempting to assess their worth."

[2] *No Critical Review.* "There is no critical examination of the 164 studies [upon which it is based]. There is [not even] a list of the studies enabling their examination…. In other words the IPCC has not carried out an evaluation of literature in the field; it has only summarized the conclusions of (a select number of) studies, with no apparent effort to check on their validity".

[3] *The Case for Biomass is Greatly Overestimated.* Trainer provides a variety of cogent reasons why the potential for biomass energy is far less that the Report indicates.

[4] *Variability, Winter, Peak Demand, etc.* Trainer points out that the Report glosses over problems of intermittency or "the crucial problem of meeting demand in mid winter". The report utilizes "annual and/or average demand and supply, whereas what matters much more are the figures for maximum demand, e.g., peak quantities, when they coincide with minimum renewable resource availability". As a result of this mismatch, particularly in winter, reserve backup capacity will be very large, adding to costs. There will occasional long gaps of several days in a row with little or no wind and sun. On the other hand, there will be considerable overcapacity during some periods when sun and wind are available in quantity.

[5] *Integration Limits.* Trainer estimates realistically that as an extreme upper limit, "… wind plus PV might contribute at best only 55% of electricity, i.e., only 14% of all energy. The Report does not deal with the question of "from which sources the other 86% is to come, apart from biomass". The Report makes it appear as if viable energy storage technology is "just around the corner" whereas in reality, energy storage remain a major hurdle for renewable energy.

Trainer also provides additional arguments regarding optimistic cost estimates and several other aspects.

In dealing with beliefs regarding the related questions of whether rising CO_2 is a very serious problem, and the extent to which renewable energy can replace fossil fuels over the next 40 years (or more), we can form analogous categories:

[1] You think that rising CO_2 is a very serious problem, and renewable energy can replace ~80% of fossil fuels by 40 years from now.

[2] You think that rising CO_2 is a very serious problem, but renewable energy cannot replace more than ~20% of fossil fuels by 40 years from now.

[3] You think that rising CO_2 is not a very serious problem; nevertheless renewable energy can replace ~80% of fossil fuels by 40 years from now.

[4] You think that rising CO_2 is not a very serious problem, and renewable energy cannot replace more than ~20% of fossil fuels by 40 years from now.

Most of us probably fit into one category or another. The IPCC and the whole community of alarmists tend to be of Type [1]. Most of those who think that CO_2 is a major problem also think that renewable energy provides the solution (they are Type [1]). Most of those who doubt that CO_2 is a major problem also doubt that renewable energy could solve the problem even if it were a problem (they are Type [4]). I am a hybrid between Type [2] and Type [4]. I am not sure about the danger presented by CO_2 but I am pretty sure that renewable energy could not solve the problem if the alarmists are right.

It is interesting that Trainer evidently subscribes to the extreme alarmist view that rising CO_2 poses an extreme danger to mankind, and that immediate draconian reductions in CO_2 emissions are needed. He is also an advocate for renewable energy development. However, unlike most advocates for renewable energy (who are Type [1]) he has analyzed the potential for renewable energy in detail, and he concluded that it is quite limited; far less than that needed to replace most of the fossil energy we expect to need by 2050. Hence, as he said:

"So what's the solution? The point is that there isn't one.... Global problems are basically due to the commitment to grossly unsustainable levels of consumption and to limitless economic growth. The problems cannot be solved on the supply side, i.e., by trying to provide the quantities of energy that a consumer-capitalist society for 10 billion would require. That kind of society is generating other major problems in addition to energy and climate, including the poverty of billions, the destruction of the ecosystems of the planet, resource conflicts, and deteriorating social cohesion. These problems cannot be solved unless there is vast and radical transition to a *Simpler Way* of some kind. This IPCC WG3 Report reinforces the dominant faith that there is no need to think about this perspective on our global situation."

In his book, Trainer describes the *Simpler Way*:

• Far less affluent living standards

• Very different economic system based on minimal production and consumption with a much lower GDP and no growth

• Shift in values away from competition, individualism and acquisition to frugality, self-sufficiency, cooperation, participation and non-material satisfaction.

It appears that Trainer's review of the IPCC Report is on target. Certainly we should push renewable energy for all it is worth, but as he says, it is not worth nearly as much as enthusiasts claim. Rising CO_2 over the next few decades may create some problems, but none that are insuperable. We'll probably muddle through, although continuing economic problems will plague us. Indeed, they are already upon us to a limited degree. Around 2040-2050, it will get worse. With rising CO_2, population growth, worldwide industrialization and declining fossil fuel availability, the world seems destined for some severe dislocations. While renewable energy will gradually take on a greater share of our energy consumption, Trainer is probably right: There is no solution on the horizon.

9.6 The Climate Debate Revisited

Politicians in developed countries tend to accept the claims by alarmists that impacts of future global warming are known to be disastrous, the only remedy is immediate draconian reduction in fossil fuel usage, and draconian reduction in fossil fuel usage can be accommodated by the renewable energy supply. In response, they have drafted legislation calling for draconian reduction in emissions over the next few decades. These reductions are technically and economically impossible. The world economy is already in dire straits in 2012; enactment of such legislation will undoutedly push the world into global economic depression.

10 FINAL REMARKS

There is evidence that the Earth was much warmer than today during much of the Phanerozoic eon (the past 540 million years). Some estimates indicate that the global average temperature may have been up to about 10°C warmer than today. The poles were free of ice. However, there were extreme glacial periods during this eon as well. In seeking a cause for this, variability of CO_2 concentration has been a prime suspect. There is evidence that the CO_2 concentration was 10 to 20 times higher than today's concentration when temperatures were highest during the Phanerozoic eon, and CO_2 concentrations plunged during the period from around 330 to 280 million years ago when the world was heavily glacial. The climate of the Earth cooled over the past 50 million years, and decreasing CO_2 is suspected as being at least partially involved as a cause. It is widely believed that over the long haul, variations in CO_2 concentration are the primary factor in climate change. However, attempts to correlate CO_2 with climate across the entire Phanerozoic eon have met with difficulty. While in general, periods of warmth tend to have higher CO_2 concentrations, there are many exceptions aqnd the correlation is ragged.

Starting about 2.7 million years ago, the climate of the Earth began to oscillate between glacial and interglacial periods. About 1 million years ago, the variations took the form of long, extended ice ages interspersed by relatively short interglacial periods. The last ice age peaked around 20,000 years before present (YBP) and the current interglacial (Holocene) commenced about 10,000 YBP. As the global average temperature rose and fell, CO_2 concentrations rose and fell in unison. However, there is no known physical reason why the CO_2 concentration should have gone through such undulations independently; hence the changing CO_2 concentrations in this instance are effects of temperature change rather than causes, although changing CO_2 concentrations would have acted in a secondary role to amplify temperature changes due to another cause. The prevailing theory is that quasi-periodic variations in the Earth's orbit about the Sun induce changes in solar input to higher latitudes, which produces the initial forcing that causes these climate changes. A variety of feedback effects amplify this forcing (Rapp, 2012).

As Figure 5.6 shows, variable CO_2 concentrations exert a forcing of the climate. Other factors being equal, higher CO_2 concentrations will produce higher temperatures, and *vice versa*. The question is how much higher, especially since other factors are rarely equal.

Climatologists have attempted to characterize the Earth's climate during the past century or so, and during the past millennium or two. This is important in (i) establishing trends that may have evolved in the past century or so from industrialization and urbanization of the Earth, and (ii) defining the range of past fluctuations in climate prior to industrialization as a conjectural baseline of expected variations independent of human intervention. If it were possible to unambiguously extract (i) and (ii) from the data, anthropogenic-induced changes could then be compared with expected fluctuations from natural causes. If such anthropogenic changes were found to be considerably greater than natural fluctuations, that would provide some experimental basis for concern about human impacts on climate.

One important factor is the question of how temperatures on Earth have varied over the past century (or more in some cases) as measured by monitoring stations dispersed around the world. While the scientists who process such data have made affirmative claims for the reliability of these data, the fact remains that the network for monitoring world temperatures suffers from a number of maladies including uneven spatial and temporal representation of large areas, poor maintenance and recording at many stations, effects of urban heating and land use of stations, and uncertainties in the measurement of sea temperatures. In addition, the number of stations reporting data decreased sharply earlier in the 20th century. Climatologists typically work in terms of a single global average temperature, or hemispheric average temperatures. In order to estimate how such averages have varied with time during the past century or so, a space–time grid is typically created from station temperature data. Unfortunately, the sparseness of spatial and temporal coverage creates considerable uncertainty in these averages— more so than the purveyors will admit to. Furthermore, such averages (i) have little physical or thermodynamic significance, (ii) tend to average out regional variations that convey more incisive information, and (iii) can be constructed in various ways to lead to different results. Nevertheless, it seems likely that the global average temperature has increased by roughly 0.6°C to 0.8°C over the past 120 years. While this warming has been far from uniform in space and time, there is no doubt that on balance the Earth is warmer today than it was 120 years ago. But 120 years ago, the Earth was emerging from the little ice age (LIA). There is considerable evidence that the Earth began pulling out of the LIA in the 19th century, well before the build-up of CO_2 concentrations in the second half of the 20th century. For example, mountain glaciers typically began their retreat well before the buildup of CO_2 in the atmosphere in the late 20th century.

There is also evidence that the temperature rise of the 20th century occurred in two steps, one from 1900 to about 1940, and the other from about 1976 to about 2000. The initial rise occurred prior to massive build-up of CO_2 in the atmosphere and was probably due partly to deposition of black carbon on Arctic snow and ice and partly to changes in ocean currents. Warming was greatest by far in the Arctic region. The dip from 1940 to 1976 has been attributed to aerosols generated by power plants. The sharp rise after 1976 coincides with a change in the Pacific Ocean in which the usual upwelling of deep cold waters seems to have suddenly diminished. In addition, the regions of greatest temperature rise appear to be near the greatest concentration of urban centers. There has been some divergence of opinion regarding the importance of the urban heating effect on measured temperatures. However, recent measurements indicate that this effect was much larger than previously thought. There has been little change in global average temperature from 2000 to 2012.

In attempting to extend temperature estimates backward in time prior to the advent of surface measurements, climatologists have relied on a variety of temperature proxies that leave remnants from an earlier time that were formed via a temperature-dependent process, from which (in principle) the temperatures at the times of formation can be extracted. Unfortunately, all proxies are subject to confounding influences of one type or another that add noise to the signal and introduce considerable uncertainty in the veracity of the derived temperatures. In the past decade, a number of climatologists attempted to integrate the results of a large number of proxies with varying spatial and temporal coverage in an effort to extract a historical global average (or hemispheric average) temperature over the past millennium or two. The accuracy of such procedures is limited by the sparse spatial and temporal coverage provided by the proxies, especially for dates prior to about 1600. However, an even more serious problem with these models has emerged. The use of principal components analysis (PCA) to discover the principal trends in the integrated data was not carried out properly, and the resultant temperature profile derived from these models had the characteristic *hockey stick* form in which the temperature of the Earth was estimated to be almost flat with little variation for a thousand years (or more). When the measured relatively steep rise in the 20th century was tacked on to the flat proxy results, the *hockey stick* resulted. This result was interpreted to mean that the human intervention of the 20th century produced an unprecedented, alarming rate of temperature rise, and this led to a number of extravagant claims that the last few years of the 20th century were the hottest in at least several millennia, with some making even wilder claims that they were the hottest years dating back millions of years. This became a rallying point for global climate alarmists and the *hockey stick* has been widely promulgated by the U.N., Al Gore, and numerous other organizations, institutions and governments. There were several things wrong with this picture. One is that when the proxy analysis is carried out to the very end of the 20th century, the result indicates that global temperatures should decrease sharply in the latter part of the 20th century. Since the experimental data show that the temperature actually rose, the scientists involved used the "trick" of replacing the downtrending model temperature with the measured temperatures after about 1980. A seocnd problem with the analysis was that McIntyre and McKitrick (M&M) and Wegman pointed out that the use of PCA was misguided, but the climatologists involved ignored the criticism and stubbornly defended their incorrect procedures. A close-knit group of co-proposing climatologists control manuscript publication in journals, and this *paleo-climatological cabal* has managed to exclude the major criticisms from the journals, and criticisms have been mainly relegated to blogs. Even if the statistical processing of the data did not contain this serious error, there is another fundamental problem with summing up large numbers of proxies. The sum of multiple noisy proxies seems to be noise. When you add an exaggerated version of the measured temperature rise in the 20th century to a flat profile from noisy data, with the late dowtrend excluded, the inevitable result is the *hockey stick*.

There is considerable anecdotal and proxy evidence to suggest that a relatively warm Medieval Warm Period (MWP) occurred in the time frame of about 900–1000 and a relatively cold Little Ice Age (LIA) occurred from perhaps 1600 to 1850. However, the magnitude and regional extent of these climate fluctuations remain uncertain.

Thus, we have a controversy. If the *hockey stick* picture were correct, that would provide support for the belief that human intervention has drastically altered a pattern of a flat temperature profile for a thousand years (or more). However, we know that the *hockey stick* result is incorrect. On the other hand, if the fluctuations in the MWP and the LIA were of comparable magnitude with the temperature rise of the 20th century, we might infer that the current warming is within the range of natural fluctuations. However, we don't have enough data to pin down the temperature changes in the MWP and the LIA. We do know that the temperatures around 1880 represented the final vestiges of the LIA, and since the temperature around 1880 was about 0.6°C to 0.8°C lower than today, we can be fairly sure that at the depth of the LIA in the

18th century, temperatures were at least 1°C cooler than they are today; yet there is no evidence that the CO_2 concentration was lower. The fact that the LIA was colder than the present climate may reflect unusual cold during the LIA rather than unusual warmth today. The real issue is how current temperatures compare with those in the MWP, for this would provide insight into how unusual the current warmth is. Unfortunately, the data are not accurate enough to answer this question.

Data taken over the past 50 years show that the CO_2 concentration has risen continuously and now well exceeds the levels found in ice cores from past interglacial periods. The pre-industrial level appears to be about 280 ppm, and it is presently about 395 ppm. Ice core data suggest that the rise in CO_2 began in the late 1800s, at about the time that temperatures started rising. The coincidence between the CO_2 increase and the temperature rise has suggested that the rise in CO_2 (together with other greenhouse gases) produced the observed rise in temperatures via the greenhouse gas effect. In order to examine this hypothesis, climatologists have developed global climate models in which the Earth is mathematically divided into cells that interact with nearest neighbors and evolve in time according to equations that attempt to represent all the physical processes that take place. The goal of these models has typically been to estimate the rise in global average temperature that will occur due to a future doubling of CO_2 concentration from the pre-industrial level of 280 ppm. The greenhouse heating effect of CO_2 is limited. Typical models predict that a doubling of CO_2 would produce (by itself, without feedback effects) a temperature rise of about 1°C. However, the warming that results from the CO_2 greenhouse effect would increase evaporation of water, and the increased water vapor content in the atmosphere would tend to amplify the warming via a water vapor greenhouse effect. Some models predict that this would add 2°C, bringing the total rise to about 3°C. However, the modeling of the entire water vapor cycle is primitive. As Lindzen and co-workers have shown, the effect of an increase in global average water vapor can have vastly different impacts on global temperature, depending on the regional distribution of the increase in water vapor. Furthermore, the effect of changes in clouds and aerosols as CO_2 rises are poorly understood. Because of uncertainties, different models predict a net temperature rise from a doubling of CO_2 that vary by a factor of 3 from one model to another. Unfortunately, it is difficult to test these models against actual data.

Another source of uncertainty in understanding the Earth's climate, past and future, is the possibility of variation in the irradiance emitted by the Sun. Solar irradiance can only be measured above the Earth's atmosphere and we only have data since 1980. Nevertheless, we do know quite a bit about the Sun. We have observations of the solar cycle and the relevance of sunspots and other surface markings to this cycle. Sunspot data go back several hundred years. Our data indicate that sunspot activity has varied considerably over the past 200 years, and we have anecdotal data to suggest that sunspots disappeared altogether from about 1645 to 1715. A number of investigators have attempted to model past solar irradiance based on sunspot indices, length of the solar cycle, comparison with Sun-like stars, and other phenomena. None of these is very credible. As a result, we simply do not know how much the solar irradiance has varied in the past or how much it might vary in the future. This adds further uncertainty to climate models.

The neat thing about predicting the climate 100 years from now is that no one can prove you wrong!

Martin (1979) wrote an interesting report in which he described the biases that inevitably creep into scientific research and reporting. According to Martin, scientists "do not disinterestedly look at the available evidence, do not make a balanced analysis, and do not present results in a neutral manner." Instead, he suggests that "from the beginning [they] support or favor a particular conclusion, and in a number of ways organize their scientific work so as to selectively support this conclusion." He labels this as "pushing the argument". Martin argued that pushing scientific arguments is inevitable, and therefore pushing should not reflect unfavorably upon the competence or integrity of the scientist. (I disagree with this).

He went on to say:

"A scientist in developing an argument to support an hypothesis draws evidence from a number of sources. In presenting evidence one must always be selective— all the evidence and arguments cannot be presented. Often different authorities support different viewpoints, present different 'facts', and offer different interpretations of evidence. Depending on the field, a scientist may draw sound support for many points of view and find some support for nearly any view. Therefore it is easy for a scientist, knowingly or unknowingly, to push an argument by selective choice and use of available evidence."

In examining the literature on SST emissions and their impact on the ozone layer, Martin concluded:

"From my point of view, the authors do not disinterestedly look at the available evidence, do not make a balanced analysis, and do not present results in a neutral manner. Rather, it appears to me that the authors from the beginning support or favor a particular conclusion, and in a number of ways organize their scientific work so as to selectively support this conclusion."

These claims made by Martin (1979) are backed up by lengthy and detailed discussions and analyses that seem quite credible to this writer.

In the 30 years that have passed since Martin wrote this report, several major changes have taken place in the way that scientific information is distributed. With the advent of the Internet, the monopoly of scientific journals has been weakened. Other cultural changes have taken place. Of some relevance is the fact that scientists are now far more prone to issue news releases on their work prior to publication, and these tend to find their way onto many websites. Other scientists, disagreeing with the orthodoxy of the consensus, have difficulty getting published in the journals. A number of so-called web blogs dealing with climate change have emerged over the past several years, and these have become foci for discussions and commentary. Some blogs are rabidly one-sided and present forums for either alarmists or naysayers to agree with one another. Any moron can voice his or her opinion. Two blogs that stand out above the others are *climateaudit.org* which has become a universal watchdog for reviewing statistical analysis of large data sets, and *judithcurry.com* which provides an even-handed forum for both sides. Unfortunately, the responses on these blogs have become so numerous that the wheat gets lost in the chaff.

Starting in the 1990s, and gradually building up in time, concern has grown amongst many climatologists that putative global warming produced by greenhouse gas emissions presents a grave danger to humanity. With the advent of Al Gore's film: *An Inconvenient Truth*, in 2006, this concern has escalated to become one of the major defining issues of our time. Governments are contemplating policies for extreme reduction of carbon emissions that are likely to cost many trillions of dollars and could produce global economic woe. Climatology, which used to be a minor science that was widely ignored, became thrust into the limelight. Prominent climatologists responded by issuing many repeated warnings of impending disaster, and now receive much attention, adulation (and funding) as a result. It appears that many climatologists routinely bias their results in such a way as to exaggerate the threat of rising CO_2. Others strive to relate almost any distant phenomenon to greenhouse gases – often to an absurd degree. Many have operated in a heavy-handed way to shut out opposing views and avoid criticism of their own work. Funding for climate research and analysis has become the goose that laid a golden egg, and human nature being what it is, many have succumbed to the temptation to seek a share of the lucre.

The subject of global climate change seems to have bifurcated into two groups, each opposed to one another, each certain that they are correct, and each predisposed to interpret everything from a one-sided viewpoint. One group, the alarmists, believe that insidious global warming will cause great havoc and suffering in the 21st century if we don't move quickly and draconically to suppress greenhouse gas emissions. The majority of climatologists are alarmists. Unfortunately, their insistence on use of the *hockey stick* result, their extravagant predictions of future CO_2 emissions, their excessive predictions of catastrophe, and their inordinate belief in climate models, provides the opposition with considerable ammunition for attack. Many alarmists are motivated by high ideals; they want to protect the planet. But it appears that the underpinnings of many are based on a similar situation that occurs in earthquake science with their mantra: "the big one is coming". If the climate is not a threat, who would want to fund their work? The skeptical opposition tends to be equally one-sided. The skeptics enjoy a number of open forums on the Internet blogs where they voice their disdain for the *cabal* and its alarmist platform. But as Steve Mosher has noted, the government estbalishment and its scientific arms (NOAA, NSF, EPA, ...) all strongly subscribe to the alarmist agenda, and government plans for cap and trade, carbon tax, regulations against coal, etc. will be enacted. The establishment pays little or no attention to the diatribes on the climate blogs.

Table 10.1. Dimensions of the climate debate.

Aspect	Conclusions
Variability of climate over past few thousand years	Climate change over recent millennia appear to have been significant prior to human generation of greenhouse gases but data are somewhat inconclusive.
Temperature rise over past century compared to natural fluctuations	Probably within past variations, but the rate of change in the 20th century was probably higher than in previous millennia.
Rising CO_2 was the cause of rising temperatures in 20th century	Rising CO_2 was probably one of several factors, Increasing CO_2 does produce a warming effect and it is likely that the temperature changes in the past century were at least partly due to greenhouse gases.
CO_2 concentration will rise further in 21st century in business-as-usual	Agree. There does not seem to be any way to provide the world with energy in the 21st century without continued dependence on fossil fuels.
Climate models provide reasonable estimates of future warming in business-as-usual scenario	Climate models cannot account for regional variability in humidity nor can they account for changes in cloudiness accompanying warming from greenhouse gases. Predictions of future climates remain dubious.
Impacts of future global warming are known to be disastrous	Impacts of future warming seem exaggerated. The one serious concern is a potential rise in sea level. This needs to be monitored carefully.
Only remedy is immediate draconian reduction in fossil fuel usage	This is true if predicted impacts by alarmists are accurate; but their predictions are dubious.
Draconian reduction in fossil fuel usage can be accommodated by renewable energy supply	The world population is growing and developing nations are using ever more energy. Providing the world with energy is the major challenge of the 21st century. Draconian reduction in fossil fuel usage will produce a worldwide depression.

The major problem in climatology is the lack of credible, precise long-term data in almost every facet of the science. Returning to the climate debate, we draw the conclusions listed in Table 10.1.

11 REFERENCES

Ackerman, Frank (2008) Hot, It's Not: Reflections on Cool It!, by Bjorn Lonborg, available at: *ase.tufts.edu/gdae/Pubs/rp/Ackerman_CoolIt.pdf*

Agee, Ernest M., Kandace Kiefer and Emily Cornett (2011) "Relationship of Lower Troposphere Cloud Cover and Cosmic Rays: An Updated Perspective" submitted to *Journal of Climate, http://journals.ametsoc.org/doi/abs/10.1175/JCLI-D-11-00169.1*

Allen, Myles R.; David J. Frame; Chris Huntingford; Chris D. Jones; Jason A. Lowe; Malte Meinshausen; and Nicolai Meinshausen (2009) "Warming caused by cumulative carbon emissions towards the trillionth tonne," Nature, 458, 1163–1166

Alley, Richard B.; Peter U. Clark; Philippe Huybrechts; and Ian Joughin (2005), "Ice-sheet and sea-level changes," *Science* 310, 456.

Andreadis, K. M. and Lettenmaier, D. P. (2006) "Trends in 20th century drought over the continental United States," *Geophysical Research Letters* 33, L10403.

Anderson, E. R. (1954) "Energy Budget Studies," from Technical Report, United States Geological Survey Professional Paper 269, *Water Loss Investigations - Lake Hefner Studies*, p. 71-117, 1954.

Andronova, N., Penner, J.E. and Wong, T. (2009) "Observed and modeled evolution of the tropical mean radiation budget at the top of the atmosphere since 1985". *Journal of Geophysical Research* 114, D14106.

Anon. (D), Ice over the Poles, online at *http://earthobservatory.nasa.gov/Library/PolarIce/*

Anon. (E), Solar Influences on Global Climate Change, National Academy Press, Washington, D.C. (1994).

Anon. (F), The Greenhouse Effect and Climate Change, Australian Government Department of Meteorology, online at *http://www.bom.gov.au/info/GreenhouseEffectAndClimate Change.pdf*

Anon. (H), Climate Change and the Greenhouse Effect, Hadley Center, U.K. Dept. of Meteorology, online at *http://www.reefrelief.org/scientificstudies/climate_greenhouse.pdf*

Anon. (J), Published e-letter responses to Hoffert et al. (2002), online at *http://intl.science mag.org/cgi/eletters/298/5595/981?ck=nck—676*

Anon. (N) Surface Temperature Reconstructions for the Last 2,000 Years, Committee on Surface Temperature Reconstructions for the Last 2,000 Years, National Research Council (2006) online at *http://www.nap.edu/catalog/11676.html*

Anon. (R), Hot Topics in Climate Change Science, Species Extinctions Will Increase Due To Global Warming, Australian Government *http://www.climatechange.gov.au/science/hottopics/pubs/topic11.pdf*

Archibald, David C. (2009), "Solar Cycle 24: Expectation sand implications," *Energy and Environment*, **20**, 1–10.

Arrak, Arno (2011) "Arctic Warming is not Greenhouse Warming" *Energy and Environment* **22**, 1069-1084.

Badalyan, O. G.; V. N. Obridko; and J. Sykora (2001), "Brightness of the coronal green line and prediction for activity cycles," *Solar Physics* **199**, 421–435.

Baldwin, R. C. (1970) "A Dispersion Model for Heated Effluent from an Ocean Outfall", Thesis, United States Naval Postgraduate School, *http://www.dtic.mil/dtic/tr/fulltext/u2/710730.pdf.*

Ball, Tim (2007), *The Science Isn't Settled: The Limitations of Global Climate Models*, Marshall Institute, online at *http://www.marshall.org/article.php?id=524*

Ballantyne, A. P., Alden, C. B., Miller, J. B., Tans, P. P. & White, J. W. C. (2012) "Earth science: The balance of the carbon budget" *Nature* **488**, 70–72.

Ban-Weiss, George A., Govindasamy Bala, Long Cao, Julia Pongratz and Ken Caldeira (2011) "Climate forcing and response to idealized changes in surface latent and sensible heat" *Environmental Research Letters* **6**, 034032.

Bard, E.; Raisbeck, G.; Yiou, F; and Jouzel, J. (2000), "Solar irradiance during the last 1200 years based on cosmogenic nuclides," *Tellus* **B52**, 985–992.

Bauer, Mike, Anthony D. Del Genio and John R. Lanzante (2002) "Observed and Simulated Temperature–Humidity Relationships: Sensitivity to Sampling and Analysis" *Journal of Climate* **15**, 203-215.

Beckman, John E. and Terence J. Mahoney (1998), *The Maunder Minimum and climate change: Have historical records aided current research?* Library and Information Services in Astronomy III ASP Conference Series 153.

Bender, Morris A., Thomas R. Knutson, Robert E. Tuleya, Joseph J. Sirutis, Gabriel A. Vecchi, Stephen T. Garner and Isaac M. Held (2010) ""Modeled Impact of Anthropogenic Warming on the Frequency of Intense Atlantic Hurricanes" *Science* **327**, 454-458.

Benestad, Rasmus E. (2005), *Solar Activity and Earth's Climate*, Praxis Publishing, 2nd edition.

Bender, Morris A., Thomas R. Knutson, Robert E. Tuleya, Joseph J. Sirutis, Gabriel A. Vecchi, Stephen T. Garner and Isaac M. Held (2010) "Modeled Impact of Anthropogenic Warming on the Frequency of Intense Atlantic Hurricanes" *Science* **327**, 454-458.

Benestad, Rasmus E. (2005), Solar Activity and Earth's Climate, Praxis Publishing, 2nd edition.

Bengsston, L., V. A. Semenov and O. M. Johannessen (2004) "The Early Twentieth-Century Warming in the Arctic—A Possible Mechanism" *J. Climate* **17**, 4045–4057.

Berliand, M. E. and T. G. Berliand (1952) "Determination of effective radiation of the earth as influenced by cloud cover" *Izvestiia Akademii Nauk S.S.S.R., Seriia Geofiziceskaya* **1**, 64-78.

Bernaerts, Arnd (2009) *Arctic Heats Up: Spitzbergen 1919-1939*, iUniverse.

Bhatt, Nipa J.; Rajmal Jain; and Malini Aggarwal (2009), "Predicting maximum sunspot number in Solar Cycle 24," *J. Astrophys. Astr.* **30**, 71–77.

Bischof, Jens (2000), *Ice Drift, Ocean Circulation and Climate Change*, Springer/Praxis, Heidelberg, Germany/Chichester, U.K.

Boening, Carmen, Josh K. Willis, Felix W. Landerer, R. Steven Nerem and John Fasullo (2012) "The 2011 La Niña: So strong, the oceans fell" *Geophysical Research Letters* **39**, L19602-L19606.

Bonan, Gordon B. (1997), "Effects of land use on the climate of the United States", *Climatic Change* **37**, 449–486.

Bond, Gerard; Bernd Kromer; Juerg Beer; Raimund Muscheler; Michael N. Evans; William Showers; Sharon Hoffmann; Rusty Lotti-Bond; Irka Hajdas; and Georges Bonani (2001), "Persistent solar influence on North Atlantic climate during the Holocene," *Science* **294**, 2130–2136.

Bond, Tami C. (2007), Testimony for the Hearing on Black Carbon and Global Warming House Committee on Oversight and Government Reform United States House of Representatives The Honorable Henry A. Waxman, Chair, October 18, 2007.

Bony, Sandrine; Robert Colman; Vladimir M. Kattsov; Richard P. Allan; Christopher S. Bretherton; Jean-Louis Dufresne; Alex Hall; Stephane Hallegatte; Marika M. Holland; William Ingram *et al.* (2006), "How well do we understand and evaluate climate change feedback processes?" *Journal of Climate* **19**, 3446–3483.

Booker, Christopher; and Richard North (2007), *Scared to Death: From BSE to GlobalWarming—Why Scares are Costing Us the Earth*, Continuum U.K., London.

Bouville, Mathieu (2008) "Plagiarism: Words and ideas" *Science and Engineering Ethics* **14**, 311-322.

Box, Jason E.: Lei Yang; David H. Bromwich, and Le-Sheng Bai (2009), "Greenland ice sheet surface air temperature variability: 1840–2007," *Journal of Climate* (submitted). Available at *http://adsabs.harvard.edu/abs/2008AGUFM.C21D..04B*

Boykoff, M. T., and J. M. Boykoff (2004) "Balance as bias: Global warming and the US prestige press" *Global Environ. Change Hum. Policy Dimensions* **14**, 125–136.

Briffa, K. R. and Osborn, T. J. (1999) "Seeing the wood from the trees" *Science* **284**, 926-7.

Briffa, K. R.; Osborn, T. J.; Schweingruber, F. H.; Harris, I. C.; Jones, P. D.; Shiyatov, S. G.; and Vaganov, E. A. (2001), "Low-frequency temperature variations from a northern tree-ring-density network," *Journal of Geophysical Research* **106**, 2929–2941.

Briffa, K. R.; Schweingruber, F. H.; Jones, P. D.; Osborn, T. J.; Shiyatov, S. G.; and Vaganov, E. A. (1998), "Reduced sensitivity of recent tree-growth to temperature at high northern latitudes" *Nature* **391**, 678-682.

Briffa, Keith R., Vladimir V. Shishov, Thomas M. Melvin, Eugene A. Vaganov, Håken Grudd, Rashit M. Hantemirov, Matti Eronen, and Muktar M. Naurzbaev (2008) "Trends in recent temperature and radial tree growth spanning 2000 years across northwest Eurasia" *Philos. Trans. Royal Soc. London B Biol Sci.* **363**, 2271–2284.

Brillinger, David, Judith Curry, Robert Jacobsen, Elizabeth Muller, Richard Muller, Saul Perlmutter, Robert Rohde, Arthur Rosenfeld, Charlotte Wickham and Jonathan Wurtele (2011) "Berkeley Earth Surface Temperature Analysis" *http://www.berkeleyearth.org/*

Brohan, P.; J. J. Kennedy; I. Harris; S. F. B. Tett; and P. D. Jones (2006), "Uncertainty estimates in regional and global observed temperature changes: A new dataset from 1850," *J. Geophys. Res.* **111**, D12106–D12140.

Brown, James H., William R. Burnside, Ana D. Davidson, John P. DeLong, William C. Dunn, Marcus J. Hamilton, Norman Mercado-Silva, Jeffrey C. Nekola, Jordan G.

Okie, William H. Woodruff, and Wenyun Zuo (2011) "Energetic Limits to Economic Growth" *BioScience* **61**, 19-26.

Callebaut, D. K. (2008), Approach of a Deep Minimum in Cycle 26 and Effect on Climate. Available at *mearim.cu.edu.eg/new/41-Mearin-Callebaut-Approach.pdf*

Cane, Mark A. and Stephen E. Zebiak (1985) "A Theory for El Nino and the Southern Oscillation," *Science* **228**, 1085-1087.

Cane, Mark A., "Oceanographic events During El Nino," *Science* **222**, 1189–1195 (1983).

Carlin, Alan (2009), Comments on Draft Technical Support Document for Endangerment Analysis for Greenhouse Gas Emissions under the Clean Air Act, March 16, 2009. Available at *http://wattsupwiththat.com/2009/06/27/released-the-censored-epa-document-final-report/*

CCSP (2007), CCSP Synthesis and Assessment Report 3.1, *Climate Models: An Assessment of Strengths and Limitations for User Applications, Public Review Draft,* U.S. Government Climate Change Science Program, May 15, online at *http://www.climatescience.gov/Library/sap/sap3-1/public-review-draft/sap3-1prd-cover.pdf*

Changnon, S. A. (Ed.) (2000), *El Niño 1997–1998: The Climate Event of the Century,* Oxford University Press, New York.

Chapman, William L.; and John E. Walsh (2007), "A synthesis of Antarctic temperatures," *Journal of Climate,* **20**, 4096–4117.

Charlson, Robert; J. Francisco; P. J. Valero; and John H. Seinfeld (2005), "In search of balance," *Science* **308**, 806–807.

Chin, Ming; Ralph A. Kahn; Lorraine A. Remer; Hongbin Yu; David Rind; Graham Feingold; Patricia K. Quinn; Stephen E. Schwartz; David G. Streets; Philip DeCola; Rangasayi Halthore (2009) *Atmospheric Aerosol Properties and Climate Impacts,* U.S. Climate Change Science Program, Synthesis and Assessment Product 2.3, January 2009.

Choudhuri, A. R.; Chatterjee, P.; and Jiang, J. (2007), "Predicting Solar Cycle 24 with a solar dynamo model," *Phys. Rev. Lett.,* **98**, 131101.

Christy, John R.; William B. Norris; Roy W. Spencer; and Justin J. Hnilo (2007), "Tropospheric temperature change since 1979 from tropical radiosonde and satellite measurements," *Journal of Geophysical Research* **112**, D06102.

Christy, J. R., B. Herman, R. Pielke, Sr., P. Klotzbach, R. T. McNider, J. J. Hnilo, R. W. Spencer, T. Chase and D. Douglass (2010) "What do observational datasets say about modeled tropospheric temperature trends since 1979?" *Remote Sensing* **2**, 2148-2169.

Church, J. A. and White, N. J. (2006) "20th century acceleration in global sea-level rise" *Geophysical Research Letters* **33**, L01602.

Church, J. A.; White, N. J.; Coleman, R.; Lambeck, K., and Mitrovica, J. X. (2004) "Estimates of the regional distribution of sea-level rise over the 1950 to 2000 period" *Journal of Climate* **17**, 2609– 2625.

Church, John A., Neil J. White, Leonard F. Konikow, Catia M. Domingues, J. Graham Cogley, Eric Rignot, Jonathan M. Gregory, Michiel R. van den Broeke, Andrew J. Monaghan, and Isabella Velicogna (2011) "Revisiting the Earth's sea level and energy budgets from 1961 to 2008" *Geophysical Research Letters* **38**, L18601.

Chylek, P. (2009) "Interactive comment on Comment on 'Aerosol radiative forcing and climate sensitivity deduced from the Last Glacial Maximum to Holocene transition'," by P. Chylek and U. Lohmann, *Geophys. Res. Lett.,* 2008' by J. C. Hargreaves and J. D. Annan" *Clim. Past Discuss.,* **4**, S752–S754.

Chylek, P.; J. E. Box; and G. Lesins (2004), "Global warming and the Greenland ice sheet," *Climatic Change* **63**, 201–221.

Chylek, P.; M. K. Dubey; and G. Lesins (2006), "Greenland warming of 1920–1930 and 1995– 2005," *Geophysical Research Letters* **33**, L11707.

Chylek, Petr and Ulrike Lohmann (2008) "Aerosol radiative forcing and climate sensitivity deduced from the Last Glacial Maximum to Holocene transition" *Geophysical Research Letters* **35**, L04804.

Chylek, Petr, Chris K. Folland, Glen Lesins, Manvendra K. Dubey, and Muyin Wang (2009a) "Arctic air temperature change amplification and the Atlantic Multidecadal Oscillation" *Geophysical Research Letters* **36**, L14801.

Chylek, Petr; Ulrike Lohmann; Manvendra Dubey; Michael Mishchenko; Ralph Kahn and Atsumu Ohmura (2007) "Limits on climate sensitivity derived from recent satellite and surface observations," Journal of Geophysical Research, **112**, D24S04-D24S12.

Clark, Peter U. and Peter Huybers (2009) "Interglacial and future sea level" *Nature* **462**, 856.

Collins, W. D., V. Ramaswamy, M. D. Schwarzkopf, Y. Sun, R. W. Portmann, Q. Fu, S. E. B. Casanova, J.-L. Dufresne, D. W. Fillmore, P. M. D. Forster, V. Y. Galin, L. K. Gohar, W. J. Ingram, D. P. Kratz, M.-P. Lefebvre, J. Li, P. Marquet, V. Oinas, Y. Tsushima, T. Uchiyama and W. Y. Zhong (2006), "Radiative forcing by well-mixed greenhouse gases: Estimates from climate models in the Intergovernmental Panel on Climate Change (IPCC) Fourth Assessment Report (AR4)" *J. Geophys. Res.*, **111**, D14317.

Comiso, Josefino C. (2000), "Variability and trends in Antarctic surface temperatures from in situ and satellite infrared measurements," *Journal of Climate* **13**, 1674–1697.

Conway, H.; B. L. Hall; G. H. Denton; A. M. Gades; and E. D. Waddington (1999), "Past and future grounding-line retreat of the West Antarctic Ice Sheet," *Science* **286**, 280–283.

Crichton, Michael (2003), *"Aliens Cause Global Warming." http://www.crichton-official.com/speech-alienscauseglobalwarming.html*

Crowley, Thomas J.; and Thomas S. Lowery (2000), "How warm was the *Medieval Warm Period?" Ambio* **29**, 51–54.

Crucifix, M. (2006) "Does the Last Glacial Maximum constrain climate sensitivity?" *Geophysical Research Letters* **33**, L18701.

Cubasch, Ulrich; E. Zorita; F. Gonzales-Rouco; H. von Storch; and I. Fast (2002), Cool stars, the sun and climate variability: Is there a connection?, online at *http://www.hs.unihamburg.de/cs13/day1/02_Cubasch.ppt*

Curry, J. A., P. J. Webster, and G. J. Holland (2006) "Mixing Politics and Science in Testing the Hypothesis That Greenhouse Warming Is Causing a Global Increase in Hurricane Intensity" *BAMS* August 2006, 1025-1037.

Curtin, Tim (2009) "Climate change and food production" *Energy and Environment* **20**, 1098-1116.

Dai, Aiguo, Thomas R. Karl, Bomin Sun, and Kevin E. Trenberth (2006) "Recent Trends in Cloudiness over the United States - A Tale of Monitoring Inadequacies" *BAMS*, May 2006, 597-606.

Dansgaard, Willi (2005), *Frozen Annals*, Niels Bohr Institute, Copenhagen, Denmark.

D'Arrigo, Rosanne, Rob Wilson and Gordon Jacoby (2006) *Further Discussion on: Tree-Ring Temperature Reconstructions for the Past Millennium* *www.ldeo.columbia.edu/res/fac/trl/.../D'AWilsJac.nrc.followup.pdf*

D'Arrigo, Rosanne, Rob Wilson and Gordon Jacoby (2006a) "On the long-term context for late twentieth century warming" *Journal of Geophysical Research* **111**, D03103.

Dasgupta, Susmita; Benoit Laplante; Craig Meisner; David Wheeler; and Jianping Yan (2007), *The impact of sea level rise on developing countries: A comparative analysis*, World Bank Policy Research Working Paper 4136, February, online at *http://www-wds.world bank.org/external/default/WDSContentServer/IW3P/IB/2007/02/09/000016406_20070 209 161430/Rendered/PDF/wps4136.pdf*

Davey, Christopher A.; and Roger A. Pielke Sr. (2005), "Microclimate exposures of surface-based weather stations implications for the assessment of long-term temperature trends," *Bulletin of the American Meteorological Society* **86**, 497–506.

Deffeyes, K. S. (2001), *Hubbert's Peak: The Impending World Oil Shortage*, Princeton University Press.

Desler, Clara; Michael A. Alexander; and Michael S. Timlin (1996), "Upper-ocean thermal variations in the North Pacific during 1970–1991," *Journal of Climate* **9**, 1840–1855.

Dessler, A. E. (2010) "A Determination of the Cloud Feedback from Climate Variations over the Past Decade" *Science* **330**, 1523-7.

Dessler, A. E. and S. M. Davis (2010) "Trends in tropospheric humidity from reanalysis systems" *Journal of Geophysical Research* 115, D19127.

Dessler, A. E. (2011) "Cloud variations and the Earth's energy budget" *Geophysical Research Letters geotest.tamu.edu/userfiles/216/Dessler2011.pdf*

Dessler, A. E., P. Yang, J. Lee, J. Solbrig, Z. Zhang, and K. Minschwaner (2008), "An analysis of the dependence of clear-sky top-of-atmosphere outgoing long-wave radiation on atmospheric temperature and water vapor," *Journal of Geophysical Research*, **113**, D17102-D17111.

Dessler, A. E., Z. Zhang, and P. Yang (2008), "Water-vapor climate feedback inferred from climate fluctuations, 2003–2008," Geophysical Research Letters, **35**, L20704-7.

Deutsche Bank Group (2008), Investing in Climate Change 2009, Necessity and Opportunity in Turbulent Times. Available at *http://dbadvisors.com/climatechange*

Dikpati, Mausumi; Giuliana de Toma; and Peter A. Gilman (2006) "Predicting the strength of Solar Cycle 24 using a flux-transport dynamo-based tool," *Geophysical Research Letters*, **33**, L05102–L05105.

DiLorenzo, E.; N. Schneider; K. M. Cobb; P. J. S. Franks; K. Chhak; A. J. Miller; J. C. McWilliams; S. J. Bograd; H. Arango; E. Curchitser *et al.* (2007), "North Pacific Gyre Oscillation links ocean climate and ecosystem change," *Geophysical Research Letters* **35**, L08607.

Divine, D. V.; and C. Dick (2006), "Historical variability of sea ice edge position in the nordic seas," *Journal of Geophysical Research* **111**, C01001.

Doherty, S. J., S. G. Warren, T. C. Grenfell, A. D. Clarke, and R. E. Brandt (2010) "Light-absorbing impurities in Arctic snow" *Atmos. Chem. Phys. Discuss.*, **10**, 18807–18878.

Domingues, Catia M., John A. Church, Neil J. White, Peter J. Gleckler, Susan E. Wijffels, Paul M. Barke and Jeff R. Dunn (2008) "Improved estimates of upper-ocean warming and multi-decadal sea-level rise" *Nature* **453**, 1090-1093.

Dorman, C. E. (1974) "Analysis of Meteorological and Oceanographic Data from Ocean Station Vessel N (30N 140W)" Ph. D. Thesis

Douglas, B.C. (1992) "Global sea level acceleration" *Journal of Geophysical Research* **97**, 12699–12706.

Douglas, B. C.; and W. R. Peltier (2002), *The puzzle of global sea-level rise*, Physics Today, March, online at *http://www.aip.org/web2/aiphome/pt/vol-55/iss-3/p35.html*

Douglass, D. H. (2005) "Observational Climate Data and Comparison with Models", *The Science and Culture Series Nuclear Strategy and Peace Technology*, Ed. Richard Ragaini, World Scientific Publishing Co. Ltd.

Douglass, D. H. (2010), "El Niño Southern Oscillation: Magnitudes and Asymmetry," *Journal of Geophysical Research*, **115**, D15111.

Douglass, D. H., John R. Christy, Benjamin D. Pearson, and S. Fred Singer (2007) "A comparison of tropical temperature trends with model predictions", *http://onlinelibrary.wiley.com/doi/10.1002/joc.1651/abstract*

Douglass, D. H.; and John R. Christy, (2009) "Limits on CO_2 Climate Forcing from Recent Temperature Data of Earth," *Energy and Environment* **20**, 177-189.

Douglass, D. H.; and R. S. Knox (2005), "Climate forcing by the volcanic eruption of Mount Pinatubo," *Geophys. Res. Lett.* **32**, L05710. Revised version available at: *http://arxiv.org/abs/physics/0509166.*

Douglass, David H. and Robert S. Knox (2005) "Climate forcing by the volcanic eruption of Mount Pinatubo" *Geophys. Res. Lett.*, **32**, L05710.

Eastman, Ryan, Stephen G. Warren, Carole J. Hahn (2011) "Variations in Cloud Cover and Cloud Types over the Ocean from Surface Observations, 1954–2008" *J. Climate*, **24**, 5914–5934.

Eddy, John (1976), "The Maunder Minimum," *Science* **192**, 1189–1202.

Eddy, John *et al.* (2004) "The Sun-Climate Connection" *http://www.ams.confex.com/ams/pdfpapers/74103.pdf*

Emanuel, K. A. (2007) "Environmental Factors Affecting Tropical Cyclone Power Dissipation," *Journal of Climate* **20**, 5497-5510.

Emanuel, Kerry (2005), "Increasing destructiveness of tropical cyclones over the past 30 years, *Nature* **436**, 686-688. *Environmental Influences on Tropical Cyclone Variability and Trends*, *http://ams.confex.com/ams/pdfpapers/107575.pdf*

Emanuel, Kerry (2010), "Tropical Cyclone Activity Downscaled from NOAA-CIRES Reanalysis, 1908-1958" *Journal of Advances in Modeling Earth Systems* **2**, 1-12.

EPA, *Report to Congress on Black Carbon*, Report No. EPA-450/R-12-001, March 2012.

Eschenbach, Willis (2012) "An Ocean of Overconfidence" http://wattsupwiththat.com/2012/04/23/an-ocean-of-overconfidence/

Esper, Jan; Edward R. Cook; and Fritz H. Schweingruber (2002), "Low-frequency signals in long tree-ring chronologies for reconstructing past temperature variability," *Science* **295**, 2249–2252.

Esper, Jan; Fritz H. Schweingruber; and Matthias Winiger (2002), "1300 years of climatic history for Western Central Asia inferred from tree-rings," *The Holocene* **12**, 267–277.

Esper, Jan; Robert J. S. Wilson; David C. Frank; Anders Moberg; Heinz Wanner; and Jurg Luterbacher (2005), "Climate: Past ranges and future changes," *Quaternary Science Reviews* **24**, 2164–2166.

Esper, Jan, David C. Frank, Mauri Timonen, Eduardo Zorita, Rob J. S. Wilson, Jürg Luterbacher, Steffen Holzkämper, Nils Fischer, Sebastian Wagner, Daniel Nievergelt, Anne Verstege and Ulf Büntgen (2012a) "Orbital forcing of tree-ring data" *Nature Climate Change* on line, 8 July 2012.

Esper, Jan, Ulf Büntgen, Mauri Timonen and David C. Frank (2012b) "Variability and extremes of northern Scandinavian summer temperatures over the past two millennia" *Global and Planetary Change* **88–89**, 1–9.

Etheridge, D. M.; L. P. Steele; R. I. Langenfels; R. J. Francey; J. M. Barnola; and V. I. Morgan (1996), "Natural and anthropogenic changes in atmospheric CO_2 over the last 1000 years from air in Antarctic ice and firn," *Journal of Geophysical Research* **101**, 4115–4128.

Evan, Amato T., Andrew K. Heidinger and Daniel J. Vimont (2007) "Arguments against a physical long-term trend in global ISCCP cloud amounts" *Geophysical Research Letters* **34**, L04701.

Fall, Souleymane, Anthony Watts, John Nielsen-Gammon, Evan Jones, Dev Niyogi, John R. Christy and Roger A. Pielke Sr. (2011) "Analysis of the impacts of station exposure on the U.S. Historical Climatology Network temperatures and temperature trends" Accepted for publication in the *Journal*

Feulner, Georg and Stefan Rahmstorf (2010) "On the effect of a new grand minimum of solar activity on the future climate on Earth" *Geophysical Research Letters* **37**, L05707.

Fjeldskaar, Willy; (2008) *Web commentaries on sea-level rise; stiger virkelig havnivaet mer enn før?* http://www.geoportalen.no/planetenjorden/klima/sealevel/; http://www.geo365.no/planetenjorden/klima/sealevel/

Flanner, M. G.; C. S. Zender; P. G. Hess; N. M. Mahowald; T. H. Painter; V. Ramanathan; and P. J. Rasch (2008) "Springtime warming and reduced snow cover from carbonaceous particles," *Atmos. Chem. Phys. Discuss.*, **8**, 19819–19859.

Flanner, Mark G.; Charles S. Zender; James T. Randerson; and Philip J. Rasch (2007) "Present- day climate forcing and response from black carbon in snow," *Journal of Geophysical Research*, **112**, D11202–D11219.

Foster, G. L., D. J. Lunt and R. R. Parrish (2009) "Mountain uplift and the threshold for sustained Northern Hemisphere Glaciation" *Clim. Past Discussions* **5**, 2439–2464.

Foster, G., J. D. Annan, P. D. Jones, M. E. Mann, J. Renwick, J. Salinger, G. A. Schmidt and K. E. Trenberth (2010) "Comment on "Influence of the Southern Oscillation on tropospheric temperature" by J. D. McLean, C. R. de Freitas, and R. M. Carter", *Journal of Geophysical Research*, **115**, D09110.

Foukal, P.; C. Frohlich; H. Spruit; and T. M. L. Wigley (2006), "Variations in solar luminosity and their effect on the Earth's climate," *Nature* **443**, 161–166.

Francis, Jennifer A. and Elias Hunter (2007) "Changes in the fabric of the Arctic's greenhouse blanket" *Environ. Res. Lett.* **2**, 045011.

Frauenfeld, Oliver W. (2005a), "Predictive skill of the ENSO and related teleconnections," in *Shattered Consensus*, edited by P. J. Michaels, Rowman and Littlefield, New York.

Frauenfeld, Oliver W.; Robert E. Davis; and Michael E. Mann (2005), "A distinctly interdecadal signal of Pacific Ocean atmosphere interaction," *Journal of Climate*, **18**, 1709–1718.

Frohlich, C. (2009) "Evidence of a long-term trend in total solar irradiance" *Astronomy and Astrophysics* **501**, L27–L30.

Frolov, Ivan; Zalman Gudkovich, Valery Karklin, Evgeny Kovalev and Vasily Smolyanitsky (2009) *Climate Change In Eurasian Arctic Shelf Seas: Centennial Ice Cover Observations*, Praxis Publishing.

Fye, F. K., Stahle, D. W. and Cook, E. R. (2003). "Paleoclimatic analogs to twentieth-century moisture regimes across the United States," *Bulletin of the American Meteorological Society* **84**, 901-909.

Gentemann, Chelle L. and Peter J. Minnett "Radiometric measurements of ocean surface thermal variability" *Journal of Geophysical Research* **113**, C08017.

Gettleman, A. and Q. Fu (2008), "Observed and Simulated Upper Tropospheric Water Vapor Feedback," *Journal of Climate*, **21**, 3282.

Ghan, Steven J.; and Stephen E. Schwartz (2007), "Aerosol properties and processes: A path from field and laboratory measurements to global climate," *Bull. Amer. Meteorological Soc.* **88**, 1059–1083.

Gill, Adrian (1982) *Atmosphere-Ocean Dynamics*, Academic Press.

Goody, Richard (1980) "Polar Process and World Climate (A Brief Overview)" *Monthly Weather Review* **108** (December 1980).

Goosse, Hughes; O. Arzel; J. Luterbacher; M. E. Mann; H. Renssen; N. Riedwyl; A. Timmermann; E. Xoplaki; and H. Wanner (2006), "The origin of the European Medieval Warm Period," *Clim. Past* **2**, 99–113.

Gorodetskaya, Irina V.; Mark A. Cane; L.-Bruno Tremblay; and Alexey Kaplan (2006), "The effects of sea-ice and land-snow concentrations on planetary albedo from the Earth Radiation Budget Experiment," *Atmosphere–Ocean* **44**, 195–205.

Graversen, R. G., S. Drijfhout, W. Hazeleger, R. Bintanja, R. van de Wal and M. Helsen (2010) "Greenland's contribution to global sea-level rise by the end of the 21st century" *Climate Dynamics, www.knmi.nl/publications/fulltexts/gravesen.pdf.*

Grenfell, Tom; Steve Warren; Tony Clarke; and Vladimir Radionov (n.d.), Black Carbon in Arctic Snow: Concentrations and Effect on Surface Albedo. Available at *http://niflheim.nilu.no/spac/nilu-workshop/summary/grenfell-bc-and-arctic-snow*

Grove, Jean M. (1988), *The Little Ice Age*, Routledge, London, UK.

Grove, J. M. (2001), "The initiation of the 'Little Ice Age' in regions around the North Atlantic," *Climatic Change* **48**, 53–82.

Guilderson, Thomas P.; and Daniel P. Schrag (1998), "Abrupt shift in subsurface temperatures in the tropical Pacific associated with changes in El Nino," *Science* **281**, 240–243.

Hadley, O. L., C. E. Corrigan, T. W. Kirchstetter, S. S. Cliff, and V. Ramanathan (2010) "Measured black carbon deposition on the Sierra Nevada snow pack and implication for snow pack retreat" *Atmos. Chem. Phys.* **10**, 7505–7513.

Hadley, O. L. and T. W. Kirchstetter (2012) "Black-carbon reduction of snow albedo" Nature Climate Change, *http://www.nature.com/nclimate/journal/vaop/ncurrent/full/nclimate1433.html*

Haerter, J. O., E. Roeckner, L. Tomassini and J.-S. von Storch (2009) "Parametric uncertainty effects on aerosol radiative forcing" *Geophysical Research Letters*, **36**, L15707.

Hall, Alex; and Syukuro Manabe (1999), "The role of water vapor feedback in unperturbed climate variability and global warming," *Journal of Climate* **12**, 2327–2346.

Hamon, M., G. Reverdin, P.-Y. Le Traon (2012) "Empirical Correction of XBT Data" *J. Atmos. Oceanic Technol.* **29**, 960–973.

Hanna, Edward; and John Cappelen (2003), "Recent cooling in coastal southern Greenland and relation with the North Atlantic Oscillation," *Geophysical Research Letters* **30**, 1132.

Hansen, James E.; Makiko Sato; Jay Glascoe; and Reto Ruedy (1998), "A common-sense climate index: Is climate changing noticeably?" *Proc. Natl. Acad. Sci.* **95**, 4113–4120.

Hansen, James E.; R. Ruedy; J. Glascoe; and M. Sato (1999), "GISS analysis of surface temperature change," *J. Geophys. Res.* **104**, 30997–31022.

Hansen, James E., Makiko Sato, Reto Ruedy, Andrew Lacis, and Valdar Oinas (2000) "Global warming in the twenty-first century: An alternative scenario" *PNAS* **97**, 9875-9880.

Hansen, James E.; R. Ruedy; M. Sato; M. Imhoff; W. Lawrence; D. Easterling; T. Peterson; and T. Karl (2001), "A closer look at United States and global surface temperature change," *Journal of Geophysical Research* **106**, 23947–23964.

Hansen, James E.; M. Sato; L. Nazarenko; R. Ruedy; A. Lacis; D. Koch; I. Tegen; T. Hall; D. Shindell; B. Santer *et al.* (2002), "Climate forcings in Goddard Institute for Space Studies SI2000 simulations," *Journal of Geophysical Research* **107**, 4347–4383.

Hansen, J., and L. Nazarenko (2003) "Soot climate forcing via snow and ice albedos" *PNAS*, **101**, 423-8.

Hansen, James E. (2004), "Defusing the global warming time bomb," *Scientific American*, 69– 77, March.

Hansen, James E. (2005), "A slippery slope: How much global warming constitutes 'dangerous anthropogenic interference'?: An editorial essay," *Climatic Change* **68**, 269–279, online at *http://www.columbia.edu/jeh1/hansen_slippery.pdf*

Hansen, James E.; L. Nazarenko; R. Ruedy; M. Sato; J. Willis; A. Del Genio; D. Koch; A. Lacis; K. Lo; S. Menon *et al.* (2005), "Earth's energy imbalance: Confirmation and implications," *Science* **308**, 1431–1434.

Hansen, J. M. Sato, R. Ruedy, L. Nazarenko, A. Lacis,, G. A. Schmidt,, G. Russell, I. Aleinov, M. Bauer, S. Bauer, N. Bell, B. Cairns, V. Canuto, M. Chandler, Y. Cheng, A. Del Genio,, G. Faluvegi, E. Fleming, A. Friend, T. Hall,, C. Jackman, M. Kelley, N. Kiang, D. Koch,, J. Lean, J. Lerner, K. Lo, S. Menon, R. Miller,, P. Minnis, T. Novakov, N. Oinas, Ja. Perlwitz, Ju. Perlwitz, D. Rind,, A. Romanou,, D. Shindell,, P. Stone, S. Sun,, N. Tausnev, D. Thresher, B. Wielicki, T. Wong, M. Yao, and S. Zhang (2005) "Efficacy of climate forcings" *Journal of Geophysical Research* **110**, D18104.

Hansen, J.; M. Sato; P. Kharecha; D. Beerling; R. Berner; V. Masson-Delmotte; M. Pagani; M. Raymo; D. L. Royer and J. C. Zachos (2008) "Target atmospheric CO2: Where should humanity aim?" *Open Atmos. Sci. J.* **2**, 217-231.

Hansen, J. E. (2008a) "Threat to the planet: Dark and bright sides of global warming" *EOS, Trans. Amer. Geophys. Union* **89** (Fall).

Hansen, J., *et al.* (2007) "Climate simulations for 1880–2003 with GISS model" *Climate Dynamics* **29**, 661–696.

Hansen, J., R. Ruedy, M. Sato and K. Lo (2010) "Global surface temperature change" "Global surface temperature change" *Reviews of Geophysics* **48**, RG4004. *http://data.giss.nasa.gov/gistemp/paper/gistemp2010_draft0601.pdf.*

Hansen, J. and Makiko Sato (2011) "Paleoclimate Implications for Human-Made Climate Change" *http://www.columbia.edu/~jeh1/mailings/2011/20110118_MilankovicPaper.pdf*

Hansen, James, Makiko Satoa and Reto Ruedyb (2012) "Perception of climate change" PNAS, published on line, *http://www.pnas.org/content/early/2012/07/30/1205276109.abstract*

Hare, Steven R.; and Nathan J. Mantua (2000) "Empirical evidence for North Pacific regime shifts in 1977 and 1989," *Progress in Oceanography* **47**, 103–145.

Harrison, D. E.; and M. Carson (2007) "Is the World Ocean Warming? Upper-Ocean Temperature Trends: 1950– 2000," *Journal of Physical Oceanography*, **37**, 174-187.

Hasse, Lutz (1971) "The Sea Surface Temperature Deviation and the Heat Flow at the Sea-Air Interface" *Boundary-Layer Meteorology* **1**, 368-379.

Hathaway, David H.; Robert M. Wilson; and Edwin J. Reichmann (2002), "Group Sunspot Numbers: Sunspot cycle characteristics," *Solar Physics* **211**, 357–370.

Hegerl, Gabriele C. and Francis Zwiers (2011) "Use of models in detection and attribution of climate change" *Wiley Interdisciplinary Reviews,May 26, 2011, http://wires.wiley.com/WileyCDA/WiresArticle/wisId-WCC121.html*

Hegerl, Gabriele C., Thomas J. Crowley, Myles Allen, William T. Hyde, Henry N. Pollack, Jason Smerdon, and Eduardo Zorita (2007) "Detection of Human Influence on a New, Validated 1500-Year Temperature Reconstruction" *Journal of Climate* **20**, 650-666.

Held, Isaac M.; and Brian J. Soden (2000), "Water vapor feedback and global warming," *Annual Rev. Energy Environ.* **25**, 441–475.

Hoffert, Martin I.; Ken Caldeira; Atul K. Jain; Erik F. Haites; L. D. Danny Harvey; Seth D. Potter; Michael E. Schlesinger; Stephen H. Schneider; Robert G. Watts; Tom M. L. Wigley et al. (1998), "Energy implications of future stabilization of atmospheric CO_2 content," *Nature* **395**, 881–884.

Hoffert, Martin I.; Ken Caldeira; Gregory Benford; David R. Criswell; Christopher Green; Howard Herzog; Atul K. Jain; Haroon S. Kheshgi; Klaus S. Lackner; John S. Lewis et al. (2002), "Advanced technology paths to global climate stability: Energy for a greenhouse planet," *Science* **298**, 981.

Hoffman, John S. (1984), *Estimates of Future Sea Level Rise*, online at *http://www.epa.gov/climatechange/effects/coastal/SLRChallenge.html*

Holden, P. B., N. R. Edwards, E. W. Wolff, N. J. Lang, J. S. Singarayer, P. J. Valdes, and T. F. Stocker (2009) "Interhemispheric coupling and warm Antarctic interglacials" *Clim. Past Discuss.*, **5**, 2555–2575.

Holgate, S. J. (2007), "On the decadal rates of sea level change during the twentieth century," *Geophysical Research Letters* **34**, L01602-5.

Holgate, S. J.; and P. L. Woodworth (2004), "Evidence for enhanced coastal sea level rise during the 1990s," *Geophysical Research Letters* **31**, L07305-9.

Holland, M. M.; and C. M. Bitz (2003), "Polar amplification of climate change in coupled models," *Climate Dynamics*, **21**, 221–232.

Holloway, Greg; and Tessa Sou (2000), *Is Arctic Sea Ice Rapidly Thinning?*; *Journal of Climate* (2002) **15**, 1691-1701. Available at *http://adaptation.nrcan.gc.ca/projdb/pdf/73_e.pdf*

Houghton, John (2004), *Global Warming, Third Edition*, Cambridge University Press.

Houston, J. R. and R. G. Dean (2011) "Sea-Level Acceleration Based on U.S. Tide Gauges and Extensions of Previous Global-Gauge Analyses" *Journal of Coastal Research* (in press). (*www.jcronline.org/doi/pdf/10.2112/JCOASTRES-D-10-00157.1*); "Sea Level Acceleration Characteristics in the 20th Century and Extrapolation to 2100" *http://www.fsbpa.com/2011TechPresentations/Dean%20Robert.pdf*

Howat, I. M.; Joughin, I.; and Scambos, T. A. (2007), "Rapid changes in ice discharge from Greenland outlet glaciers," *Science* **315**, 1559–1561.

Hoyt, Douglas (2006), *A Critical Examination of Climate Change*, online at *http://www.warwickhughes.com/hoyt/climate-change.htm*

Huang, J. C. K. and J. M. Park (1975) "Effective Cloudiness Derived from Ocean Buoy Data" *Journal of Applied Meteorology* **14**, 240-245.

Humlum, Ole (2011) *http://www.climate4you.com/*

Humlum, Ole, Kjell Stordahl, Jan-Erik Solheim (2012) "The phase relation between atmospheric carbon dioxide and global temperature", accepted for publication in: *Global and Planetary Change*.

Hunt, B. G. (1981) "An examination of some feedback mechanisms in the carbon dioxide climate problem" *Tellus* **33**, 78-88.

Huybers, Peter (2010) "Compensation between Model Feedbacks and Curtailment of Climate Sensitivity" *Journal of Climate* **23**, 3009-3018.

Iacono, M. J., J. S. Delamere, E. J. Mlawer, M. W. Shephard, S. A. Clough, and W. D. Collins (2008) "Radiative forcing by long-lived greenhouse gases: Calculations with the AER radiative transfer models" *Journal of Geophysical Research* **113**, D13103.

Idso, Craig D. (2008), *Public Comment in Response to EPA's proposed Rulemaking on Regulating Greenhouse Gas Emissions Under the Clean Air Act*, 24 November 2008. Available at: *http://www.co2science.org/*

Idso, Craig, Robert Carter and S. Fred Singer (2011) *Climate Change Reconsidered*, NIPCC Report, Heartland Institute, Chicago, IL.

Idso, Sherwood B.; and Craig D. Idso (2007), *Carbon Dioxide and Global Change: Separating Scientific Fact from Personal Opinion*, a critique of the April 26, 2007 testimony of James E. Hansen made to the Select Committee of Energy Independence and Global Warming of the United States House of Representatives entitled "Dangerous Human-Made Interference with Climate," Center for the Study of Carbon Dioxide and Global Change, June 6, online at *http://www.co2science.org*

Idso, Sherwood B.; Craig D. Idso; and Keith E. Idso (2003), *The Specter of Species Extinction*, online at *http://www.co2science.org*

Levin, Ingeborg (2012) "The balance of the carbon budget" *Nature* **488**, 35-36.

IPCC (2007), *Fourth Assessment Report*, Intergovernmental Panel on Climate Change.

IPCC (2011) "IPCC Special Report on Renewable Energy Sources and Climate Change Mitigation" Intergovernmental Panel on Climate Change.

IPCC (2011a) Intergovernmental Panel on Climate Change, Working Group 111, Mitigation of Climate Change, Special Report on Renewable Energy Sources and Climate Mitigation. June, 2011.

Jacobson, Mark Z. (2002), "Control of fossil-fuel particulate black carbon and organic matter, possibly the most effective method of slowing global warming," *Journal of Geophysical Research*, **107**, 4410–4431.

Jacobson, Mark Z. (2007), Testimony for the Hearing on Black Carbon and Global Warming House Committee on Oversight and Government Reform United States House of Representatives The Honorable Henry A. Waxman, Chair October 18, 2007.

Jacoby Gordon C., Nikolai V. Lovelius, Oleg I. Shumilov, Oleg M. Raspopov, Juri M. Karbainov and David C. Frank (2000) "Long-Term Temperature Trends and Tree Growth in the Taymir Region of Northern Siberia" *Quaternary Research* **53**, 312–318.

James, R. W. (1966) *Ocean Thermal Structures Forecasting*, U. S. Naval Oceanographic Office.

Jevrejeva, S,. J. C. Moore, A. Grinsted and P. L. Woodworth (2008) "Recent global sea level acceleration started over 200 years ago?" *Geophysical Research Letters* **35**, L08715.

Jevrejeva, S.; Grinsted, A.; Moore, J. C.; and Holgate, S. (2006), "Nonlinear trends and multiyear cycles in sea level records," *Journal of Geophysical Research* **111**, C09012-22.

Jevrejeva, S.; Moore, J. C., and Grinsted, A. (2010) "How will sea level respond to changes in natural and anthropogenic forcings by 2100?" *Geophysical Research Letters* **37**, L07703.

Johannessen, O. M., L. Bengtsson, M.W. Miles, S.I. Kuzmina, V.A. Semenov, G.V. Alekseev, A.P. Nagurnyi, V.F. Zakharov, L.P. Bobylev, L.H. Pettersson, K. Hasselmann and H.P. Cattle (2004) "Arctic climate change: observed and modelled temperature and sea-ice variability" *Tellus* **56A**, 328-341.

Jones, P. D., M. New, D. E. Parker, S. Martin and I. G. Rigor, "Surface Air Temperature and its Changes over the Past 150 Years," *Reviews of Geophysics* **37**, 173-199 (1999).

Jones, C. D., P. M. Cox, Peter Simmonds and Alistair Manning (2005) "On the significance of atmospheric CO_2 growth rate anomalies in 2002– 2003", *Geophys. Res. Lett.*, **32**, L14816.

Jones, P. D.; K. R. Briffa; T. J. Osborn; J. M. Lough; T. D. van Ommen; B. M. Vinther; J. Luterbacher; E. R. Wahl; F. W. Zwiers; M. E. Mann; G. A. Schmidt; C. M. Ammann; B. M. Buckley; K. M. Cobb; J. Esper; H. Goosse; N. Graham; E. Jansen; T. Kiefer; C. Kull; M. Küttel; E. Mosley-Thompson; J. T. Overpeck; N. Riedwy; M. Schulz; A. W. Tudhope; R. Villalba; H. Wanner; E. Wolff; and E. Xoplaki (2009) "High-resolution palaeoclimatology of the last millennium: a review of current status and future prospects," *The Holocene* **19**, (2009) 3–49.

Jones, P. D.; K. R. Briffa; T. P. Barnett; and S. F. B. Tett (1998), "High-resolution palaeoclimatic records for the last millennium: Interpretation, integration and comparison with General Circulation Model control-run temperatures," *The Holocene* **8**, 455–471.

Jones, P. D.; T. J. Osborn; and K. R. Briffa (2001), "Evolution of climate over the last millennium," *Science* **292**, 662–668.

Joughin, I.; Sarah B. Das; Matt A. King; Ben E. Smith; Ian M. Howat; and Twila Moon (2008), "Seasonal speedup along the western flank of the Greenland Ice Sheet," *Science*, **320**, 781–783.

Juckes, M. N.; M. R. Allen; K. R. Briffa; J. Esper; G. C. Hegerl; A. Moberg; T. J. Osborn; S. L. Weber; and E. Zorita (2006), "Millennial temperature reconstruction intercomparison and evaluation," *Clim. Past Discussions* **2**, 1001–1049; (2007) "Millennial temperature reconstruction intercomparison and evaluation," *Clim. Past Discussions* **3**, 591-609.

Kane, R. P. (2008), "Prediction of Solar Cycle 24 based on the Gnevyshev–Ohl– Kopecky rule and the three-cycle periodicity scheme," *Ann. Geophys.*, **26**, 3329–3339.

Kaufman, Darrell S., David P. Schneider, Nicholas P. McKay, Caspar M. Ammann, Raymond S. Bradley, Keith R. Briffa, Gifford H. Miller, Bette L. Otto-Bliesner, Jonathan T. Overpeck, Bo M. Vinther andArctic Lakes 2k Project Members (2009) "Recent Warming Reverses Long-Term Arctic Cooling" *Science* **325**, 1236-9.

Kemp, A.C., Horton, B.P., Donnelly, J.P., Mann, M.E., Vermeer, M. and Rahmstorf, S. (2011) "Climate related sea-level variations over the past two millennia" *Proc. Nat. Acad. Sci.* (*http://www.pnas.org/content/early/2011/06/13/1015619108.abstract*)

Kernthaler, S. C., Toumi, R. and Haigh. J. D. (1999), "Some doubts concerning a link between cosmic ray fluxes and global cloudiness," *Geophysical Research Letters* **26**, 863– 865.

Khalil, M. A. K.; C. L. Butenhoff; and R. A. Rasmussen (2007), "Atmospheric methane: Trends and cycles of sources and sinks," *Environmental Science and Technology* **41**, 2131– 2137.

Kiehl, J. T.; and Kevin E. Trenberth (1997), "Earth's annual global mean energy budget," *Bull. Amer. Meteorological Soc.* **78**, 197–208.

Kim, Hey-Jin and Arthur J. Miller (2007) "Did the Thermocline Deepen in the California Current after the 1976/77 Climate Regime Shift?" *Journal of Physical Oceanography* **37**, 1733-1739.

Kirkby, J.; A. Mangini; and R. A. Muller (2004), *The Glacial Cycles and Cosmic Rays*, online at *http://arxiv.org/pdf/physics/0407005*

Kirkby, Jasper (2008) "Cosmic Rays and Climate" *Surveys in Geophysics* **28**, 333–375.

Klotzbach, P. J. (2006), "Trends in global tropical cyclone activity over the past twenty years (1986–2005)," *Geophysical Research Letters* **33**, L010805.

Klotzbach, P. J. (2011) "El Niño – Southern Oscillation's Impact on Atlantic Basin Hurricanes and U. S. Landfalls" *Journal of Climate* **24**, 1252-1263.

Klotzbach, P. J., R. A. Pielke Sr., R. A. Pielke Jr., J. R. Christy, and R. T. McNider (2009), "An alternative explanation for differential temperature trends at the surface and in the lower troposphere", *J. Geophys. Res.*, **114**, D21102.

Klotzbach, Philip J. and William M. Gray (2011) "Extended Range Forecast Of Atlantic Seasonal Hurricane Activity And Landfall Strike Probability For 2011" *http://hurricane.atmos.colostate.edu/Forecasts*

Kniveton, Dominic R. and Martin C. Todd (2001), "On the relationship of cosmic ray flux and precipitation," *Geophysical Research Letters*, **28**, 1527–1530.

Knorr, Wolfgang (2009)) "Is the airborne fraction of anthropogenic CO_2 emissions increasing?" *Geophysical Research Letters* **36**, L21710.

Knox, Robert S. and David H. Douglass (2010) "Recent Energy Balance of Earth" *International Journal of Geosciences* **1**, 99-101.

Knutson, Thomas R. (2008) *Global Warming and Hurricanes: An Overview of Current Research Results, http://www.gfdl.noaa.gov/~tk/glob_warm_hurr_webpage.html*

Knutson, Thomas R., John L. McBride, Johnny Chan, Kerry Emanuel, Greg Holland, Chris Landsea, Isaac Held, James P. Kossin, A. K. Srivastava and Masato Sugi (2010) "Tropical cyclones and climate change" *Nature Geoscience* **3**, 157-163.

Koberle, C.; and R. Gerdes (2003), "Mechanisms determining the variability of Arctic sea ice conditions and export," *Journal of Climate* **16**, 2843–2858.

Kohler, P., Richard Bintanja, Hubertus Fischer, Fortunat Joos, Reto Knutti, Gerrit Lohmann, and Valerie Masson-Delmotte (2009) "What caused Earth's temperature variations during the last 800,000 years? Data-based evidence on radiative forcing and constraints on climate sensitivity" *Quaternary Science Reviews* **29** ,129-145.

Kolker, A. S., and S. Hameed (2007); "Meteorologically driven trends in sea level rise," Geophys. Res. Lett., **34**, L23616.

Konikow, L. F. (2011) "Contribution of global groundwater depletion since 1900 to sea level rise", *Geophys. Res. Lett.*, **38**, L17401.

Kopp, Greg and Judith L. Lean (2011) "A new, lower value of total solar irradiance: Evidence and climate significance" *Geophysical Research Letters* **38**, L01706.

Kopp, Robert E., Frederik J. Simons, Adam C. Maloof and Michael Oppenheimer (2009) "Probabilistic assessment of sea level during the last interglacial stage" *Nature* **462**, 863-867.

Koto, Hideto (1966) "Stagnation, Mixing and Renewal of the Water of the Funka Bay" *Memoirs of the Faculty of Fisheries Hokkaido University*, **13**, 65-78.

Kraus, Eric B. and Claes Rooth (1961) "Temperature and Steady State Vertical Heat Flux in the Ocean Surface Layers" *Tellus* **13**, 231-8, *http://www.tellusb.net/index.php/tellusb/article/download/12986/14753*

Krey, Volker and Leon Clarke (2011) "Role of renewable energy in climate mitigation: a synthesis of recent scenarios" *Climate Policy* **11**, 1-28.

Krivova, N. A., S. K. Solankia and Y. C. Unruh (2009) "Towards a long-term record of solar total and spectral irradiance" *J. Atmospheric and Solar-Terrestrial Physics* **73**, 223-234.

Landsea, C. W., G. A. Vecchi, L. Bengtsson and T. T. Knutson (2010) "Impact of duration thresholds on Atlantic tropical cyclone counts" *Journal of Climate* **23**, 2508-2518.

Lane, A. (1989) "The Heat Balance of the North Sea" Report No. 8 Proudman Oceanographic Laboratory, *http://nora.nerc.ac.uk/3872/*

Lefohn, Allen S.; Janja D. Husar; and Rudolf B. Husar (1999), "Estimating historical anthropogenic global sulfur emission patterns for the period 1850–1990," *Atmospheric Environment* **33**, 3435–3444.

Leroy, M. (2010) "Siting Classification for Surface Observing Stations on Land, Climate, and Upper-air Observations", JMA/WMO Workshop on Quality Management in Surface, Tokyo, Japan 27-30 July 2010.

Levitus, Sydney, John I. Antonov, Tim P. Boyer, Olga K. Baranova, Hernan Eduardo Garcia, Ricardo Alejandro Locarnini, Alexey V. Mishonov, James Reagan, Dan Seidov, Evgeney S. Yarosh, Melissa Marie Zweng (2012) "World ocean heat content and thermosteric sea level change (0-2000), 1955-2010" *Geophys. Res. Lett.*, **39**, L10603. *http://www.agu.org/pubs/crossref/pip/2012GL051106.shtml*

Lightfoot, H. Douglas; and Christopher Green (2002), *Energy Intensity Decline Implications for Stabilization of Atmospheric CO₂ Content*, Dept. of Economics, McGill University, McGill Center for Climate and Global Change Research (C2GCR) Report No. 2001-7, October 2001 (January 2002), online at *http://people.mcgill.ca/files/christopher.green/energyintensity decline.pdf*

Lindsay, R. W.; and J. Zhang (2005), "The thinning of Arctic sea ice, 1988–2003: Have we passed a tipping point?" *Journal of Climate* **18**, 4879–4894.

Lindsay, R. W.; J. Zhang; A. Schweiger, M. Steele; and H. Stern (2009), "Arctic sea ice retreat in 2007 follows thinning trend," *Journal of Climate,* **22**, 165–176.

Lindzen, R. S., A. Y. Hou and B. F. Farrell (1982), "The role of convective model choice in calculating the climate impact of doubling CO₂," *Journal of the Atmospheric Sciences* **39**, 1189-2005.

Lindzen, Richard S. (1997), "Can increasing carbon dioxide cause climate change?" *Proc. Nat. Acad. Science USA* **94**, 8335-8342.

Lindzen, Richard S, Ming-Dah Chou, and Arthur Y. Hou (2001), "Does the Earth Have an Adaptive Infrared Iris?" *Bulletin of the American Meteorological Society* **82**, 417-432.

Lindzen, Richard S. and Constantine Giannitsis (2002) "Reconciling observations of global temperature change" *Geophysical Research Letters* **29**, 1583-5.

Lindzen, Richard S. (2007),"Taking Greenhouse Warming Seriously," *Energy and Environment* **18**, 937-950.

Lindzen, Richard S. (2008), "Climate Science: Is it currently designed to answer questions?" Creativity and Creative Inspiration in Mathematics, Science, and Engineering: Developing a Vision for the Future, San Marino 29-31 August 2008.

Lindzen, Richard S. and Yong-Sang Choi (2009) "On the determination of climate feedbacks from ERBE data" *Geophysical Research Letters* **36**, L16705.

Lindzen, Richard S. and Yong-Sang Choi (2011) "On the observational determination of climate sensitivity and its implications" *Asia-Pacific J. Atmos. Sci.* **47**, 377-390. *http://www.pensee-unique.fr/Erice-ERBEpaper-2009finalL.pdf*

Lobell, David B., Wolfram Schlenker and Justin Costa-Roberts (2011) "Climate trends and global crop production since 1980" *http://www.sciencemag.org/content/early/2011/05/04/science.1204531*

Lockwood, Michael and Claus Frohlich (2007) "Recent oppositely directed trends in solar climate forcings and the global mean surface air temperature," *Proc. Roy. Soc. A* **463**, 2447–2460.

Loeb, Norman G., John M. Lyman, Gregory C. Johnson, Richard P. Allan, David R. Doelling, Takmeng Wong, Brian J. Soden and Graeme L. Stephens (2012) "Observed changes in top-of-the-atmosphere radiation and upper-ocean heating consistent within uncertainty", Nature Geoscience, published on line 22 January 2012, *http://www.nature.com/ngeo/journal/vaop/ncurrent/full/ngeo1375.html*

Loehle, Craig (2007), "A 2000-year global temperature reconstruction based on non-tree ring proxies," *Energy and Environment*, **18**, 1049–1058.

Lombard, A.; Cazenave, A.; Le Traon, P.-Y.; and Ishii, M. (2005), "Contribution of thermal expansion to present-day sea-level change revisited," *Global and Planetary Change* **47**, 116.

Lonborg, Bjorn (2007) Perspective on Climate Change, *http://www.climatechangefacts.info/ClimateChangeDocuments/lomborg_testimony.pdf*; See also Lonborg, Bjorn (2001), The Skeptical Environmentalist, Cambridge University Press; and Lonborg, Bjorn (2004), *Global Crises, Global Solutions*, Cambridge University Press.

Lyman, J. M., Simon A. Good, Viktor V. Gouretski, Masayoshi Ishii, Gregory C. Johnson, Matthew D. Palmer, Doug M. Smith and Josh K. Willis (2010) "Robust warming of the global upper ocean" *Nature* **465**, 334–337.

Maasch, Kirk Allen (2009) "El Niño and Interannual Variability of Climate in the Western Hemisphere" *http://climatechange.umaine.edu/people/profile/kirk_allen_maasch*

Mackenzie, F. T., A. Lerman and L. M. B. Ver (2001) "Recent past and future of the global carbon cycle," in L. C. Gerhard, W. E. Harrison, and B. M. Hanson, eds., *Geological Perspectives of Global Climate Change*, p. 51-82.

Mackey, Richard (2007), *The Climate Dynamics of Total Solar Variability*. Available at *www.coastalconference.com/2007/papers2007/Richard%20Mackey.doc*

Mann, Michael E., Zhihua Zhang, Malcolm K. Hughes, Raymond S. Bradley, Sonya K. Miller, Scott Rutherford, and Fenbiao Ni (2008), "Proxy-based reconstructions of hemispheric and global surface temperature variations over the past two millennia," *PNAS* **105**, 13252-7.

Mann, Michael E.; and Philip D. Jones (2003), "Global surface temperatures over the past two millennia," *Geophysical Research Letters* **30**, 1820.

Mann, Michael E.; R. S. Bradley; and M. K. Hughes (2004), "Corrigendum: Global-scale temperature patterns and climate forcing over the past six centuries," *Nature* **430**, 105.

Mann, Michael E.; Raymond S. Bradley; and Malcolm K. Hughes (1998), "Global-scale temperature patterns and climate forcing over the past six centuries," *Nature* **392**, 779–807.

Mann, M. E.; R. S. Bradley; and M. K. Hughes (1999), "Northern Hemisphere temperatures during the past millennium: Inferences, uncertainties, and limitations," *Geophys. Res. Letters* **26**, 759–762.

Mann, Michael E.; Scott Rutherford; Eugene Wahl; and Caspar Ammann (2005), "Testing the fidelity of methods used in proxy-based reconstructions of past climate," *Journal of Climate* **18**, 4097–4107.

Mann, Michael E.; Scott Rutherford; Eugene Wahl; and Caspar Ammann (2007), "Robustness of proxy-based climate field reconstruction methods," *Journal of Geophysical Research* **112**, D12109.

Marsh, Gerald E. (2002), *Global Warming Primer, National Center for Public Policy Research*, Report No. 420 (July), online at *http://www.nationalcenter.org/NPA420.pdf*

Marsh, N. D., and H. Svensmark (2000) "Low cloud properties influenced by cosmic rays" *Phys. Rev. Lett.*, **85**, 5004-5007.

Martin, Brian (1979) *The Bias of Science*. Available at *http://www.uow.edu.au/arts/sts/bmartin/pubs/79bias/index.html*

Maughan, P. M. (1966) "Measurement of Radiant Energy Over s Mixed Water Body", Ph. D. Thesis, Oregon State University.

McConnell, Joseph R.; Ross Edwards; Gregory L. Kok; Mark G. Flanner; Charles S. Zender; Eric S. Saltzman; J. Ryan Banta; Daniel R. Pasteris; Megan M. Carter; and Jonathan D. W. Kahl (2007), "20th-century industrial black carbon emissions altered Arctic climate forcing," *Science*, **317**, 1381–1384.

McIntyre, S.; and R. McKitrick (2003), "Corrections to the Mann et al. (1998) 'Proxy data based and Northern Hemispheric average temperature series'," *Energy and Environment* **14**, 751–771.

McIntyre, S.; and R. McKitrick (2005), *Hockey Sticks, Principal Components and Spurious Significance, Geophysical Research Letters* **32**, L03710.

McIntyre, S.; and R. McKitrick (2006), *Surface Temperature Reconstructions for the Past 1,000–2,000 Years*, presentation to the National Academy of Sciences Expert Panel, Washington, D.C. (March 2).

McIntyre, S. (2007), Climate Audit Website, online at *http://www.climateaudit.org*

McIntyre, S.; and R. McKitrick (2007), *The M&M Critique of the MBH98 Northern Hemisphere Climate Index: Update and Implications*, Informal report, online at *http://www.uoguelph.ca/rmckitri/research/trc.html*

McKitrick, Ross (2005), *What is the 'Hockey Stick' Debate About?*, online at *http://www. climatechangeissues.com/files/PDF/conf05mckitrick.pdf*

McKitrick, Ross (2008) *Atmospheric Oscillations do not Explain the Temperature-Industrialization Correlation* (July 21, 2008). Available at SSRN: http://ssrn.com/abstract=1166424

McKitrick, Ross T. (2005a), *The Mann et al. Northern Hemisphere Climate Index*, edited by P. J. Michaels, Rowman and Littlefield, New York.

McKitrick, Ross T.; and Patrick J. Michaels (2007) "Quantifying the influence of anthropogenic surface processes and inhomogeneities on gridded global climate data," *Journal of Geophysical Research*, **112**, D24S09-D24S22,

McLean, J. D. (2010) "Global Warming Issues" *http://mclean.ch/climate/ENSO_paper.htm, http://mclean.ch/climate/global_warming.htm*

McLean, J. D., C. R. de Freitas, and R. M. Carter (2009) "Influence of the Southern Oscillation on tropospheric temperature" *Journal of Geophysical Research*, **114**, D14104.

McLean, John (2007a), *Ignoring a Natural Event to Blame Humans (October)*, online at *http://mclean.ch/climate/global_warming.htm*

McLean, John (2007b), *Fallacies about Global Warming* (September), online at *http://mclean.ch/climate/global_warming.htm*

McNider, R. T., G. J. Steeneveld, B. Holtslag, R. Pielke Sr, S. Mackaro, A. Pour Biazar, J. T. Walters, U. S. Nair, and J. R. Christy (2012) "Response and sensitivity of the nocturnal boundary layer over land to added longwave radiative forcing", *J. Geophys. Res.*, in press.

McWilliams, James C. (2007), "Irreducible imprecision in atmospheric and oceanic simulations," *Proc. Natl. Academy Sci.*, **104**, 8709–8713.

Meehl, Gerald A.; and Warren M. Washington (1996), "El Nino-like climate change in a model with increased atmospheric CO_2 concentrations," *Nature* **382**, 56–60.

Meehl, Gerald A.; Warren M. Washington; T. M. L. Wigley; Julie M. Arblaster; and Aiguo Dai (2002), "Solar and greenhouse gas forcing and climate response in the twentieth century," *Journal of Climate* **16**, 426–444.

Meinshausen, Malte; Nicolai Meinshausen; William Hare1; Sarah C. B. Raper; Katja Frieler; Reto Knutti; David J. Frame; and Myles R. Allen (2009), "Greenhouse-gas emission targets for limiting global warming to 2°C," *Nature*, **458**, 1158–1163.

Menne, M.J., C.N. Williams, Jr., and M.A. Palecki (2010) "On the reliability of the U.S. surface temperature record" J. Geophys. Res. 115, D11108.

Michaels, P. J.; P. C. Knappenberger; and R. E. Davis (2006), "Sea-surface temperatures and tropical cyclonesin the Atlantic basin," *Geophys. Res. Lett.*, **33**, L09708–L09711.

Michaels, Patrick J.; and R. C. Balling, Jr. (2009) *Climate of Extremes*, Cato Institute, Washington, D.C.

Miller, G. H., J. Brigham-Grette, R. B. Alley, L. (2010a) "Arctic amplification: can the past constrain the future?" *Quaternary Science Reviews* **29**, 1779-1790.

Miller, G. H., J. Brigham-Grette, R. B. Alley, L. Anderson, H. A. Bauch, M. S. V. Douglas, M. E. Edwards, S. A. Elias, B. P. Finney, J. J. Fitzpatrick, S. V. Funder, T. D. Herbert, L. D. Hinzman, D. S. Kaufmanm, G. M. MacDonald, L. Polyak, A. Robock, M. C. Serreze, J. P. Smol, R. Spielhagen, J. W. C. White, A. P. Wolfe and E. W. Wolff (2010) "Temperature and precipitation history of the Arctic" *Quaternary Science Reviews* **29**, 1679-1715.

Mills, A. F. (1999) *Basic Heat and Mass Transfer*, 2nd ed., Prentice-Hall.

Mills, A. F. (2001) *Mass Transfer*, 2nd ed., Prentice-Hall.

Min, S-K., Zhang, X., Zwiers, F.W., Hegerl, G.C. (2011) "Human contribution to more-intense precipitation extremes" *Nature* **470**, 378-381.

Minschwaner, Ken; and Andrew E. Dessler (2004), "Water vapor feedback in the tropical Upper Troposphere: Model results and observations," *Journal of Climate* **17**, 1272–1282.

Minschwaner, Ken; Andrew E. Dessler; and Parnchai Sawaengphokhai (2006), "Multimodel analysis of the water vapor feedback in the tropical Upper Troposphere," *Journal of Climate* **19**, 5455–5464.

Moberg, A.; D. M. Sonechkin; K. Holmgren; N. M. Datsenko; W. Karlen; and S. E. Lauritzen (2005), "Highly variable Northern Hemisphere temperatures reconstructed from low-and high-resolution proxy data," *Nature* **433**, 613–617.

Monaghan, A. J., D. H. Bromwich, W. Chapman, and J. C. Comiso (2008), "Recent variability and trends of Antarctic near-surface temperature," *Journal of Geophysical Research Atmospheres*, **113**, D04105.

Montford, A. W. (2010) *The Hockey Stick Illusion: Climategate and the Corruption of Science*, Stacey International, London, UK.

Morice, Colin P., John J. Kennedy, Nick A. Rayner and Phil D. Jones (2012) "Quantifying uncertainties in global and regional temperature change using an ensemble of observational estimates: the HadCRUT4 data set" *Journal of Geophysical Research* **117**, D08101-D08122.

Mörner, Nils-Axel (1973), "Eustatic changes during the last 300 years," *Palaeogeography, Palaeoclimatology, Palaeoecology* **9**, 153–181.

Mörner, Nils-Axel (2004) "Estimating future sea level changes from past records," Global and Planetary Change **40**, 49–54, online at *http://gsa.confex.com/gsa/inqu/finalprogram/abstract_54461.htm*

Mosher, Stephen and Thomas W. Fuller (2010) *Climategate: The CRU tape letters*, available at amazon.com.

Mudelsee, M.; M. Borngen; G. Tetzlaff; and U. Grunewald (2003), "No upward trends in the occurrence of extreme floods in Central Europe," *Nature* **425**, 166-169.

Muller, Richard A., Judith Curry, Donald Groom, Robert Jacobsen, Saul Perlmutter, Robert Rohde, Arthur Rosenfeld, Charlotte Wickham and Jonathan Wurtele (2011a) "Decadal Variations in the Global Atmospheric Land Temperatures" *http://www.berkeleyearth.org/*

Muller, Richard A., Judith Curry, Donald Groom, Robert Jacobsen, Saul Perlmutter, Robert Rohde, Arthur Rosenfeld, Charlotte Wickham and Jonathan Wurtele (2011b)

"Earth Atmospheric Land Surface Temperature and Station Quality in the United States" *http://www.berkeleyearth.org/*

Murphy, D. M., S. Solomon, R. W. Portmann, K. H. Rosenlof, P. M. Forster, and T. Wong (2009), An observationally based energy balance for the Earth since 1950, *J. Geophys. Res.* **114**, D17107.

Murray, T., Scharrer, K., James, T. D., Dye, S. R., Hanna, E., Booth, A. D., Selmes, N., Luckman, A., Hughes, A. L. C., Cook, S., and Huybrechts, P. (2010) "Ocean regulation hypothesis for glacier dynamics in southeast Greenland and implications for ice sheet mass changes" *Journal of Geophysical Research* **115**, F03026.

Muscheler, Raimund; Fortunat Joos; Simon A. Müller; and Ian Snowball (2005), "How unusual is today's solar activity?" *Nature* **436**, E3–E4.

Nagashima, Tatsuya; Hideo Shiogama; Tokuta Yokohata; Simon A. Crooks; and Toru

Newell, R. E. and T. G. Dopplick (1979) "Questions Concerning the Possible Influence of Anthropogenic CO_2 on Atmospheric Temperature" *Journal of Applied Meteorology* **18**, 822-5.

Newell, R. E. and T. G. Dopplick (1981) "Reply to Robert G. Watts Discussion of Questions Concerning the Possible Influence of Anthropogenic CO_2 on Atmospheric Temperature" *Journal of Applied Meteorology* **20**, 114-117.

NIPCC (2011) *Climate Change Reconsidered*, Center for the Study of Carbon Dioxide and Global Change, Tempe, AZ.

Nordhaus, William (2006), Review of Stern Review (November 17), online at *http://www. carbontax.org/wp-content/uploads/2006/12/nordhaus-_-stern-review-_-dec-2006.pdf*

Nordhaus, William (2007), The Stern Review on the Economics ofClimate Change. Available at *nordhaus.econ.yale.edu/stern_050307.pdf*

Norris, J. R. (2005) "Multidecadal changes in near-global cloud cover and estimated cloud cover radiative forcing" *J. Geophys. Res.* **110**, D08206.

Norris, Joel R. and Anthony Slingo (2009) "Trends in Observed Cloudiness and Earth's Radiation Budget- What Do We Not Know and What Do We Need to Know?" *meteora.ucsd.edu/~jnorris/.../02_Norris_and_Slingo.pdf*

Northrup, Amy (2004), *The Global Carbon Cycle in the Ocean and the Threat of Climate Change Because of Anthropogenic Carbon Emissions*, BISC 419 (April 22), online at *http:// bioweb.usc.edu/courses/2004-spring/documents/bisc419-carboncycle.pdf*

Nozawa (2006), "The effect of carbonaceous aerosols on surface temperature in the mid twentieth century," *Geophysical Research Letters* **33**, L04702.

Nozawa, T.; T. Nagashima; T. Ogura; T. Yokohata; N. Okada; and H. Shiogama (2007), *Climate Change Simulations with a Coupled Ocean–Atmosphere GCM (MIROC)*, Center for Global Environmental Research, Japan, Report No. CGER 1073.

Nozawa, Toru; Tatsuya Nagashima; Hideo Shiogama; and Simon A. Crooks (2005), "Detecting natural influence on surface air temperature change in the early twentieth century," *Geophys. Res. Letters* **32**, L20719.

O'Donnell, Ryan, Nicholas Lewis, Steve McIntyre and Jeff Condon (2011) "Improved methods for PCA-based reconstructions: case study using the Steig *et al.* (2009) Antarctic temperature reconstruction" *Journal of Climate* **24**, 2099-2115.

Ogilvie, A. E. J.; and T. Jonsson (2001), "Little Ice Age research: A perspective from Iceland," *Climatic Change* **48**, 9–52.

Oppenheimer, Michael; and R. B. Alley (2005), *Ice sheets, global warming, and Article 2 of the UNFCCC: An editorial essay*, *Climatic Change* **68**, 257–267, online at *http:// www.princeton.edu/step/people/Oppenheimer%oand%Alley%II%published.pdf*

Oreskes, N. (2004), "The scientific consensus on climate change," Science **306**, 1686.

Oreskes, N. (2007a), "The long consensus on climate change," Washington Post, February 1, 2007. Available at *http://www.washingtonpost.com/wp-dyn/content/article/2007/01/31/AR200701301808_pf.html*

Oreskes, N. (2007b), *The Science of Global Warming: How Do We Know We're Not Wrong?* AMS Environmental Science Seminar Series, American Meteorological Society, Washington, D.C. Available at
www.lpl.arizona.edu/resources/globalwarming/documents/oreskeson-science-consensus.pdf

Oreskes, Naomi, Erik M. Conway, and Matthew Shindell (2008), *From Chicken Little to Dr. Pangloss: William Nierenberg, Global Warming, and the Social Deconstruction of Scientific Knowledge,*
http://www.lse.ac.uk/collections/CPNSS/projects/ContingencyDissentInScience/DP/DPOreske setalChickenLittleOnlinev2.pdf

Overpeck, J.; K. Hughen; D. Hardy; R. Bradley; R. Case; M. Douglas; B. Finney; K. Gajewski; G. Jacoby; A. Jennings *et al.* (1997), "Arctic environment change of the last four centuries," *Science* **278**, 1251.

B. Ozsoy-Cicek, H. Xie, S. F. Ackley, and K. Ye "Antarctic summer sea ice concentration and extent: comparison of ODEN 2006 ship observations, satellite passive microwave and NIC sea ice charts" *The Cryosphere* **3**, 1-9, 2009.

Pall, P., Aina, T., Stone, D.A., Stott, P.A., Nozawa, T., Hilberts, A.G.J., Lohmann, D and Allen, M.R. (2011) "Anthropogenic greenhouse gas contribution to flood risk in England and Wales in autumn 2000" *Nature* **470**, 382-386.

Palmer, M. D. and K. Haines (2009) "Estimating Oceanic Heat Content Change Using Isotherms" *Journal of Climate* **22**, 4953-4969.

Paltsev, Sergey, John M. Reilly, Henry D. Jacoby, and Jennifer F. Morris (2009) "The Cost of Climate Policy in the United States " *MIT Joint Program on the Science and Policy of Global Change* Report No. 173 April 2009.

Patterson, R. Timothy (2007), *Read the Sunspots* (June 20, 2007), online at *http://www.canada.com/nationalpost/financialpost/comment/story.html?id = 597d0677-2a05-47b4-b34fb84068db11f4*

Perovich, Donald K., Jacqueline A. Richter-Menge, Kathleen F. Jones, Bonnie Light, Bruce C. Elder, Christopher Polashenski, Daniel Laroche, Thorsten Markus and Ronald Lindsay (2011) "Arctic sea-ice melt in 2008 and the role of solar heating" *Annals of Glaciology* **52**, 355-9.

Peterson, Thomas C.; and Russell S. Vose (1997), "An overview of the global historical climatology network temperature database," *Bulletin of the American Meteorological Society* **78**, 2837–2849.

Pielke, Sr., Roger (2003) "Heat Storage within the Earth System" *Bull. Amer. Meteorological Soc.*, March, 2003, 331-5.

Pielke Jr., Roger A.; Joel Gratz; Christopher W. Landsea; Douglas Collins; Mark A. Saunders; and Rade Musulin (2008) "Normalized Hurricane Damage in the United States: 1900–2005" *Natural Hazards Review*, February 2008, pp. 29-42.

Pielke Jr., Roger, Tom Wigley and Christopher Green (2008), "*Dangerous assumptions: How big is the energy challenge of climate change?*" *Nature Commentary* **452**, 531-2.

Pielke Sr., Roger A. (2009) "Reply to comment by David E. Parker *et al.* on "Unresolved issues with the assessment of multidecadal global land surface temperature trends" *Journal of Geophysical Research*, **114**, D05105.

Pielke Sr., Roger A.; C. A. Davey; Dev Niyogi; Souleymane Fall; Jesse Steinweg-Woods; Ken Hubbard; Xiaomao Lin; Ming Cai; Young-Kwon Lim; Hong Li et al. (2007a), *Unresolved Issues with the Assessment of Multi-Decadal Global Land-Surface Temperature Trends* (February 5), online at *http://climatesci.colorado.edu/publications/pdf/R-321.pdf*

Pielke Sr., Roger A.; Jimmy O. Adegoke; Thomas N. Chase; Curtis H. Marshall; Toshihisa Matsui; and Dev Niyogi (2007b), "A new paradigm for assessing the role of agriculture in the climate system and in climate change," *Agricultural and Forest Meteorology* **142**, 234– 254.

Pielke Sr., Roger; John Nielsen-Gammon; Christopher Davey; Jim Angel; Odie Bliss; Nolan Doesken; Ming Cai; Souleymane Fall; Dev Niyogi; Kevin Gallo et al. (2007c), "Documentation of uncertainties and biases associated with surface temperature measurement sites for climate change assessment," *Bull. Amer. Meteor. Soc.* **88**, 913–928.

Pielke, Jr., R. A.; Landsea, C.; Mayfield, M.; Laver, J.; and Pasch, R. (2005). "Hurricanes and global warming," *Bulletin of the American Meteorological Society* **86**, 1571-1575.

Pielke, Sr., Roger (2008) "A Broader View of the Role of Humans in the Climate System" *Physics Today*, November, 2008, 54-55.

Pielke, Sr., Roger A. (2009), "Comments on the new paper 'The United States His torical Climatology Network Monthly Temperature Data—Version 2' by Menne *et al.* 2009." Available at *http://climatesci.org/2009/05/12/comments-on-the-new-paper-the-unitedstates- historical-climatology-network-monthly-temperature-data-%E2%80%93-version -2-by-menne-et-al-2009/*

Pielke Sr., R.A. (2012) "Summary Of Several Climate Science Issues – October 2012", http://pielkeclimatesci.wordpress.com/2012/04/22/comment-on-ocean-heat-content-world-ocean-heat-content-and-thermosteric-sea-level-change-0-2000-1955-2010-by-levitus-et-al-2012/

Pierrehumbert, R. T. (2009) *Principles of planetary climate*, available online at *http://geosci.uchicago.edu/~rtp1/ClimateBook/ClimateBook.html*

Pokhrel, Yadu N., Naota Hanasaki, Pat J-F. Yeh, Tomohito J. Yamada, Shinjiro Kanae and Taikan Oki (2012) "Model estimates of sea-level change due to anthropogenic impacts on terrestrial water storage" *Nature Geoscience*, published on line 20 May 2012, *http://www.nature.com/ngeo/journal/vaop/ncurrent/abs/ngeo1476.html*

Polyak, Leonid, Richard B. Alley, John T. Andrews, Julie Brigham-Grette, Thomas M. Cronin, Dennis A. Darby, Arthur S. Dyke, Joan J. Fitzpatrick, Svend Funder, Marika Holland, Anne E. Jennings, Gifford H. Miller, Matt O'Regan, James Savelle, Mark Serreze, Kristen St. John, James W. C. White and Eric Wolff (2010) "History of sea ice in the Arctic" *Quaternary Science Reviews* **29**, 1757–1778.

Polyakov, I. G. V. Alekseev, R. V. Bekryaev, U. Bhatt, R. Colony, M. Johnson, V. P. Karklin, D. Walsh, and A. V. Yulin (2003) "Long-Term Ice Variability in Arctic Marginal Seas" *Journal of Climate* **16**, 2078-2085.

Polyakov, I. V.; and M. A. Johnson (2000), "Arctic decadal and interdecadal variability," *Geophysical Research Letters* **27**, 4097–4100.

Polyakov, I. V.; R. V. Bekryaev; G. V. Alekseev; U. S. Bhatt; R. L. Colony; M. A. Johnson; A. P. Maskshtas; and D. Walsh (2003b), "Variability and trends of air temperature and pressure in the maritime Arctic, 1875–2000," *Journal of Climate* **16**, 2067–2077.

Polyakov, I.; D. Walsh; I. Dmitrenko; R. L. Colony; and L. A. Timokhov (2003a), "Arctic Ocean variability derived from historical observations," *Geographic Research Letters* **30**, 31–34.

Power, Scott B.; and Ian N. Smith, "Weakening of the Walker Circulation and apparent dominance of El Nino both reach record levels, but has ENSO really changed?" *Geophysical Research Letters* **34**, L18702-5 (2007).

Przybylak, Rajmund (2002), "Changes in seasonal and annual high-frequency air temperature variability in the Arctic from 1951 to 1990," *Int. J. Climatology* **22**, 1017–1032.

Quiggan, John (2006), Stern and the Critics Discounting. Available at *johnquiggin.com/wpcontent/uploads/2006/12/sternreviewed06121.pdf*

Quirk, Tom (2012) "Did the global temperature trend change at the end of the 1990s?" accepted for publication, *Asia-Pacific Journal of Atmospheric Sciences.*

Radic, Valentina (2008) "Modeling Future Sea Level Rise From Melting Glaciers" Ph. D. Dissertation, University of Alaska at Fairbanks.

Rahmstorf, Stefan; (2007) "A Semi-Empirical Approach to Projecting Future Sea-Level Rise," *Science* **315**, 368-370.

Ramanathan, V. (1988), "The greenhouse theory of climate change: A test by an inadvertent global experiment," *Science* **240**, 293–299.

Ramanathan, V. (1981) "The Role of Ocean-Atmosphere Interactions in the CO_2 Climate Problem" *Journal of the Atmospheric Sciences* **38**, 918-930.

Ramanathan, V. (1998) "Trace Gas Greenhouse Effect and Global Warming" *Ambio* **27**, 187-197.

Randall, D. A., R. D. Cess, J. P. Blanchet, G. J. Boer, D. A. Dazlich, A. D. Del Genio, M. Deque, V. Dymnikov, V. Galin, S. J. Ghan, A. A. Lacis, H. Le Treut, Z.-X. Li, X.-Z. Liang, B. J. Mcavaney, V. P. Meleshko, F. B. Mitchell, J.-J. Morcrette, G. L. Potter, L. Rikus, E. Roeckner, J. F. Royer, U. Schlese, D. M. Sheinin, J. Slingo, A. P. Sokolov, K . E. Taylor, W. M. Washington, R. T. Wetherald, I. Yagai and M.-H. Zhang (1992) "Intercomparison and Interpretation of Surface Energy Fluxes in Atmospheric General Circulation Models" *Journal of Geophysical Research* **97**, 3711-3724.

Rapp, Donald (2012), *Ice Ages: Measurements, Interpretation and Models*, 2nd edition, Springer-Praxis Publishing.

Rapp, Donald (2010), *Assessing Climate Change*, 2nd ed., Springer-Praxis Publishing.

Rapp, D. (2011) "The Relation Between Ancient Climates and CO_2 Concentration" *http://www.spaceclimate.net/*.

Richey, J. N.; R. Z. Poore; B. P. Hower; and T. M. Quinn (2007), "1400 yr multi-proxy record of climate variability from the northern Gulf of Mexico," *Geology*, **35**, 423.

Rignot, E., I. Velicogna, M. R. van den Broeke, A. Monaghan, and J. Lenaerts (2011) "Acceleration of the contribution of the Greenland and Antarctic ice sheets to sea level rise" *Geophysical Research Letters*, **38**, L05503.

Rignot, E.; and P. Kanagaratnam (2006), "Changes in the velocity structure of the Greenland Ice Sheet," *Science* **311**, 986–990.

Rind, David; Drew Shindell; Judith Perlwitz; Jean Lerner; Patrick Lonergan; Judith Lean; and ChrisMclinden (2004), "The relative importance of solar and anthropogenic forcing of climate change between the Maunder Minimum and the Present," *Journal of Climate* , **17**, 906–929.

Roe, Gerard H. and Marcia B. Baker (2007) "Why Is Climate Sensitivity So Unpredictable?" *Science* **318**, 629-632.

Roe, Gerard H. (2009) "Feedbacks, Timescales, and Seeing Red" *Annual Rev. Earth Planet. Sci.* **37**, 93–115.

Roe, Gerard H. and Yoram Bauman (2011) "Should the climate tail wag the policy dog?" *earthweb.ess.washington.edu/roe/GerardWeb/.../RoeBauman_FatTail.pdf.*

Roe, Gerard H. and K. C. Armour (2011) "How sensitive is climate sensitivity?" *Geophysical Research Letters* submitted 2011.

Rohde, Robert, Judith Curry, Donald Groom, Robert Jacobsen, Richard A. Muller, Saul Perlmutter, Arthur Rosenfeld, Charlotte Wickham and Jonathan Wurtele (2011) "Berkeley Earth Temperature Averaging Process" *http://www.berkeleyearth.org/*

Rohde, Robert, Richard A. Muller, Robert Jacobsen, Elizabeth Muller, Saul Perlmutter, Arthur Rosenfeld, Jonathan Wurtele, Donald Groom, Charlotte Wickham (2012) "A New Estimate of the Average Earth Surface Land Temperature Spanning 1753 to 2011" Submitted to JGR, *The Third Santa Fe Conference on Global and Regional Climate Change*, Manuscript # 2012JD018202.

Rohling, E. J., K. Grant, Ch. Hemleben, M. Siddall, B. A. A. Hoogakker, M. Bolshaw and M. Kucera (2008) "High rates of sea-level rise during the last interglacial period" *Nature Geoscience* **1**, 38-42.

Royer, Dana L. (2010) "Fossil soils constrain ancient climate sensitivity" *PNAS* **107**, 517-518.

Rutledge, D. (2007), *Hubbert's Peak and Climate Change*, California Inst. of Technology, presentation at Jet Propulsion Laboratory, Pasadena, CA (May), online at *http://www.rutledge.caltech.edu/*

Sabine, Christopher L.; Richard A. Feely; Nicolas Gruber; Robert M. Key; Kitack Lee; John L. Bullister; Rik Wanninkhof; C. S. Wong; Douglas W. R. Wallace; Bronte Tilbrook *et al.* (2004), "The oceanic sink for anthropogenic CO_2," *Science* **305**, 367–371.

Santer, B. D., C. Mears, F. J. Wentz, K. E. Taylor, P. J. Gleckler, T. M. L. Wigley, T. P. Barnett, J. S. Boyle, W. Bruggemann, N. P. Gillett, S. A. Klein, G. A. Meehl, T. Nozawa, D. W. Pierce, P. A. Stott, W. M. Washington and M. F. Wehner (2007) "Identification of human-induced changes in atmospheric moisture content" *PNAS* **104**, 15248-15253.

Santer, B. D., C. Mears, C. Doutriaux, P. Caldwell, P. J. Gleckler, T. M. L. Wigley, S. Solomon, N. P. Gillett, D. Ivanova, T. R. Karl, J. R. Lanzante, G. A. Meehl, P. A. Stott, K. E. Taylor, P. W. Thorne, M. F. Wehner, and F. J. Wentz (2011) "Separating Signal and Noise in Atmospheric Temperature Changes: The Importance of Timescale" *Journal of Geophysical Research (Atmospheres)* **116**, D22105.

Schatten, Kenneth; and W. Dean Presnell (2007), "Solar Cycle 24 and the Solar Dynamo". Available at *ntrs.nasa.gov/archive/nasa/casi.ntrs.nasa.gov/20070032658_2007033016.pdf*

Schewe, J., A. Levermann and M. Meinshausen (2011) "Climate change under a scenario near 1.5°C of global warming: monsoon intensification, ocean warming and steric sea level rise" *Earth Syst. Dynam.* **2**, 25–35.

Schmidt, Gavin; and David Archer (2009), "Too much of a bad thing," *Nature*, **458**, 1117–1118.

Schmitt, C. and D. A. Randall (1991) "Effects of Surface Temperature and Clouds on the CO_2 Forcing" *Journal of Geophysical Research* **96**, 9159-9168.

Schulte, Klaus-Martin (2008) "Scientific consensus on climate change?" *Energy and Environment* **19**, 281-6.

Schuster, Ute; and Andrew J. Watson, *A variable and decreasing sink for atmospheric CO_2 in the North Atlantic*, to be published in the *Journal of Geophysical Research* (2007), online at *http://lgmacweb.env.uea.ac.uk/ajw/Reprints/Schuster_Watson_JGR_in_press.pdf*

Schwartz, Stephen E. (2003), *Requirements for empirical determination of Earth's climate sensitivity*, AAAS Annual Meeting, Denver, CO, February 14–18, online at *http://www.ecd.bnl.gov/steve/abstracts/Empirical.html*

Schwartz, Stephen E. (2004), "Uncertainty requirements in radiative forcing of climate change," *J. Air and Waste Manage. Assoc.* **54**, 1351–1359.

Schwartz, Stephen E. (2007), *Heat capacity, time constant, and sensitivity of Earth's climate system, Journal of Geophysical Research* **112**, D24S05, online at *http://www.ecd.bnl.gov/pubs/BNL-76939-2006-AB.pdf*

Schwartz, Stephen E., Robert J. Charlson, Ralph A. Kahn, John A. Ogren and Henning Rodhe (2010) "Why Hasn't Earth Warmed as Much as Expected?" *Journal of Climate* **23**, 2453-2464.

Schwartz, Stephen E.; Robert J. Charlson; and Henning Rodhe (2007), "Quantifying climate change: Too rosy a picture?" *Nature Reports, Climate Change* **2**, 23–24.

Sedlacek, Jan and Lawrence A. Mysak (2008) "Sensitivity of sea ice to wind-stress and radiative forcing since 1500: a model study of the Little Ice Age and beyond" *Climate Dynamics* **32**, 817-831.

Serreze, M. C.; A. P. Barrett; J. C. Stroeve; D. N. Kindig; and M. M. Holland (2008), "The emergence of surface-based Arctic amplification," *The Cryosphere Discussions*, **2**, 601–622. Available at *http://www.the-cryosphere-discuss.net/2/601/2008/*

Serreze, Mark C. and Jennifer A. Francis (2006) "The Arctic Amplification Debate" *Climatic Change* **76**, 241–264.

Serreze, Mark C. and Roger G. Barry (2011) "Processes and impacts of Arctic amplification: A research synthesis" *Global and Planetary Change* **77**, 85-96.

Serreze, Mark C., Marika M. Holland and Julienne Stroeve (2007) "Perspectives on the Arctic's Shrinking Sea-Ice Cover" *Science* **315**, 1533-1537.

Shakun, J.D. and A.E. Carlson (2010) "A global perspective on Last Glacial Maximum to Holocene climate change", *Quaternary Sci. Rev.* **29**, 1801-1816.

Shapiro, A. I., W. Schmutz, E. Rozanov, M. Schoell, M. Haberreiter, A. V. Shapiro, and S. Nyeki (2011) "A new approach to long-term reconstruction of the solar irradiance leads to large historical solar forcing" *arxiv.org/pdf/1102.4763*.

Sheffield, Justin, Eric F.Wood and Michael L. Roderick (2012) "Little change in global drought over the past 60 years" *Nature* **491**, 435-440.

Shepherd, A.; and Wingham, D. (2007), "Recent sea-level contributions of the Antarctic and Greenland Ice Sheets," *Science* **315**, 1529–1532.

Shindell, Drew; and Greg Faluvegi (2009), "Climate response to regional radiative forcing during the twentieth century," *Nature Geoscience*, **2**, 294–300.

Siegel, D. A. and T. D. Dickey (1986) "Variability of Net Long Wave Radiation Over the Eastern North Pacific Ocean" *Journal of Geophysical Research* **91**, 7657-7666.

Silverman, Jacob, Boaz Lazar, Long Cao, Ken Caldeira, and Jonathan Erez (2009) "Coral reefs may start dissolving when atmospheric CO_2 doubles" *Geophysical Research Letters* **36**, L05606.

Sime, L. C., E. W. Wolff, K. I. C. Oliver and J. C. Tindall (2009) "Evidence for warmer interglacials in East Antarctic ice cores" *Nature* **462**, 342-346.

Singer, S. Fred; and Dennis T. Avery (2007), *Unstoppable Global Warming—Every 1500 years*, Rowman & Littlefield.

Smith, Doug M., Rosie Eade, Nick J. Dunstone, David Fereday, James M. Murphy, Holger Pohlmann and Adam A. Scaife (2010) "Skillful multi-year predictions of Atlantic hurricane frequency" *Nature Geoscience* **3**, 846-849.

Smith, S. J.; J. van Aardenne, Z. Klimont, R. J. Andres, A. Volke, and S. Delgado Arias (2011) "Anthropogenic sulfur dioxide emissions: 1850–2005" *Atmos. Chem. Phys.*, **11**, 1101–1116.

Soden, Brian J., Richard T. Wetherald, Georgiy L. Stenchikov and Alan Robock (2002) "Global Cooling After the Eruption of Mount Pinatubo: A Test of Climate Feedback by Water Vapor" *Science* **296**, 727-730.

Soden, Brian J. Darren L. Jackson, V. Ramaswamy, M. D. Schwarzkopf and Xianglei Huang (2005) "The Radiative Signature of Upper Tropospheric Moistening" *Science* **310**, 841-4.

Sokolov, Andrei P., P.H. Stone, C.E. Forest, R. Prinn, M.C. Sarofim, M. Webster, S. Paltsev, C.A. Schlosser, D. Kicklighter, S. Dutkiewicz, J. Reilly, C. Wang, B. Felzer, J. Melillo, and H.D. Jacoby (2009) "Probabilistic Forecast for 21st Century Climate Based on Uncertainties in Emissions (without Policy) and Climate Parameters" *MIT Joint Program on the Science and Policy of Global Change* Report No. 169, January 2009.

Solanki, S. K.; and M. Fligge (1998), "Solar irradiance since 1874 revisited," *Geophysical Research Letters* **25**, 341–344.

Solanki, S. K.; and M. Fligge (1999), "A reconstruction of total solar irradiance since 1700," *Geophys. Res. Lett.* **26**, 2465–2468.

Solanki, S. K.; I. G. Usoskin; B. Kromer; M. Schussler; and J. Beer (2004), "Unusual activity of the Sun during recent decades compared to the previous 11,000 years," *Nature* **431**, 1084–1087.

Solanki, S. K.; I. G. Usoskin; B. Kromer; M. Schussler; and J. Beer (2004), "Unusual activity of the Sun during recent decades compared to the previous 11,000 years," *Nature* **431**, 1084–1087.

Solanki, S.; N. Krivova; and T. Wenzler (2005), "Irradiance models," *Adv. in Space Research* **35**, 376–383.

Solanki, S.; N. Krivova; and T. Wenzler (2005), "Irradiance models," *Adv. in Space Research* **35**, 376–383.

Solanki, Sami K.; and Natalie A. Krivova (2004), "Solar irradiance variations: From current measurements to long-term estimates," *Solar Physics* **224**, 197–208.

Solomon, Susan, Karen H. Rosenlof, Robert W. Portmann, John S. Daniel, Sean M. Davis, Todd J. Sanford and Gian-Kasper Plattner (2010) "Contributions of Stratospheric Water Vapor to Decadal Changes in the Rate of Global Warming" *Science* **327**, 1219-1223.

Solomon, Susan; Gian-Kasper Plattner, Reto Knutti and Pierre Friedlingstein (2009) "Irreversible climate change due to carbon dioxide emissions," *PNAS* **106**, 1704-1709.

Soon, Willie; and Sallie Baliunas (2003a), *Lessons and Limits of Climate History: Was the 20th Century Climate Unusual?*, informal report, George C. Marshall Institute, Washington,

Soon, Willie; and Sallie Baliunas (2003b), "Reconstructing climatic and environmental changes of the past 1000 years: A reappraisal," *Energy and Environment* **14**, 233–299.

Sopkin, K. L. (2008) "Heat fluxes in Tampa Bay, Florida" Master of Science Thesis, College of Marine Science, University of South Florida.

Spencer, Roy W. and William D. Braswell (2007) "Potential Biases in Feedback Diagnosis from Observational Data: A Simple Model Demonstration" *Journal of Climate* **21**, 5624-5628.

Spencer, Roy W. and William D. Braswell (2010) "On the diagnosis of radiative feedback in the presence of unknown radiative forcing" *Journal of Geophysical Research*, **115**, D16109.

Spencer, Roy W. and William D. Braswell (2011) "On the Misdiagnosis of Climate Feedbacks from Variations in Earth's Radiant Energy Balance" *Remote Sensing* **3**, 1603-1613.

Spencer, Roy W., William D. Braswell, John R. Christy, and Justin Hnilo (2007), "Cloud and radiation budget changes associated with tropical intraseasonal oscillations," *Geophysical Research Letters*, **34**, L15707-L15711.

Spielhagen, Robert F. Kirstin Werner, Steffen Aagaard Sørensen, Katarzyna Zamelczyk, Evguenia Kandiano, Gereon Budeus, Katrine Husum, Thomas M. Marchitto, Morten Hald (2011) "Enhanced Modern Heat Transfer to the Arctic by Warm Atlantic Water" *Science* **331**, 450-453.

Stahle, D. W., Cook, E. R., Cleaveland, M. K, Therrell, M. D., Meko, D. M., Grissino-Mayer, H. D., Watson, E. and Luckman, B. H. (2000), "Tree-ring data document 16th century megadrought over North America," *EOS Transactions, American Geophysical Union* **81**, 121-125.

Stainforth, D. A.; M. R. Allen; E. R. Tredger; and L. A. Smith (2007), "Confidence, uncertainty and decision-support relevance in climate predictions," Phil. Trans. R. Soc., **A365**, 2145–2161.

Steig, Eric J., David P. Schneider, Scott D. Rutherford, Michael E. Mann, Josefino C. Comiso and Drew T. Shindell (2009) "Warming of the Antarctic ice-sheet surface since the 1957 International Geophysical Year," Nature **457**, 459-453.

Stephens, Graeme L., Juilin Li, Martin Wild, Carol Anne Clayson, Norman Loeb, Seiji Kato, Tristan L'Ecuyer, Paul W. Stackhouse Jr, Matthew Lebsock and Timothy Andrews (2012) "An update on Earth's energy balance in light of the latest global observations" *Nature Geoscience* published on line 23 September 2012.

Stern, David I. (2005a), Reversal in the Trend of Global Anthropogenic Sulfur Emissions, Rensselaer working papers in economics, No. 0504 (May), online at *http://www.rpi.edu/sternd/GEC2006.pdf*

Stern, David I. (2005b), "Global sulfur emissions from 1850 to 2000," *Chemosphere* **58**, 163–175.

Svalgaard, Leif , Edward W. Cliver, and Yohsuke Kamide (2005) "Sunspot cycle 24: Smallest cycle in 100 years?" *Geophysical Research Letters* **32**, L01104.

Svensmark H. and Friis-Christensen, E. (1997), "Variation of cosmic ray flux and global cloud coverage—a missing link in solar-climate relationships," *Journal of Atmospheric and Solar-Terrestrial Physics* **59**, 1225–1232.

Svensmark, H. (2000), "Cosmic Rays and Earth's Climate," Space Science Reviews 93, 155-166. *http://www.junkscience.com/Greenhouse/influence-of-cosmic-rays-on-the-earth.pdf*

Svensmark, H. and E. Friis-Christensen (2007), Reply to Lockwood and Fröhlich – The persistent role of the Sun in climate forcing, Danish National Space Center Scientific Report, March 2007. *http://www.spacecenter.dk/publications/scientific-report-series/Scient_No._3.pdf/view*

Sverdrup, H. U., Martin W. Johnson and Richard H. Fleming (1942) "The Oceans Their Physics, Chemistry, and General Biology" The Oceans, Their Physics, Chemistry, and General Biology. New York: Prentice-Hall, *publishing.cdlib.org/ucpressebooks/data/.../kt167nb66r_ch04.pdf*

Tapping, K. F.; D. Boteler; A. Crouch; P. Charbonneau; A. Manson; and H. Paquette (2006), *Modelling Solar Magnetic Flux and Irradiance during and since the Maunder Minimum*, online at *http://www.lps.umontreal.ca/paquetteh/Maunder_SP.pdf*

Taylor, K. E., C. D. Hewitt, P. Braconnot, A. J. Broccoli, C. Doutriaux, J. F. B. Mitchell (2001) "Analysis of Forcing Response, and Feedbacks in a Paleoclimate Modeling Experiment" Lawrence Livermore National Laboratory report UCRL-JC-143363

Taylor, K. E., M. Crucifix, P. Braconnot, C. D. Hewitt, C. Doutriaux, A. J. Broccoli, J. F. B. Mitchell, and M. J. Webb (2000) "Estimating Shortwave Radiative Forcing and Response in Climate Models" *Journal of Climate* **20**, 2530-2544.

Tedesco , M. , X. Fettweis, M. R. van den Broeke, R. S. W. van de Wal, C. J. P. P. Smeets, W. J. van de Berg, M. C. Serreze and J. E. Box (2011) "The role of albedo and accumulation in the 2010 melting record in Greenland" *Environ. Res. Lett.* **6**, 014005.

Tedesco M., X. Fettweis, M. R. van den Broeke, R. S. W. van de Wal, C. J. P. P. Smeets, W. J. van de Berg, M. C. Serreze, and J. E. Box (2010) "Record summer melt in Greenland in 2010", *EOS AGU* **92**, 126.

Tedesco, M. and Monaghan, A. J. (2010) "Climate and melting variability in Antarctica" *EOS, Transactions, American Geophysical Union* **91**, L18502.

Tedesco, M., X. Fettweis, T. Mote, N. Steiner and J. E. Box (2011a) "Year 2011 Greenland melting remains well above the (1979 – 2010) average; close-to-record mass loss" *http://greenland2011.cryocity.org/*

Thejll, Peter (2009), Update of the Solar Cycle Length Curve, and the Relationship to the Global Mean Temperature, Danish Climate Centre Report 09-01. Available at *http://www.dmi.dk/dmi/dkc09-01.pdf*

Thompson, D. W.; Kennedy, J. J.; Wallace, J. M.; and Jones, P. D. (2008), "A large discontinuity in the mid-twentieth century in observed global-mean surface temperature," *Nature*, **453**, 646–649.

Thorsteinsson, Thorsteinn "Stable isotopes and climate history from polar ice cores." *http://www.ifa.hawaii.edu/UHNAI/NAIweb/presentations/26-Thorsteinsson-isotopeclimate.pdf*

Tisdale, Bob (2009, 2012), *Who turned on the heat? http://bobtisdale.wordpress.com/*

Titus, James G. (1990), *Greenhouse Effect, Sea Level Rise, and Barrier Islands: Case Study of Long Beach Island, New Jersey*, online at *http://yosemite.epa.gov/oar/globalwarming.nsf/content/ResourceCenterPublicationsSLRBarrier_Islands.html*

Trainer, Ted (2011) "A critique of the 2011 IPCC Report on Renewable Energy", *http://bravenewclimate.com/2011/08/09/ipcc-renewables-critique/*

Trenberth, Kevin E. and Timothy J. Hoar (1997) "El Nino and Climate Change," *Geophysical Research Letters* **24**, 3057-3060.

Trenberth, Kevin E. (2004), "Comment on 'Impact of urbanization and land-use change on climate' by E. Kalnay and M. Cai in *Nature* **423**, 528–531 (2003)," *Nature* **427**, 214.

Trenberth, Kevin E., John T. Fasullo and Jeffrey Kiehl (2009) "Earth's Global Energy Budget" *Bull. Amer. Meteorological Soc.* March 2009, 311-324.

Trenberth, K. (2010) "The ocean is warming, isn't it?" Nature 465, 304.

Trenberth, K.E., J.T. Fasullo, Chris O'Dell, and T. Wong (2010) "Relationships between tropical sea surface temperature and top-of-atmosphere radiation", *Geophys. Res. Lett.*, **37**, L03702.

Turner, John; Steve R. Colwell; Gareth J. Marshall; Tom A. Lachlan-Cope; Andrew M. Carleton; Phil D. Jones; Victor Lagun; Phil A. Reide; and Svetlana Iagovkina (2005), "Antarctic climate change during the last 50 Years," *Int. J. Climatology* **25**, 279–294.

Turner, John; Josefino Comiso; Gareth J. Marshall; Tom A. Lachlan-Cope; Tom Bracegirdle; Tim Maksym; Michael Meredith; Zhaomin Wang; and Andrew Orr (2009), "Non annular atmospheric circulation change induced by stratospheric ozone depletion and its role in the recent increase of Antarctic sea ice extent," *Geophysical Research Letters* **36**, L08052.

Van de Wal, R. S.; W. Boot; M. R. van den Broeke; C. J. P. P. Smeets; C. H. Reijmer; J. J. A. Donker; and J. Oerlemans (2008), "Large and rapid Greenland Ice Sheet," *Science*, **321**, 111–113.

Van Dorland, Rob (1998) "Radiation and Climate from Radiative Transfer Modelling to Global Temperature Response" Ph. D. Dissertation, University of Utrecht.

Vaquero, J. M.; M. C. Gallego; R. M. Trigo; F. Sanchez-Bajo; M. L. Cancillo; and J. A. Garcia (2006), "A new reconstruction of total solar irradiance since 1832," Atmosfera **19**, 267– 274.

Vecchi, G. A.; B. J. Soden; A. T. Wittenberg; I. M. Held; A. Leetma; and M. J. Harrison (2006), "Weakening of tropical Pacific atmospheric circulation due to anthropogenic forcing," *Nature* **441**, 73–76.

Vermeer, M. and Rahmsdorf, S. (2009) "Global sea level linked to global temperature" *Proceedings of the National Academy of Sciences* **106**, 21527–21532.

Vieira, L. E. A., S. K. Solanki, N. A. Krivova, and I. Usoskin "Evolution of the solar irradiance during the Holocene" *Astronomy and Astrophysics* **531**, A6-A25.

Villarini, G,, G. A. Vecchi, T. R. Knutson and J. A. Smith (2011) "Is the recorded increase in short-duration North Atlantic tropical storms spurious?" *Journal of Geophysical Research* **116**, D10114.

Villarini, G,, G. A. Vecchi, T. R. Knutson, Ming Zhao and J. A. Smith (2011a) "North Atlantic Tropical Storm Frequency Response to Anthropogenic Forcing: Projections and Sources of Uncertainty" *Journal of Climate* **24**, 3224-3238.

Vinther, B. M., Jones, P.D., Briffa, K.R., Clausen, H.B., Andersen, K.K., Dahl-Jensen, D., and Johnsen, S.J. (2010) "Climatic signals in multiple highly resolved stable isotope records from Greenland" *Quaternary Science Reviews* **29**, 522–538.

von Storch, H. E.; E. Zorita; J. Jones; Y. Dimitriev; F. Gonzalez-Rouco; and S. Tett (2004), "Reconstructing past climate from noisy data," *Science* **306**, 679–682.

Wada, Y., L. P. H. van Beek, C. M. van Kempen, J. W. T. M. Reckman, S. Vasak, and M. F. P. Bierkens (2010), "Global depletion of groundwater resources", *Geophys. Res. Lett.*, **37**, L20402.

Wada, Y., L. P. H. van Beek, F. C. Sperna Weiland, B. F. Chao, Y.-H. Wu, and M. F. P. Bierkens (2012) "Past and future contribution of global groundwater depletion to sea-level rise" *Geophys. Res. Lett.*, **39**, L09402.

Wahl, Eugene R.; Ammann, Caspar M. (2007). "Robustness of the Mann, Bradley, Hughes reconstruction of Northern Hemisphere surface temperatures: Examination of criticisms based on the nature and processing of proxy climate evidence" *Climatic Change* **85**, 33–69.

Wang, Chien (2002) "A Modeling Study on the Climate Impacts of Black Carbon Aerosols" *MIT Joint Program on the Science and Policy of Global Change* Report No. 84, March 2002.

Wang, M.; and J. E. Overland (2009), "A sea ice free summer Arctic within 30 years?" *Geophys. Res. Lett.*, **36**, L07502.

Wang, K., R. E. Dickinson, M. Wild, and S. Liang (2012) "Atmospheric impacts on climatic variability of surface incident solar radiation" *Atmos. Chem. Phys. Discuss.*, **12**, 14009–14042.

Ward, Bud (2008), *Communicating on Climate Change: An Essential Resource for Journalists, Scientists, and Educators*, Metcalf Institute for Marine & Environmental Reporting. Available at *http://www.metcalfinstitute.org/Communicating_ClimateChange.htm*

Watson, P. J. (2011) "Is There Evidence Yet of Acceleration in Mean Sea Level Rise around Mainland Australia?" *Journal of Coastal Research* **27**, 36377.

Watts, Robert G. (1980) "Discussion of Questions Concerning the Possible Influence of Anthropogenic CO_2 on Atmospheric Temperature" *Journal of Applied Meteorology* **19**, 494-5.

Watts, Robert G. (1982) "Further Discussion of Questions Concerning the Possible Influence of Anthropogenic CO_2 on Atmospheric Temperature" *Journal of Applied Meteorology* **21**, 243-7.

Webster, Mort, Sergey Paltsev, John Parsons, John Reilly and Henry Jacoby (2008) "Uncertainty in Greenhouse Gas Emissions and Costs of Atmospheric Stabilization" *MIT Joint Program on the Science and Policy of Global Change* Report No. 165, November 2008.

Webster, P. J.; G. J. Holland; J. A. Curry; and H-R. Chang (2005), "Changes in tropical cyclone number, duration, and intensity in a warming environment," *Science* **309**, 1844–1846.

Wegman, Edward J.; David W. Scott; and Yasmin H. Said (2006), *Ad Hoc Committee Report on the hockey stick Global Climate Reconstruction*, The Congressional Committee on Energy and Commerce, Washington, D.C. (July 14), online at *http://republicans.energy commerce.house.gov/108/home/07142006_Wegman_Report.pdf*

Weiler, C. Susan, Jason K. Keller and Christina Olex (2011) "Personality type differences between Ph.D. climate researchers and the general public: implications for effective communication" Climatic Change, *disccrs.org/files/WeilerEtAl_2011_ClimaticChange_MBTI.pdf*

Weitzman, Martin (2007), "A review of The Stern Review of the Economics of Climate Change," *Journal of Economic Literature*, **45**, 703–724.

Weitzman, Martin (2009), "Additive Damages, Fat-Tailed Climate Dynamics, and Uncertain Discounting" Economics: The Open-Access, Open-Assessment E-Journal 3, 2009-39, *http://www.economics-ejournal.org/economics/journalarticles/2009-39*.

Wickham, Charlotte, Judith Curry, Don Groom, Robert Jacobsen, Richard Muller, Saul Perlmutter, Robert Rohde, Arthur Rosenfeld and Jonathan Wurtele (2011) "Influence of Urban Heating on the Global Temperature Land Average Using Rural Sites Identified from MODIS Classifications" *http://www.berkeleyearth.org/*

Wilmking, M. and J. Singh (2008) "Eliminating the "divergence problem" at Alaska's northern treeline" *Clim. Past Discuss.* **4**, 741–759.

Wingham, D. J.; A. Shepherd; A. Muir; and G. J. Marshall (2006), "Mass balance of the Antarctic Ice Sheet," *Transactions of the Royal Society* **364**, 1627–1635.

Woodworth, P.L. (1990) "A search for accelerations in records of European mean sea level". *International Journal of Climatology*, **10**, 129–143.

Woppelmann, G.; B. Martin Miguez; M.-N. Bouin; and Z. Altamimi (2007), "Geocentric sea level trend estimates from GPS analyses at relevant tide gauges world-wide," *Global and Planetary Change*, **57**, 396–406.

Wopplemann G., N. Pouvreau, A. Coulomb, B. Simon and P. L. Woodworth (2008) "Tide gauge datum continuity at Brest since 1711: France's longest sea-level record" *Geophysical Research Letters* **35**, L22605.

Wu, Lixin; Dong Eun Lee; and Zhengyu Liu (2005), "The 1976/77 North Pacific climate regime shift: The role of subtropical ocean adjustment and coupled ocean–atmosphere feedbacks," *Journal of Climate* **18**, 5125–5140.

Wunsch C., R. M. Ponte and P. Heimbach (2007) "Decadal trends in sea level patterns: 1993–2004", *J. of Climate*, **20**, 5889–5911.

Wunsch, Carl (1999), "The interpretation of short climate records, with comments on the North Atlantic and Southern Oscillations, *Bulletin of the American Meteorological Society* **80**, 245–255.

Wunsch, Carl (2003), "Greenland—Antarctic phase relations and millennial time-scale climate fluctuations in the Greenland ice-cores," *Quaternary Science Reviews* **22**, 1631–1646.

Wunsch, Carl (2005), "The Total Meridional Heat Flux and Its Oceanic and Atmospheric Partition," *Journal of Climate* **18**, 4374-4380.

Wunsch, Carl (2006), "Abrupt climate change: an alternative view," *Quaternary Research* **65**, 191–203.

Yang, Bao, Jinsong Wang, Achim Brauning, Zhibao Dong, and Jan Esper (2009) "Late Holocene climatic and environmental changes in arid central Asia" *Quaternary International* **194**, 68–78.

Zapadka, Tomasz, Bogdan Woźniak and Jerzy Dera (2007) "A more accurate formula for calculating the net long wave radiation flux in the Baltic Sea" *Oceanologia* **49**, 449–470.

Zwally, H. J.; Giovinetto, M. B.; Li, J.; Cornejo, H. G.; Beckley, M. A.; Brenner, A. C.; Saba, J. L.; and Yi, D. (2005), "Mass changes of the Greenland and Antarctic ice sheets and shelves and contributions to sea-level rise: 1992–2002," *Journal of Glaciology* **51**, 509–527.

Donald Rapp

ABOUT THE AUTHOR

Donald Rapp received his M.S. degree in chemical engineering from Princeton in 1956 and Ph. D. in chemical physics at Berkeley in January 1960. He worked as a researcher in chemical physics for a number of years, amassing a considerable number of publications. He was on the faculty of the University of Texas and was promoted to full professor in 1973 at the age of 39. While at the University of Texas, he published textbooks on quantum mechanics, statistical mechanics and solar energy.

He came to JPL/Caltech in 1979 to take a position as the Division Chief Technologist (senior technical person) in the Mechanical and Chemical Systems Division (staff of 700 including 100 Ph.Ds). At JPL, he was a pioneer in pointing the institution toward new technologies. He was Proposal Manager on the Genesis Discovery Project to return samples of the solar wind to Earth. His proposal won in a field of about 25 competitors, being funded at ~ $220M in the Discovery 5 competition. Genesis carried out its mission in space from 2001 to 2004. After that, he played a major role in putting together the OMEGA MIDEX proposal ($139M) in 1998. Subsequently, he acted as Proposal Manager for the Deep Impact Discovery proposal to hit a comet with a projectile to allow examination of the interior, which won, bringing in about $320M to JPL. Deep Impact was a spectacular success in 2005.

He was manager of the Mars Exploration Technology Program for a period, and he was manager of the In Situ Propellant Production (ISPP) task in this Program. He wrote a landmark report on converting Mars resources into usable propellants for return to Earth. He wrote the Mars Technology Program Plan in 2001.

During the period 2001-2002, he played an important role in NASA assessments of technology for radioisotope power conversion, energy storage, and photovoltaic power conversion. In 2002 he wrote the NASA Office of Space Science Technology Blueprint for Harley Thronson, NASA Technology Director, a 100-page assessment of technology needs and capabilities for future missions.

In the period 2003-2006, he prepared a revised and expanded version of the Technology Blueprint for Harley Thronson at NASA HQ. In 2004, he was Proposal Manager for a proposal for a ground-penetrating radar experiment for the Mars Science Laboratory.

In the period 2004-2006, he concentrated on mission design for Mars and lunar human missions. This work led to his writing the book *Human Missions to Mars* that was published by Praxis/Springer in 2007. This is a major work, comprising 520 pages with over 200 figures.

Starting in 2006, he devoted most of his time to the study of climate change, reading hundreds of published papers and about 20 books. This effort culminated in publication of two books: *Assessing Climate Change* (2008) and *Ice Ages* (2009). A significantly expanded version of *Assessing Climate Change* was published as a second edition in 2010. The book on climate change is comprehensive, containing 1,348 specific citations to references in the field of climate change and 411 specific quotations of authors with their own words. Throughout the period 2009-2012 he has continued to diligently read every significant new paper in climate change and has monitored the discussions of climate change on the good blogs: climateaudit.org and judithcurry.com. On a daily basis, he takes notes and integrates new data and theories into his comprehensive overview of the subject.

During 2012 he also prepared a second edition to his book on ice ages, and wrote a book on use of extraterrestrial resources for space missions to the Moon and Mars.

Honors:

Elected Fellow of the American Physical Society, 1974

Referee for the Journal of Chemical Physics, the Physical Review, the American Journal of Physics, the Journal of Physical Chemistry, and other journals on over 300 occasions.

Book reviewer for Physics Today and other journals.

Two articles were chosen as Citation Classics with over 450 citations.

Listed in Who's Who in the West

Listed in Who's Who in Frontiers of Science and Technology

Listed in Who's Who in America

Listed in Men of Achievement

Listed in International Who's Who of Contemporary Achievement

Listed in International Who's Who of Professionals

Listed in Personalities of the Americas

Listed in Who's Who in Technology Today

Listed in Who's Who in Technology

Listed in Who's Who in California

Listed in Who's Who of Professionals

Listed in Two Thousand Notable Americans

Listed in Dictionary of International Biography

Listed in Strathmore's Who's Who

Received Exceptional Service Award from NASA October, 2002

Associate Editor of the Mars Journal 2006 - present

Published Books:

Quantum Mechanics, 672 pages, published 1971 by Holt, Rinehart and Winston

Statistical Mechanics, 330 pages, published in 1972 by Holt, Rinehart and Winston; translated into Japanese 1977. Reissued in 2012 as a new, updated version on amazon.com.

Solar Energy, 516 pages, published in 1981 by Prentice-Hall

Human Missions to Mars: Enabling Technologies for Exploring the Red Planet

Hardback, October, 2007

552 pages, two 8-page color sections

Assessing Climate Change – Temperatures, Solar Radiation and Heat Balance

Series: Springer Praxis Books – Environmental Sciences

410 p. 130 illus., Hardcover, January, 2008; second edition 2010

Ice Ages and Interglacials

Series: Springer Praxis Books – Environmental Sciences

263 p. Hardcover, 2009; second edition, summer, 2012.

Use of Extraterrestrial Resources in Human Space Missions to the Moon and Mars, Springer Praxis books, publication date November 2012.

Donald Rapp

www.ingramcontent.com/pod-product-compliance
Lightning Source LLC
Chambersburg PA
CBHW051437170526
45166CB00001B/21